ADVANCES IN INTERNATIONAL RICE RESEARCH

Edited by **Jinquan Li**

Advances in International Rice Research
http://dx.doi.org/10.5772/63007
Edited by Jinquan Li

Contributors

André Andres, Germani Concenço, Giovani Theisen, Jose M. B. Parfitt, Gustavo Mack Teló, Leandro Galon, Matheus Martins, Cristian Rossi, Ombretta Marconi, Valeria Sileoni, Giuseppe Perretti, Dayana Ceccaroni, Min Huang, Shuanglü Shan, Jiana Chen, Fangbo Cao, Ligeng Jiang, Yingbin Zou, Chang-Sheng Wang, Da-Gin Lin, Masayuki Fujita, Mirza Hasanuzzaman, Kamrun Nahar, Anisur Rahman, Md. Shahadat Hossain, Jubayer-Al- Mahmud, Yogesh Vikal, Dharminder Bhatia, Maxwell Darko Asante, Ken\'Ichi Ohtsubo, Mario Luiz Mesquita, Germani Concenco, Walkyria B. Scivittaro, Jaqueline T. Silva, Marilia A. B. Pinto, Jimmy Lamo, Pangirayi Tongoona, Sie Moussa, Semon Mande, Geoffrey Onaga, Patrick Okori, Ana Paula Santos, Liliana Ferreira, Sebastião Ravasco, Duarte Figueiredo, Christoph Peterhansel, Nelson Saibo, Maria Margarida Oliveira, Ibemhal Asem

Published by InTech
Janeza Trdine 9, 51000 Rijeka, Croatia

Publishing Process Manager Romina Rovan
Technical Editor SPi Global
Cover InTech Design team

First published March, 2017
Printed in Croatia
Legal deposit, Croatia: National and University Library in Zagreb

Additional hard copies can be obtained from orders@intechopen.com

Advances in International Rice Research, Edited by Jinquan Li
p. cm.
Print ISBN 978-953-51-3009-3
Online ISBN 978-953-51-3010-9

For EU product safety concerns:
IN TECH d.o.o., Prolaz Marije Krucifikse Kozulić 3, 51000 Rijeka, Croatia,
info@intechopen.com or visit our website at intechopen.com.

Meet the editor

Dr. Jinquan Li obtained his master's and PhD degrees on Crop Science at Shenyang Agricultural University and South China Agricultural University in 2000 and 2003, respectively. During July 2003 and 2011, he worked at South China Agricultural University as an assistant professor (an associate professor since December 2008) in Plant Genetics and Breeding. During 2009 and 2011, he worked at the Max Planck Institutes for plant breeding research, Germany, as a postdoc and later as a project scientist between October 2011 and October 2014. From November 2014 to present, he works as a scientist in the German Plant Phenotyping Network (DPPN) project at the Institute of Plant Sciences (IBG-2), Institute of Bio- and Geosciences, Research Center, Jülich, Germany.

Contents

Preface

Rice (*Oryza sativa* L.) provides staple food for more than 50% of the world's population and is one of the most important crops in the world. It is a treasure for human being. It provides 30% of total energy to rice eaters and serves as the main food with various forms including cooked rice and rice noodles as well as the source of beer, wine, and vinegar. Moreover, it produces many by-products, such as rice husks, rice bran, broken rice, rice flour, rice milk, rice pudding, rice starch, rice straw, rice cakes, etc. The straw and bran can be used as cattle and poultry feed. Nowadays, the rice straw can be used to produce biochar, rice hulls can be used in manufacturing insulation materials, and rice bran oil is an important edible oil.

Recently, great advancements have been made on rice researches, especially with the technologies of high-throughput genome sequencing, high-throughput phenotyping, integrated "-omis" methods, as well as the improvement of cultivation methods, which ensures rice to provide increasing food as the world's population increases. To have a glance of the international rice researches, this book was organized to show some advancements in rice researches with the rice researchers from all over the world.

Different to the traditional rice book, the book was aimed to introduce some new advances in the rice researches. It first introduced cultivation, management, and breeding of upland rice in Latin America (Brazil; Chapters 1 and 2) and the weed management in rice fields (Chapter 3). Then, in Chapter 4, the use of rice in beer production was introduced. As more and more importance was paid attention to rice breeding for good quality and biotic and abiotic stress resistance, two core sections (the second and the third section) of the book were focused on these two topics, respectively. The second section introduced a breeding program for improved grain quality (Chapter 5), the evaluation of palatability of cooked rice (Chapter 6), the comparison of grain quality between super hybrid rice and popular inbred rice cultivars (Chapter 7), as well as the characteristics of scent and color of a special "Chakhao" rice (aromatic black rice) in India (Chapter 8). The third section introduced the mechanisms of salt-stress tolerance (Chapter 9), the genetic and genomic mechanisms of bacterial blight resistance (Chapter 10), and breeding for drought tolerance in Africa (Chapter 11). Furthermore, the fourth section introduced the new technologies in rice research. In this section, it reported the chromatin immunoprecipitation (ChIP) technology on epigenetics research in rice (Chapter 12), the integrated "-omis" methods in studying a bacterial blight-resistant rice mutant (Chapter 13), and the bistatic interferometry technology on monitoring plant height and production from space (Chapter 14).

As more and more new technologies appear and are applied in rice research, it is no doubt that these technologies will make rice research a bright future and make this important crop as a cornerstone for food safety to human being.

Dr. Jinquan Li
Institute of Plant Science, Research Center,
Jülich, Germany

State Key Laboratory for Conservation and Utilization of Subtropical Agro-Bioresources,
South China Agricultural University, China

Section 1

Cultivation, Production and Management

Chapter 1

Soil and Water Management for Sprinkler Irrigated Rice in Southern Brazil

José Maria Barbat Parfitt, Germani Concenço,
Walkyria Bueno Scivittaro, André Andres,
Jaqueline Trombetta da Silva and
Marília Alves Brito Pinto

Additional information is available at the end of the chapter

http://dx.doi.org/10.5772/66024

Abstract

Rice is grown in lowland paddies, which is flood irrigated. In the most undulating areas, continuous flooding is difficult and some farmers seek alternative irrigation methods. Grain yield in sprinkler irrigated rice ranges between 80 and 100% of that obtained under flooding, but for this, fertilizer and water should be properly managed. For sprinkler irrigated rice, fertilizer should be corrected by adding 10 kg/ha of P_2O_5 and 15 kg/ha of K_2O for every expected additional ton of grains, over the standard recommendation. Regarding nitrogen fertilizer, it is recommended to be applied about 20 kg/ha of N at planting and the rest as topdressing. This can be done via soil, split into two applications: 50–60% of the topdressing dose at tillering start and the rest at panicle initiation. When N is applied by fertigation, 25% of the recommended topdressing N should be applied at tillering start; the remainder of the dose may be partitioned into four to six weekly applications through irrigation water. For water management, soil water tension should be kept below 10 kPa. At the vegetative stage, irrigation can be applied aiming to avoid water tensions in soil above 30 kPa at any moment.

Keywords: *Oryza sativa*, soil moisture, water application, fertilizer, center pivots

1. Introduction

Rice is among the most consumed cereals in the world, constituting a staple food. The Brazilian *per capita* consumption of rice is 25 kg/year, and it is the ninth largest producer, with production concentrated in the Southern region, particularly in the state of Rio Grande do Sul, which accounts for about 60% of national production. The projections for rice production and consumption show that Brazil will harvest 14.12 million tons of rice in the 2019/2020 cropping season, which is equivalent to an annual increase in production of 1.15% [1].

In the state of Rio Grande do Sul, rice is grown in flood irrigated lowland paddies. Cropping systems include conventional system, where soil tillage is accomplished in the spring; minimum tillage, with advanced soil preparation in the fall / winter and later direct planting in the spring; and water-seeded rice, restricted to leveled flat areas.

The lowland areas of Rio Grande do Sul present diversified relief, ranging from very flat areas (slope <0.2%) to gently undulating areas (slope >2%). The latter occur more frequently in the region called *Fronteira Oeste* (West Border), although they can occur in all rice-growing regions of the state. In the most undulating areas, the method of continuous flood irrigation hinders crop management, particularly planting, harvesting, and irrigation, due to the large amount of levees necessary for keeping the water layer into the field. This has led some producers of that region to seek alternative irrigation methods for rice, among which the sprinkler irrigation under center pivots stands out.

The rice grain yield obtained in sprinkler irrigated system, in some cases, has proved comparable to those under flood irrigation, and in other cases, varied between 80 and 90% of that obtained under continuous flooding. Research data show that, even when water management and other practices are appropriate, some soil types can promote yield drops when rice is sprinkler irrigated, which is directly related to the fraction of the micropores in soil [2].

It is estimated that for every kilogram of rice produced under flood irrigation, 1300 L of water is required, which is not much compared to crops such as soybeans, which demands about 2300 L per kg of grain produced. However, the difference in water demand between these crops lies in the fact that for rice, nearly 100% of the water comes from irrigation, while for soybean the water demand is supplied primarily by rain.

According to Mota et al. [3], rice evapotranspiration averages 650 mm per cropping season. However, when flood irrigated, the total amount of water required to meet rice demand include other components such as soil saturation, keeping the layer of standing water, as well as losses by percolation or lateral runoff [4]. In this system, there is also greater loss by evaporation from the free standing water surface mainly in the early crop stages. Thus, sprinkler irrigation in rice is of great importance, especially for regions with scarce water resources, which is the current condition in many regions of Brazil and other countries.

Another important aspect in rice production is that the no-till planting system is not easily possible, due mainly to the physical damage inflicted to the flooded soil by machinery tires during crop management and harvest. Such damage practically require soil disturbance at least for soil leveling, preceding the next planting [5]. With sprinkler irrigation, soil physics is not strongly impacted, being dispensable a new tillage operation before the next planting.

The perspectives for sprinkler irrigated rice can be classified into two groups: (a) *needs*: imposed by climatic conditions, increased demand for food, and the requirement of environmental agencies and (b) *expectations*: increasingly search for positive results following adoption of the technology. The perspectives of the first group (needs) consider the increasing limitations on water availability on the planet [6], requiring the adoption of technologies that provide a more rational use of this natural resource. Moreover, the required increase in food production due to the exponential growth of the world population will be a result of both increased productivity and production area expansion [7]. The sprinkler irrigated rice (mainly by center pivots and lateral-movement irrigation equipment [linears]) will help in these two aspects, since the technology is adopted with proper technical and scientific basis. The multiple use of water requires that all water consuming sectors contribute to the more efficient use of water in their activities [8], which is the only alternative capable of accommodating the various needs of use of this renewable but finite resource.

The expectations comprise the equalization of rice productivity levels under sprinkler irrigation to those obtained in paddy rice, with the development of research aimed at overcoming the current grain yields obtained under sprinkler irrigation. Such superiority will be achieved by adapting cultivars to new water regime plus improvement in management techniques, resulting in lower environmental impact of rice production, with social benefits that include better quality of the final product, rational use of pesticides, and cost reduction. In addition, improvement in working conditions of rural staff, who no longer would work in flooded paddy, unstable, and irregular environment, would work on a stable and less laborious soil.

In rice fields established in areas with slopes higher than 2%, water is saved by sprinkler irrigating the rice. Research reports that the cost for growing rice under center pivot is smaller than flooded paddy rice cultivation, with higher net profit. The management of the flooded rice paddies demand greater number of machines per area due to the lower speed in agricultural operations, the presence of levees, and the wheel drive slipping on the muddy soil; at the same time, there is need for more powerful machines and often adjustments are required in the equipment to specific operation conditions for areas with great number of levees. This results in increased costs with fuel and maintenance. The cost for rice production under center pivot in the West Border region of Rio Grande do Sul was about 20–25% lower than the average cost of nearby flooded rice paddies. The main economy factors under center pivot were fuel for machinery and irrigation costs, which included electricity, machinery repair and maintenance, and human labor.

The evolution of crop production systems requires that at a given time, new technologies should be adopted to ensure the achievement of further increases in the efficiency of use of natural resources and production levels. The sprinkler irrigation stands out among the alternatives studied to save water in rice production due to its flexibility, high productivity potential, and ease of adoption. Evidently, only research and continuous improvement will keep the technology ahead of other alternatives. Currently, the opportunities associated to sprinkler irrigation in rice are promising, including the possibility of full adoption of no-till and crop rotation practices in rice production; but for its success, there are many aspects that need to be improved.

This chapter is aimed at presenting the basic aspects for soil and water management in sprinkler irrigated rice, based on research results carried out for over 5 years, as well as through experiences of the productive sector in the Brazilian state of Rio Grande do Sul.

2. Cropping system

In Rio Grande do Sul, although rice cultivars recommended for sprinkler irrigation are those developed for flooding, crop management practices differ. The main reason is the lack of standing water in the sprinkler irrigation, eliminating the levees. This feature, on the one hand, gives some advantages to the system, such as the possibility of no-till adoption and application of all practices by ground, with no need for fertilizer and pesticide applications with aircrafts; on the other hand, there is need for more attention to soil fertility and integrated pest management.

The Brazilian rice cultivars developed to be grown in flooded paddies present high grain yield potential, but they are very susceptible to water stress, especially during the reproductive stages. Research results by Embrapa Clima Temperado conducted under conventional soil tillage, reported severe damage to rice plants in several spots into the experimental fields, even when adopting the recommended water management, e.g., by irrigating back to saturation when water tension in soil reached a maximum of 10 kPa. This was mainly due to the absence of mulching on soil at the experimental areas; in this situation, the water droplets caused disruption of topsoil, resulting in the formation of a crust and making water infiltration into soil difficult. This behavior is hardly observed when rice is grown on mulch and especially under consolidated no-till systems, which were established for some years. Therefore, it is clear that sprinkler irrigated rice should be grown in a production system involving both no-till and crop rotation.

Suggestions for a possible production system include rice, corn, and soybeans in rotation in summer, succeeding winter cover crops (pasture species), with or without cattle grazing in winter. Regardless of the established cropping system, an essential practice to be applied is the need for burndown several weeks preceding rice planting. This is because soil needs to be warm as rice seeds require a minimum temperature of 11°C to start germination [9], but the emergence is not quick and effective at temperatures below 18–20°C [10]. Thus, due to the cool climatic conditions of Rio Grande do Sul during spring, when too much mulching rests on soil, there is the risk of not reaching proper crop stand.

Research data held in Typic Albaqualf [11] cultivated with rice cv. BRS-Pampa for five consecutive seasons, with and without rotation with soybeans, showed that the rice-soybean rotation increases rice productivity (8671 kg/ha) compared to the monocrop (7464 kg/ha). This effect was independent of the presence of ryegrass as ground cover in winter.

3. Soils of Rio Grande do Sul, Brazil

The cultivation of paddy rice in Rio Grande do Sul is done in lowlands, comprising the floodplain soils and soils located at higher levels. Lowland soils are found in river, lakes, and lagoons, presenting as common characteristic its formation in various conditions of drainage deficiency (hydromorphism). In the state of Rio Grande do Sul, they cover large areas with relief ranging from flat to gently undulated, being found in the South Coast, Inner Coastal

Plain, Outer Coastal Plain, Central Depression, Campaign, and West Border regions. They occur usually at low altitudes (0–200 m) and cover an area of about 4,395,000 ha.

Floodplain soils have developed from fluvial, lagoons, and marine sediments from coastal plains and alluvial sediments derived from sedimentary, igneous, and metamorphic rocks of the depressions, plateaus, and mountains of Rio Grande do Sul; thus, source materials are very distinct.

Poor drainage or natural hydromorphism is usually motivated by the predominantly flat terrain, often associated with a profile with shallow surface layer and more impermeable subsurface layer. This characteristic is identified in maximum intensity by the gley soil feature—greyish or blue-grey colors, and less accentuated intensity of red/orange mottling dispersed in a gray background. In the landscape, this trait is less present in higher level soils, and may even be absent in the case of sandy soils.

Paddy rice cultivation in Rio Grande do Sul is also developed in higher levels floodplains or lowlands, which are adjacent to the floodplains, with undulated to plain relief. Such areas are preferred for sprinkler irrigation, facilitating management operations, particularly planting, harvesting, and irrigation, and eliminating the use of levees. These soils are found in the West Border and Campaign regions, developing from basalts and its sediments, or from silt or clay sedimentary rocks (siltstones, shales, and mudstones), respectively.

4. Fertilizer and liming management

In flood conditions, soils undergo profound chemical transformations resulting from the reduction process caused by anaerobic microorganisms, which use the oxygen of oxidized substances for their metabolism [12]. The changes resulting from flooding increase the availability of soil nutrients, both native and supplied through fertilizers, especially phosphorus (P) and potassium (K). Raising of the pH of acidic soils to between 6.0 and 6.5 is also reported, with subsequent neutralization of toxic aluminum [13].

The changes resulting from soil submersion have direct influence on the response of rice to soil liming, P and K fertilizers, which is smaller than that is observed in cultivations on aerated conditions [14], including sprinkler irrigated rice. This fact, along with low to moderate fertility of Rio Grande do Sul soils [15–17], make the adequacy of fertilization and liming essential to meet nutritional demand of sprinkler irrigated rice, enabling it to achieve yields consistent with the potential of the rice cultivars, which were developed for the flooded system, making the sprinkler irrigated rice economically viable.

The main aspects related to the management of soil fertility, fertilizer recommendations, and liming for sprinkler irrigated rice in lowlands in Rio Grande do Sul are discussed below. We consider information contained in the *Fertilization and Liming Manual for the States of Rio Grande do Sul and Santa Catarina* [18] and also the *Technical Recommendations of the Research for Southern Brazil* [10], as well as the results of research on nutrition and fertilization of rice produced in sprinkler irrigated rice system [19–21].

4.1. Fertilizer and liming management in sprinkler irrigated rice

The management of fertilization and liming for sprinkler irrigated rice should be based on the diagnosis of soil fertility and the nutritional requirements of rice into production systems involving rotations and crop sequences. The adequacy of these practices is critical to the performance of rice crop, as well as other species, which take part into the production system.

Recommendations for rice Based on soil analysis are basic instrument to determine the need of using liming and fertilizers. In the case of cultivation on drained soils, or on saturated soils with no water layer, rice should be treated as a component of a "dry system." Thus, soil samples for fertility evaluation should be performed at least every two crops.

The success of the recommendations depends on the adequacy of the collection and analysis of soil samples and the interpretation of analytical results and other production factors involved, in particular climatic conditions, rice cultivar, planting time and density, water, and integrated pest management.

Nutritional requirement varies among rice cultivars, particularly with its productivity potential and genetic background, which depends on the adequacy of production factors. Thus, crops with higher yield potential and expected response to fertilization require greater supply of nutrients compared to those less productive, regardless of the irrigation method, be it by flooding or sprinkling.

4.1.1. Liming

In general, lowland soils of Rio Grande do Sul are acidic, prevailing pH in water between 4.5 and 5.4, which correspond to the interpretation as "very low" ($4.5 < pH < 4.9$) and "low" ($5.0 < pH < 5.4$). In this pH range, the availability of many essential nutrients to crops is low, except for some micronutrient (copper, iron, manganese, and zinc). The acidity, in many cases, leads to high saturation by aluminum ($m > 20\%$), affecting root development and, consequently, absorption of water and nutrients by plants.

As rice and other components of lowland crop production systems grow best in soils with pH near neutral ($pH > 6.0$), special attention should be given to liming, making it a major issue in sprinkler irrigation. The water layer in the flooded paddies automatically increase the soil pH to between 6.0 and 6.5 with subsequent neutralization of toxic aluminum due to reduction processes, which does not occur under sprinkler irrigation. Liming reduces or eliminates also the toxic effects of manganese. Other benefits associated with this practice are improved root environment for absorption of nutrients favoring microbial activity and increasing the availability of nutrients and the supply of calcium (Ca) and magnesium (Mg).

The amount of limestone to be used varies with the pH to be reached and soil characteristics, in particular aluminum, clay, and organic matter content, which are the main sources of acidity and pH buffering. Larger liming will be required in soils where these attributes have higher values. In practice, in Brazil, the need for liming is estimated by the SMP index, supplied by soil analysis. The indication of general liming for rice as well as for the major component species of sprinkler irrigated production systems, particularly soybeans, corn, and cover crops, aims at increasing soil water pH to 6.0.

Liming has an average persistence of 3–5 years, depending on the amount and type of corrective used, the intensity of cultivation, soil and crop management, etc. In the field, the efficiency of liming depends on the amount and type of corrective, on the homogeneity of the mixture, soil moisture, and period allowed for the reactions after application. Soil pH reaches its maximum value between 3 and 12 months after liming. Thus, this operation should be accomplished at least 3 months prior to planting rice or other crops, which will comprise the production system.

4.1.2. Fertilization

Fertilization indications for rice aim a rational use of inputs in order to increase and maintain soil nutrient content and to optimize the economic return for each crop. They assume, also, that the use of the fertilizer will be accomplished under adequate correct soil acidity levels and the application of proper soil and crop management practices. Furthermore, indications are related to different crop response expectations to fertilizer. This is because the production factors (genetic potential of cultivars, soil and climate conditions, as well as management practices) determine different crop yield potential and therefore response to fertilizers.

Current fertilizer indications for rice consider two levels of expected yield levels: "average" and "high," and can be extrapolated to the respective response expectations to fertilization: "low" and "very high," adjusting the doses recommended for less or more, respectively. Thus, fertilization indications for rice are flexible and adaptable to the diversity of cultivars, environmental conditions, and crop management, as well as the availability of farmer's financial resources.

4.2. Soil analysis interpretation for fertilizers

Fertilizer recommendations for rice are based on soil analysis, using the contents of organic matter, phosphorus and potassium extracted by the Mehlich-I method, to estimate the availability of nitrogen, phosphorus, and potassium, respectively. For phosphorus and potassium, interpretation classes are set (**Tables 1** and **2**). Although the interpretation of phosphorus content in soil has been established for the flood irrigation, this can also be used for sprinkler system, with no changes being demanded.

Interpretation of potassium fertilizer for rice considers the cation exchange capacity (CTC) of the soil.

P content in soil	P extracted (mg/dm^3)[1]
Low	≤3
Mean	3.1–6.0
High	6.1–12.0
Very high	>12

[1] Extractor Mehlich-I.
Source: Adapted from Ref. [10].

Table 1. Interpretation of phosphorus (P) content in soil for phosphorus recommendation in rice (flood and sprinkler irrigation).

K content in soil		$CTC_{pH\,7.0}$ ($cmol_c/dm^3$)	
	<5	5–15	>15
		K extracted[1] (mg/dm^3)	
Low	≤30	≤40	≤60
Mean	31–45	41–60	61–90
High	46–90	61–120	91–180
Very high	>90	>120	>180

[1]Extractor Mehlich-I.
Source: Adapted from Ref. [10].

Table 2. Interpretation of potassium (K) content in soil for potassium recommendation in rice (flood and sprinkler irrigation).

4.2.1. Indications for phosphorus and potassium fertilization

As mentioned, it is necessary to increase phosphorus and potassium fertilization indications for sprinkler irrigated rice, compared to the doses recommended for the flooded system.

Accordingly, an approach that has provided good results, especially when the system is first installed, consists of adding 10 kg/ha of P_2O_5 and 15 kg/ha of K_2O for every expected additional ton of grains compared to that obtained with the recommendation established for the flooded system, considering a high expectation to fertilization response (**Tables 3** and **4**). As an example, considering a sprinkler irrigated area that used fertilizer recommendation established for the flooded rice and obtained an average yield of 7000 kg/ha, it is indicated to be added to the recommendation of the flooded system, 10 kg/ha of P_2O_5 and 15 kg/ha of K_2O for every expected additional ton of grain. It is limited, however, to a maximum of 90 kg/ha of P_2O_5 and 120 kg/ha of K_2O per cropping cycle.

It is recommended to supply phosphorus and potassium during rice planting. However, when the recommended dose of potassium is higher than 80 kg/ha of K_2O, its partitioning is

P content in soil[1]	Response expectation to fertilization	
	Mean	**High[2]**
	kg/ha of P_2O_5	
Low	50	60
Mean	40	50
High	30	40
Very high	≤30	≤40

[1]Mehlich-I method.
[2]The doses of P_2O_5 recommended for the high response expectation should be added with 10 kg/ha of P_2O_5, per each ton of additional expected grain yield, compared to the obtained with the recommendation for the flooded rice.
Source: Adapted from Ref. [10].

Table 3. Recommendation for phosphorus (P) fertilization for irrigated rice, considering the expected response to fertilization.

K content in soil[1]	Response expectation to fertilization	
	Mean	High[2]
	kg/ha of K_2O	
Low	75	90
Mean	55	70
High	35	50
Very high	≤35	≤50

[1]Mehlich-I method.
[2]The doses of K_2O recommended for the high response expectation should be added with 15 kg/ha of K_2O, per each ton of additional expected grain yield, compared to the obtained with the recommendation for the flooded rice.
Source: Adapted from Ref. [10].

Table 4. Recommendation for potassium (K) fertilization for irrigated rice, considering the expected response to fertilization.

especially necessary for sandy soils where the risk of nutrient losses is greater. In this case, it is indicated that half the recommended dose of potassium be applied at planting and the rest in topdressing associated to the nitrogen at the beginning of the reproductive stage.

4.2.2. Indications for nitrogen fertilization

The level of nitrogen fertilization, which leads to maximum economic yield of rice grains, depends on the interaction of several factors, especially the availability of N in the soil, plant type, and climatic conditions, particularly the temperature and solar radiation.

Also, nitrogen management for sprinkler irrigated rice requires changes compared to flooding. The sprinkler irrigation method is most prone to redox conditions, favoring nitrogen losses from soil-plant system. On the other hand, it allows flexibility in the topdressing fertilization, which can be accomplished directly to soil or through the irrigation water. When opting for fertilizing the soil, those in which the organic matter content is low (<2.5%), demand of N doses range from 120 to 150 kg/ha for crops with response expected equal to mean and high fertilization levels, respectively. This depends on the previous crop, climate, and crop management level [10]. N rates should be reduced by 10 and 20 kg/ha of N, for soils with mean (2.6–5.0%) and high (>5.0%) levels of organic matter.

Regarding timing and partitioning of nitrogen fertilizer, it is indicated to apply about 20 kg/ha of N at planting and the rest as topdressing. This can be done via soil, split into two applications: 50–60% of the topdressing dose at the start of tillering (*V4* stage) and the rest at the beginning of the reproductive stage, corresponding to panicle initiation (*R0*). In rainy years, the topdressing can be split into three applications, being the first at *V4* (about 30% of the dose); the second, *after 10 to 15 days*, when the plants reached the stage of six to eight leaves (about 20–30%), and the rest at panicle initiation (*R0*).

For sprinkler irrigated rice, nitrogen topdressing by fertigation is being studied. Recent data identified the following indication as with potential to maximize the grain yield of sprinkler irrigated rice: application of 25% of the recommended topdressing N dose to soil at the beginning of tillering, corresponding to the four-leaf stage (*V4*); the remainder of the dose may be partitioned into four to six weekly applications through irrigation water, according to the cycle of the cultivar, being the largest number of applications suitable for longer cycle cultivars.

5. Irrigation management

In the Brazilian state of Rio Grande do Sul, rice is grown in paddies under continuous flooding. However, this irrigation method provides, in many situations, water consumption exceeding 1500 mm per crop cycle, in areas with undulated relief. At the same locations, farmers who adopted the sprinkler irrigation (through central pivot) are using between 400 and 700 mm, depending on climatic conditions throughout the cropping season.

Research data about rice under sprinkler irrigation, by using rice cultivars developed for continuous flooding, report good crop performance when water is properly managed. The results indicated also that the susceptibility of the crop to drought varies with the phenological stage.

5.1. Susceptibility of rice crop to water deficit and water demand

In order to know the effect of soil water deficit on rice, experiments were conducted by Embrapa Clima Temperado with the cultivar BRS Pampa (early cycle) in the 2011/2012 and 2012/2013 cropping seasons [2].

To analyze the results of the effect of soil water deficit on rice productivity, the water-yield model proposed by Jensen [22] was adopted. For application of the model, the crop cycle was divided into two periods: L1—vegetative stage (from emergence to panicle initiation) and L2—reproductive stage (panicle initiation to maturation). When using the model, the variable "evapotranspiration" was replaced by the "water tension" in soil. The adaptation of the Jensen's model is shown as follows:

$$\frac{Y}{Ym}=\prod_{i=1}^{n}\left(\frac{ET}{ETm}\right)^{\lambda}\rightarrow\frac{Y}{Ym}=\prod_{i=1}^{n}\left(\frac{Tmín}{Tobs}\right)^{\lambda}$$

where:

Y = reported grain yield (kg/ha);

Ym = maximum grain yield reported in absence of water deficit (kg/ha);

n = number of stages in the phenological cycle;

λ = susceptibility of rice to drought per phenological stage in the relative productivity;

ET = true evapotranspiration (mm);

ETm = maximal evapotranspiration (mm);

$Tmín$ = minimal water tension in soil (kPa); and

$Tobs$ = reported water tension in soil (kPa).

Crop stage	Parameter	Estimation	Pr > ǀtǀ
2011/2012 Cropping season			
Vegetative	$\lambda1$	0.15	0.0001
Reproductive	$\lambda2$	0.25	<0.0001
2012/2013 Cropping season			
Vegetative	$\lambda1$	0.16	0.0095
Reproductive	$\lambda2$	0.29	<0.0001

Table 5. Estimation of the parameters for the proposed model of Jensen [22], for each phenological stage of rice under sprinkler irrigation.

Table 5 shows the values of the parameter λ in the model for both vegetative and reproductive stages of rice. For both cropping seasons, the parameters were significant at 0.1% in both stages. As higher values of λ indicate greater susceptibility to drought, it can be inferred that the vegetative and reproductive stages have different susceptibility levels, being the reproductive stage the most sensitive period.

Figure 1 is a graphical representation of the model fitting proposed by Jensen to the two monitored agricultural crops. The coefficients of determination (R^2) of 0.70 and 0.54 indicate

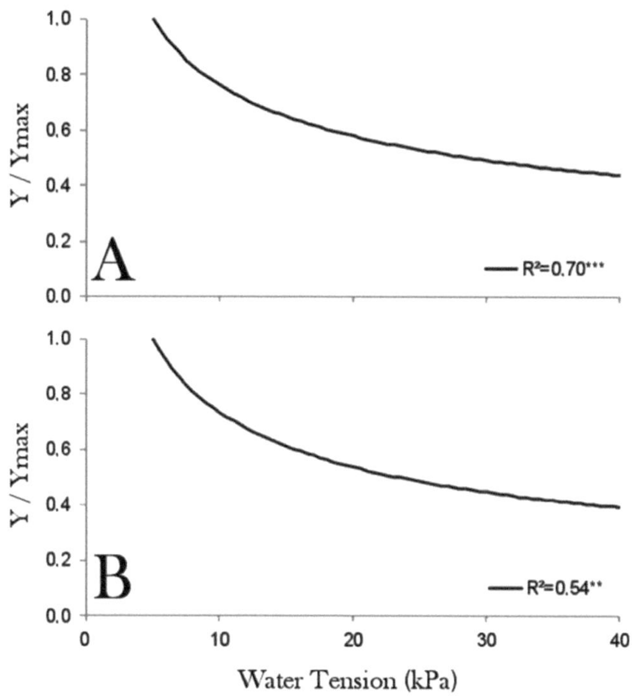

Figure 1. Relation between rice grain yield when grown under sprinkler irrigation and the mean water tension in soil, adjusted by the model of Jensen [22] for the 2011/2012 (A) and 2012/2013 (B) cropping seasons. ***Significant model at 0.1%; **significant model at 1%.

that, among the various agronomic factors affecting rice grain yield, such as soil fertility, pests incidence, seed and planting operation quality, among others, water availability influenced by 70 and 54% rice grain yield in the first and second cropping seasons, respectively. These data illustrate that rice is highly susceptible to water stress, thus proper water management is essential for grain yield.

Figure 1 also shows that the closer to saturation the soil (soil water tension equal to zero), the higher the rice yield, with remarkable decrease in grain yield in water tensions beyond 15 kPa. However, from this tension, the decrease in rice productivity due to increases in soil water tension is smaller, stabilizing in about 50% of the maximum productivity. Thus, when using sprinkler irrigation in rice, the soil water tension should be kept as close as possible to saturation, since there are no significant water losses by runoff during water application.

Table 6 shows the total irrigation applied to rice both in the vegetative and reproductive stages, under different water managements. It can be seen that the higher the water tension in soil, that is, the drier the soil, the lower the applied water and the lower the rice grain yield.

In the 2011/2012 and 2012/2013 cropping seasons, the total irrigation and rainfall in the managements of 20 kPa were 676 and 716 mm, respectively. Although the values of the applied amounts are similar, there is variation in the total irrigation between the distinct stages of the cycle (**Table 6**). In 2012/2013, water demand was lower in the vegetative and higher in the reproductive stages, compared to the previous season. This is due to the distribution of rainfall in both periods; in 2012/2013, there was higher rainfall during vegetative stage (170 mm) and less precipitation in the reproductive stage (231 mm), compared to 2011/2012, when rainfall was 102 and 283 mm in the vegetative and reproductive stages, respectively.

For the management of 20 kPa, the total amount of water (irrigation + rainfall) in 2011/2012 was 210 and 466 mm in vegetative and reproductive stages, respectively, while in the 2012/2013, the

Cropping season	Management	Applied water (mm)		
		Vegetative stage	Reproductive stage	Total
2011/2012	20 kPa	108	183	291
	40/20* kPa	81	192	273
	40 kPa	72	159	231
2012/2013	10 kPa	138	396	534
	20 kPa	63	252	315
	40 kPa	30	156	186
	40/10* kPa	30	369	399

*The first and the second numbers regard to the soil water tension at the vegetative and the reproductive stages, respectively.

Table 6. Total amounts of water applied in each phenological stage of sprinkler irrigated rice, for distinct water managements, which were established based on the water tension in soil, for the 2011/2012 and 2012/2013 cropping seasons.

http://dx.doi.org/10.5772/66024

total amounts were 233 and 483 mm at the same stages. The total rainfall accounted for almost 50% of water supply to crop. In the management of 10 kPa, in 2012/2013, the applied water depth was greater. The results indicate that systems and/or irrigation managements that allow the maximum use of rainfall constitute an important alternative to reduce water demand by rice crop.

5.2. Irrigation timing

Programming the time to irrigate is a matter of fundamental importance in the management of irrigation of any crop. This aspect assumes, however, more relevance to sprinkler irrigated rice due to its susceptibility to drought. Irrigation control can be performed by monitoring climatic or soil-related variables.

For control via climate, there is need to estimate the potential evapotranspiration (ETo) and know its relationship to the actual crop evapotranspiration (ETr), which is called crop coefficient (Kc). The Kc varies depending on soil water content, being known for flood irrigated rice, or to saturated soil conditions. This parameter has not been established for sprinkler irrigated rice, in which cultivation soil is not saturated. In this production system, Kc will vary also with the water management adopted. For this reason, it is still not possible to control irrigation via climate properly, leading one to plan sprinkler irrigation for rice based on soil-related parameters.

For water management based on soil variables, sensors are used to monitor the water tension, which is directly related to soil moisture, so that the drier the soil, the higher the measure. Knowing the water tension in soil which is ideal for proper rice development, irrigation is done in order to keep it below that value. Based on the research results discussed above, it is recommended to keep soil water tension up to 10 kPa for security reasons (time needed from measure to the irrigation to start). At the vegetative stage, if there is need to save water, irrigation can be applied aiming to avoid water tensions in soil above 30 kPa in any moment.

Currently, there are several sensor types for monitoring soil moisture, being the TDR and FDR sensor types most used for research, but tensiometers and indirect reading devices by means of electrical resistance are also vastly used. Soil sensors based on electric resistance are most accurate in higher water tensions and may present some limitations for lower water tensions.

The sensors must be installed in a representative depth of the effective plant root zone, where about 80% of the root system is located. For rice grown in lowlands of Rio Grande do Sul, the first research results indicate that the sensors can be installed up to 10 cm depth. If soil moisture is read at greater depths, there is a risk of irrigating when high humidity is still on soil surface, reflecting in increased water losses by runoff. An example would be after a rain, when infiltration rates are smaller because the rate of infiltration depends on soil moisture content.

To choose the proper installation locations of the sensors into the field, it is important to identify homogeneous areas, considering at least topography and soil type, but also fertility and other traits. In the case of undulated areas, it is important that some sensors are installed in the upper, middle, and bottom parts. In the case of rice, which is very susceptible to drought, it is better that irrigation is done according to soil moisture read in the upper areas, as these tend to be drier. Sensor should be read on a daily basis.

In the same way, as the time to start irrigation, the time to stop the process is also essential. An indication to prevent losses in milling yield in sprinkler irrigated rice is to maintain irrigation throughout the growing season, suspending it only one day before harvest.

Acknowledgements

The authors would like to thank Valmont Industries—Irrigation Division, Omaha, NE, USA, and Valmont Indústria e Comércio, Uberaba-MG, Brazil, for the research cooperation with Embrapa Clima Temperado, Pelotas-RS, Brazil, which made possible to install part of the experiments whose results are included into the present work.

Author details

José Maria Barbat Parfitt[1], Germani Concenço[1*], Walkyria Bueno Scivittaro[1], André Andres[1], Jaqueline Trombetta da Silva[2] and Marília Alves Brito Pinto[2]

*Address all correspondence to: gconcenco@yahoo.com.br

1 Embrapa—Temperate Agriculture Research Center, Pelotas, Brazil

2 Federal University of Pelotas, Pelotas, Brazil

References

[1] Ministry of Agriculture, Livestock and Supply – Mapa. (2013). Projections of the Brazilian Agribusiness. Brasil 2012/2013 a 2022-2023 – Long-Term projections. Brasília, DF. Available at: http://www.agricultura.gov.br/arq_editor/projecoes%20-20versao%20 atualizada.pdf. Accessed: 13 Nov. 2014.

[2] Pinto, M.A.B. (2015). Sprinkler irrigation in rice as function of soil water tension. 68 p. Thesis (Doctorate in Agronomy), Federasl University of Pelotas, Pelotas.

[3] Mota, F.S., Alves, E.G.P. & Becker, C.T. (1990). Climatic data for planning water demand and irrigation in rice in Rio Grande do Sul. Lavoura Arrozeira, Porto Alegre, Vol. 43, No. 392, pp. 3–6.

[4] Fietz, C.R., Cauduro, F.A. & Beltrame, L.S. (1986). Model for estimation of water demand in irrigated rice (Oryza sativa L) fields. In: CONIRD, 7. Brasília: ABID. Vol. 1, pp. 155–167.

[5] Nunes, C.D.M., Ribeiro, A.S. & Terres, A.L. (2004). Principais doenças em arroz irrigado e seu controle. In: Gones, A.S. & Magalhães Jr., A.M. *Arroz irrigado no sul do Brasil. Brasília,* DF: Embrapa Informação Tecnológica. pp. 579–621.

[6] Engelman, R., Cincotta, R.P., Dye, B., Gardner-Outlaw, T. & Wisnewski, J. (2000). *People in the Balance: Population and Natural Resources at the Turn of the Millennium*. Washington, DC: Population Action International, 1. CD-ROM.

[7] Christofoletti, J.C. (2007). Food Production. Brasília: CTNBio. Available at: http://www.ctnbio.gov.br/upd_blob/0001/1058.pdf. Accessed: 14 Mar. 2010.

[8] Piolli, A.L. (2009). Public participation and new skills: a case study of the technical council of hidrological basins of Piracicaba, Capivari and Jundiai rivers. 110 p. Dissertation (Master in Scientific and Tecnological Policy), Geosciences Institute/Federal University of Campinas, Campinas.

[9] Floss, E.L. (2004). Physiology of crop plants: studying beyond what we see. Passo Fundo: UPF. 528 p.

[10] Southern Brazilian Rice Society (SOSBAI) (2014). Irrigated Rice: Technical Recommendations of research for Southern Brazil. Itajaí: SOSBAI. 179 p.

[11] Natural Resources Conservation Service (NRCS), United States Department of Agriculture (USDA). (2009). *Soil Taxonomy*. Available at: http://www.nrcs.usda.gov/ Accessed 20 Jun. 2016.

[12] Sousa, R.O., Camargo, F.A.O. & Vahl, L.C. (2006). Flooded areas (Redox Reactions). In: Meurer, E.J. (Ed.). Fundamentals of soil chemistry. 3rd ed. Porto Alegre: Evangraf. pp. 185–211.

[13] Scivittaro, W.B. & Machado, M.O. (2004). Feetilization and liming for rice crop. In: Gomes, A.S. & Magalhães-Junior, A.M. (Eds.). Irrigated Rice in Southern Brazil. 1st ed. Brasília: Embrapa Technological Information. pp. 259–303.

[14] Scivittaro, W.B. & Gomes, A.S. (2004). Management of fertilizer and liming for irrigated rice. In: Gomes, A.S., Magalhães-Júnior, A.M. & Santos, A.B. (Eds.). Irrigated Rice Cropping System in Southern Brazil. 1st ed. Pelotas: Embrapa Temperate Agriculture. pp. 73–87.

[15] Anghinoni, I., Genro-Junior, S.A., Silva, L.S., Bohnen, H., Rheinheimer, D.S., Osório-Filho, B.D. & Macedo, V.R.M. (2004). Fertility of soils grown with rice in Rio Grande do Sul. Cachoeirinha: IRGA. Research Department, 52 p. (Technical Bulletin, 1).

[16] Boeni, M., Anghinoni, I., Genro-Junior, S.A. & Osório-Filho, B.D. (2007). Evolution of soil fertility under rice cultivation in Rio Grande do Sul. Cachoeirinha: Irga. 52 p.

[17] Vedelago, A., Carmona, F.C., Boeni, M., Lange, C.E. & Anghinoni, I. (2012). Fertility and best uses of soils for soybean cultivation in rice regions of Rio Grande do Sul. Cachoeirinha: IRGA. 48 p. (Technical Bulletin, 12).

[18] Brazilian Society of Soil Science. Commission on soil chemistry and fertility. (2004). Fertilizer and liming handbook for Rio Grande do Sul and Santa Catarina. 10th ed. Porto Alegre: SBCS-CQFS. 400 p.

[19] Scivittaro, W.B., Parfitt, J.M.B., Silva, P.S. & Silveira, A.D. (2012). Management of nitrogen fertilizer for sprinkler irrigated rice. Pelotas: Embrapa Temperate Agriculture. 27 p. (Embrapa Temperate Agriculture. Research Bulletin, 177).

[20] Scivittaro, W.B., Parfitt, J.M.B., Klumb, E.K., Silva, P.S. & Mattos, G.S. (2013a). Productive performance and nitrogen accumulation in flood- and sprinkler-irrigated rice. Pelotas: Embrapa Temperate Agriculture. 24 p. (Embrapa Temperate Agriculture. Research Bulletin, 182).

[21] Scivittaro, W.B., Parfitt, J.M.B., Klumb, E.K. & Silva, P. S. (2013b). Nutrient absorption by sprinkler irrigated rice. Pelotas: Embrapa Temperate Agriculture. 29 p. (Embrapa Temperate Agriculture. Research Bulletin, 176).

[22] Jensen, M.E. (1968). Water consumption by agricultural plants. In: Kozlowsky, T.T. (Ed.). *Water Deficits and Plant Growth*. Vol. 2. New York: Academic Press. pp. 1–22.

Chapter 2

Weed Management in Sprinkler-Irrigated Rice: Experiences from Southern Brazil

André Andres, Giovani Theisen, Gustavo Mack Teló, Germani Concenço, José Maria Barbat Parfitt, Leandro Galon and Matheus Bastos Martins

Additional information is available at the end of the chapter

http://dx.doi.org/10.5772/67146

Abstract

Sprinkler rice saves water compared to paddy rice. However, in paddy fields, the water table is efficient for weed suppression. In sprinkler rice, there is no water table on soil; thus, weed management used in paddy rice may not be suitable for sprinkler rice, since herbicides and water table are expected to interact. Weed pressure in sprinkler rice is higher than in paddy rice; annual grasses are the main weeds in both paddy and sprinkler rice. Barnyardgrass, goosegrass, crabgrass and Alexandergrass show vigorous growth in sprinkler rice. A 3-year study shows that weeds in sprinkler rice reduce grain yield between 11 and 95%. Herbicides used in conventional and Clearfield® rice (clomazone, imazethapyr + imazapic, imazapyr + imazapic, pendimethalin and penoxsulam) were tested, contrasting paddy and sprinkler rice. Additionally, the technique locally called "needle-point" (glyphosate applied over the first-day emerging rice) was combined with pre- and postemergence herbicides. When using only pre- or postemergence, weeds reduced rice grain yield; a combination of products was the best option for sprinkler-irrigated rice. The Clearfield technology was efficient in controlling most weeds. However, using it combined to the needle-point promoted the best results. The main approaches for weed management in sprinkler-irrigated rice were summarized.

Keywords: weed control, herbicides, management strategies, needle point

1. Introduction

Paddy rice is one of the most water demanding cropping systems in agriculture. Changing from surface to sprinkler irrigation can contribute to optimize water use in rice production. Under sprinkler irrigation, rice grain yield has reached similar levels as obtained in the traditional flooded system [1]. Sprinkler irrigation currently represents one of the best alternatives to improve water use efficiency in rice production.

However, weeds represent one of the main difficulties in sprinkler-irrigated rice. In the traditional surface-irrigation system, a layer of water (5–30 cm deep) remains permanently on soil during all rice cycle; the layer of water reduces free O_2 in soil, thereby suppressing the germination of most weeds. For the sprinkler-irrigated rice, the soil is maintained humid but not flooded, and oxygen levels are suitable for weed seed germination. In a practical sense, sprinkler irrigation facilitates weed establishment. This characteristic implies that weed management strategies that are successful in paddy rice usually not succeed under sprinkler irrigation [2].

In the roll of the main weed species in lowlands, annual grasses represent the most important group affecting rice, either in paddy as in sprinkler-irrigated [3]. The weeds barnyardgrass (the *Echinochloa* complex), goosegrass (*Eleusine indica*), southern crabgrass (*Digitaria ciliaris*) and Alexandergrass (*Urochloa plantaginea*) present vigorous growing under adequate soil humidity and high temperatures, and they reduce rice grain yield if not controlled efficiently. In **Table 1**, the most important weeds in rice fields of south Brazilian lowlands are listed. These weeds, besides red rice (the weedy *Oryza sativa*), are the most important species occurring in paddy rice fields of all sub-tropical South America [4].

Family/common name	Scientific name	Life cycle	Reproduction	Inf. level[2]
Poaceae				
Weedy rice (red rice)	*Oryza sativa*	Annual	Seeds	H
Barnyardgrass	*Echinochloa*[3]	Annual	Seeds	H
Goosegrass	*Eleusine indica*	Annual	Seeds	H
Alexandergrass	*Urochloa plantaginea*	Annual	Seeds	H
Crabgrass	*Digitaria ciliaris*	Annual	Seeds	H
	Digitaria sanguinalis			
German grass	*Echinochloa polystachya*	Perennial	Seeds, rhizomes	L
Cupgrass	*Eriochloa punctata*	Annual, perennial	Seeds	L
Marsh grass	*Hymenachne amplexicaulis*	Perennial	Seeds, stolons, rhizomes	L
Saramollagrass	*Ischaemum rugosum*	Annual	Seeds	L/I
Fall panicgrass	*Panicum dichotomiflorum*	Annual, perennial	Seeds, stolons	L
Brook crowngrass	*Paspalum acuminatum*	Perennial	Seeds, stolons	L
Knotgrass	*Paspalum distichum*	Perennial	Seeds, stolons, rhizomes	L
Water paspalum	*Paspalum modestum*	Perennial	Seeds, stolons	L
Mexican sprangletop	*Leptochloa uninervia*	Annual	Seeds	L
Southern cutgrass	*Leersia hexandra*	Perennial	Seeds, stolons	L
Peruvian watergrass	*Luziola peruviana*			

http://dx.doi.org/10.5772/67146

Family/common name	Scientific name	Life cycle	Reproduction	Inf. level[2]
Cyperaceae				
Sedges	*Cyperus*[4]	Annual	Seeds	I/H
–	*Cyperus laetus*	Annual	Seeds, rhizomes	I
Yellow nutsedge	*Cyperus esculentus*	Perennial	Seeds, tubers	I
Fringerush	*Fimbristylis miliacea*	Annual	Seeds	I/H
Pontederiaceae				
Kidneyleaf mudplantain	*Heteranthera reniformis*	Perennial	Seeds, stolons	L
Alimastaceae				
Arrowhead	*Sagittaria guyanensis*	Perennial	Seeds, rhizomes, tubers	L
Giant arrowhead	*Sagittaria montevidensis*			
Fabaceae				
Jointvetches	*Aeschynomene denticulata*	Annual	Seeds	I
	Aeschynomene. indica			
	Aeschynomene sensitiva			
Amaranthaceae				
Alligator weed	*Alternanthera philoxeroides*	Perennial	Seeds, vegetative parts	I
Convolvulaceae				
Morning glory	*Ipomoea grandifolia*	Annual	Seeds	L
Onagraceae				
Waterprimrose	*Ludwigia elegans*	Annual, perennial	Seeds	L
	Ludwigia longifolia			
	Ludwigia octovalvis	Perennial		
Polygonaceae				
Smartweed	*Polygonum hydropiperoides*	Annual	Seeds	L/I

[1]Adapted from Ref. [5].

[2]Infestation level varies according to region, cropping system, crop rotation and herbicides (L = low; I = intermediary; H = high).

[3]Several barnyardgrass species are found, as *Echinochloa colona, Echinochloa crus-galli, Echinochloa mitis* and *Echinochloa helodes.*

[4]Several sedges species are found, as *C. difformis, Cyperus ferax, Cyperus iria* and *Cyperus brevifolius.*

Table 1. Main weeds found in lowlands cultivated under sprinkler-irrigated rice in southern Brazil [1].

The main paddy rice weeds in Brazil are commonly classified into narrow- and broad-leaved weeds. Main narrow leaves are weedy rice (*Oryza sativa*), barnyardgrass (*Echinochloa* sp.), Alexandergrass (*U. plantaginea*), crabgrass (*Digitaria horizontalis*), goosegrass (*E. indica*), the aquatic grasses (*Leersia hexandra* and *Luziola peruviana*) and the sedges (*Cyperus difformis, Cyperus esculentus, Cyperus ferax,* and *Cyperus laetus*). Some monocotyledonous weeds that are common in corn, sorghum and soybeans are expanding due both to the increase in crop diversification in lowland areas, and to the continued use of Acetolactate Synthase (ALS)-inhibitors herbicides with abandonment of propanil. Some locations also reported the presence of perennial weeds such as Olive hymenachne (*Hymenachne amplexicaulis*), ribbed murainagrass (*Ischaemum rugosum*), Mexican sprangletop (*Leptochloa uninervia*), Fall panicum (*Panicum dichotomiflorum*), Knotgrass (*Paspalum distichum*) and *Paspalum modestum*. As broad-leaved weed representatives, there are the jointvetches (*Aeschynomene* spp.) and in some areas some species of morning glory (*Ipomoea* spp.), water pepper (*Polygonum hydropiperoides*) and alligator weed (*Alternanthera philoxeroides*). The aquatic weeds, associated mainly with fields grown in the water-seeded system (with pre-germinated seeds) are globe fringerush (*Fimbristylis miliacea*), arrowheads (*Sagittaria montevidensis* and *Sagittaria guyanensis*), water hyacinth (*Eichhornia crassipes*), kidneyleaf mudplantain (*Heteranthera reniformis*) and the Ludwigia complex (*Ludwigia elegans, Ludwigia longifolia* and *Ludwigia octovalvis*).

2. Weed management in the traditional flooded-irrigated (paddy) rice in southern Brazil

Integrated weed management in paddy rice is characterized by the association of agronomic practices to minimize the negative effect of weeds [5]. Besides the layer of water on soil, which naturally reduces germination and establishment of various weed species, other measures such as the use of vigorous genotypes and an adequate rice plant density, provide a more competitive crop against weeds. Early soil preparation, which stimulates weed germination out of rice growing season is used, concomitantly to minimum tillage. Minimum tillage is an effective way to reduce the presence of some annual *Poaceae* species. The early seeding of rice, the application of nonselective herbicides using the needle-point technique, early weed control and early irrigation are a set of commonly used measures to reduce weed impact in flooded rice areas. Moreover, it is important to highlight that in the RS state (the larger rice producer in Brazil, supplying about 70% of Brazilian production), almost 75% of the rice is tolerant to imidazolinones (Clearfield® rice). This technology first started to be used by farmers in the cropping season 2003/2004.

3. Weed management in sprinkler-irrigated rice

Despite the several alternatives for weed management available for farmers [5], the most used method for weed control in flooded-irrigated rice is the association of chemical control (herbicides) with an early formation of water layer in the soil surface, provided by irrigation [6]. However, in sprinkler-irrigated rice there is not such water layer, and weed seed germination is, in fact, mostly stimulated by irrigation in sprinkler systems [2, 7]. The strategies for weed management in sprinkler-irrigated rice are, in this way, more complex than for the flooded system.

Studies conducted at EMBRAPA—Terras Baixas Experimental Station, in Pelotas, southern Brazil, show that in sprinkler-irrigated rice weeds can reduce rice grain yield in up to 95% [3], depending on the weed control provided by herbicides (**Table 2**). In such condition, rice yield

http://dx.doi.org/10.5772/67146

			Dry mass		Weed ctrl 70DAE		Grain yield	
			Npoint[2]	Norm	NPoint	Norm	NPoint	Norm
Herbicide	Doses (g ha^{-1})	Appl. time[1]	(g m^{-2})		(%)		(t ha^{-1})	
Clomazone[3]	400	PRE	174bc	182b	29d	31c	4.38bc	1.92cde
Clomazone[3]	700	PRE	69cd	112b	56bc	36bc	6.59ab	3.46bc
Pendimethalin	1250	PRE	306a	629a	3e	4e	3.55cd	0.67de
Pendimethalin	1750	PRE	322a	763a	6e	7de	0.98d	0.44e
Penoxsulam	24	POS	104bc	194b	65b	29cd	5.77b	3.85bcd
Penoxsulam	60	POS	182b	155b	41cd	24cd	4.23bc	2.32cde
Imazethapyr + imazapic	37.5 + 12.5	PRE	66cd	70b	68b	57b	6.14ab	4.61bc
Imazethapyr + imazapic	37.5 + 12.5	POS						
Imazethapyr + imazapic	56.25 + 18.75	PRE	2d	8c	95a	91a	9.58a	8.14ab
Imazethapyr + imazapic	56.25 + 18.75	POS						
Imazapyr + imazapic	73.5 + 24.5	POS	29d	35bc	95a	88a	9.70a	9.16a

Adapted from Ref. [3].

[1]PRE = preemergent; POS = postemergent.

[2]NPoint = treatment associated to the needle-point technique; Norm = treatment not associated to the needle-point technique.

[3]Seeds treated with dietholate (Permit®).

[4]Sequential applications. Means of columns followed by same letter are not significant different (Duncan test, $P \leq 0.05$).

Table 2. Herbicides, doses, application time, weed dry mass, weed control and rice yield in sprinkler-irrigated rice.

can be reduced to zero if weeds are not controlled. The main herbicides registered for weed control in rice in Brazil are listed in **Table 3**.

Herbicide[1]	Dose (a.i.) g ha^{-1}	Time/mode of application[2]
Bentazon	960	Post
Bispyribac-sodium	100–125 mL	Post
Clomazone	360–612	Pre
Cyhalofop-butyl	360–630	Post
2,4-D	240	Post
Fenoxaprop-P-ethyl	69	Post
Glyphosate	2.160	Pre (NPoint[3])
Imazapyr + imazapic	725 + 175/725 + 175	Pre/post
Metsulfuron-methyl	2	Post
Pendimethalin	1500	Pre
Penoxsulam	48–54	Pre/post
Propanil	2800	Post
Propanil + thiobencarb	1200 + 2400–600 + 3200	Post
Pyrazosulfuron-ethyl	15–20	Post
Quinclorac	375	Post

[1]Herbicides named by their technical names, not by their commercial names.
[2]Pre = preemergent; Post = postemergent.
[3]NPoint = treatment associated to the needle-point technique.

Table 3. Main herbicides registered for weed control in paddy rice in Brazil.

Recent studies with the most used herbicides (>90% of the Brazilian lowland rice area) clomazone, imazethapyr + imazapic, imazapyr + imazapic, pendimethalin, penoxsulam and glyphosate applied at the "needle point" [8] were evaluated under a range of weed species in sprinkler-irrigated and flooded-irrigated rice (**Figure 1**). The result of these experiments, conducted between 2011 and 2015, showed that using only conventional preemergent herbicides (clomazone, pendimethalin and penoxsulam) was not sufficient to fully control weeds, and consequently rice yield was affected (**Table 2**). However, associating the preemergent clomazone (700 g ha^{-1}), to the needle-point technique [8], was an effective way to reduce weeds, and this treatment resulted in high grain yield.

Penoxsulam, applied alone in preemergence, was efficient to control sedges like *Cyperus difformis* and *Cyperus iria*. However, for grasses, after two consecutive seasons the same rates of herbicides (penoxsulam (36 and 48 g ha^{-1}), clomazone (up to 360 g ha^{-1}) and pendimethalin

http://dx.doi.org/10.5772/67146

Figure 1. Seeds of rice at S_3 stage (commonly called the "needle-point"). Source: Refs. [7, 8].

(1250 and 1750 g ha^{-1}) were not able to fully control species like *E. indica*, *Digitaria* sp. and *U. plantaginea*. In these cases, there was the need for a complimentary postemergent application of the ACCase-inhibitor cyhalofop-butyl (360 g ha^{-1}).

When the option for weed control was based on the Clearfield® technology, the sequential application of imazethapyr + imazapic (56.25 + 18.75 g ha^{-1}) in preemergence ($^1/_2$ dose) and post emergence ($^1/_2$ dose), provided adequate results. Smaller doses of imazethapyr + imazapic, applied without other supplementary herbicide, were not effective and resulted in poor weed control and in a reduced rice production. However, when the reduced dose was associated to the needle-point technique, the weed biomass was reduced and rice grain yield was improved. The commercial mix of imazapyr + imazapic (73.5 + 24.5 g ha^{-1}), applied either in pre- or postemergence, was efficient to reduce weed biomass, which allowed rice to express a high grain yield (**Table 2**). In the fields using the Clearfield technology, however, also occurred some uncontrolled plants of *E. indica*, *Digitaria* sp. and *U. plantaginea*. In these cases, an additional application of the ACCase-inhibitor cyhalofop-butyl was needed.

Another option for chemical weed control evaluated was the split of clomazone application. To test this treatment, the first application of clomazone was at preemergence (360 g ha^{-1}), at the beginning of rice germination (the needle-point stage); the second application was in postemergence (360 g ha^{-1}). Clomazone was supplemented by the ACCase inhibitor cyhalofop-butyl, applied at early postemergence (17 days after rice emergence). This combination presented a high level of control for annual grasses. However, as neither clomazone nor cyhalofop control efficiently sedges, fields with sedge infestation can require different strategies for weed management.

In a general overview, we observed benefits for weed control when we associated preemergence herbicides to the needle-point technique in sprinkler-irrigated rice. Glyphosate, applied at very early rice emergence, is very effective against most of annual grasses occurring in rice, like *Echinochloa*, *Urochloa*, *Digitaria* and others. This association effectively reduced weed biomass, enabling to attain high levels of control even at a late rice stage, as 70 days after emergence (**Figure 2**).

From these experiments, we elaborated **Table 4** with some strategies for chemical weed control in sprinkler-irrigated rice. Surely, in fields with a low weed infestation it is possible to use less herbicides than are here presented. A scale of colors was used to highlight the efficiency

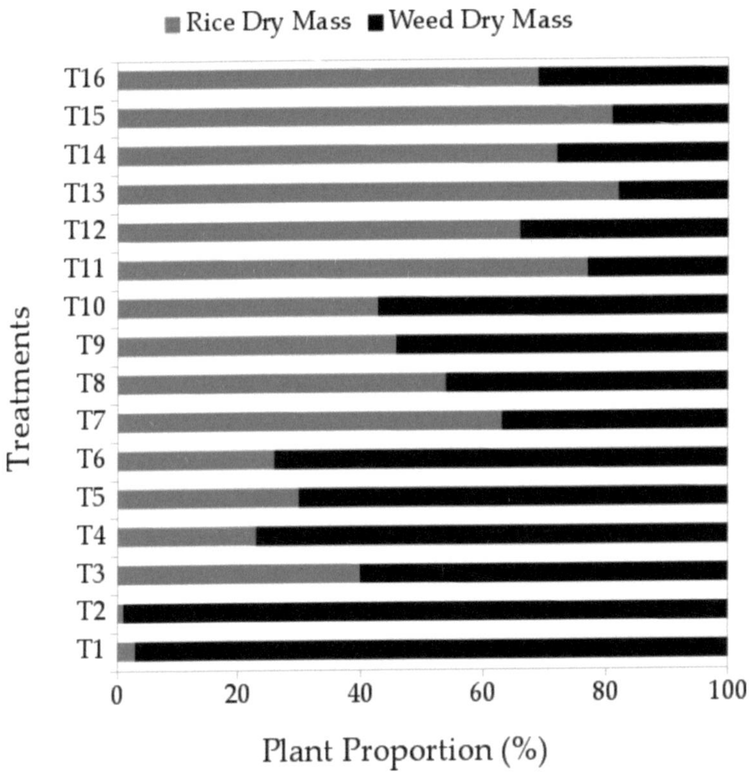

Figure 2. Dry mass of rice and of weeds as a function of chemical treatments for weed control in sprinkler-irrigated rice. (T1) glyphosate, applied at the needle-point of rice, (T2) no weed control, (T3) glyphosate*/clomazone 360 g a.i. ha^{-1} preemergence, (T4) clomazone 360 g a.i. ha^{-1} preemergence, (T5) glyphosate*/clomazone 450 g a.i. ha^{-1} preemergence, (T6) clomazone 450 g a.i. ha^{-1} preemergence, (T7) glyphosate*/penoxsulam 48 g a.i. ha^{-1} pre and postemergence, (T8) penoxsulam 48 g a.i. ha^{-1} pre- and postemergence, (T9) glyphosate*/penoxsulam 36 g a.i. ha^{-1} pre- and postemergence, (T10) penoxsulam 36 g a.i. ha^{-1} pre- and postemergence, (T11) glyphosate*/imazethapyr+imazapic (75 + 25 g a.i. ha^{-1}) preemergence, (T12) imazethapyr + imazapic (75 + 25 g a.i. ha^{-1}) preemergence, (T13) glyphosate*/imazethapyr + imazapic (75 + 25 g a.i. ha^{-1}) pre- and postemergence, (T14) imazethapyr + imazapic (37.5 + 12.5 g a.i. ha^{-1}) pre- and postemergence, (T15) glyphosate*/imazapyr + imazapic (73.5 + 3.5 g a.i. ha^{-1}) preemergence, (T16) imazapyr + imazapic (73.5 + 3.5 g a.i. ha^{-1}) preemergence. *Glyphosate at 720 g e.a. ha^{-1} applied at the needle-point of rice.

http://dx.doi.org/10.5772/67146

		Applic. time	Herbicide	Label dose (ge.a/i.a ha^{-1})	Main weeds and herbicide efficiency#							Estim. Cost (US$ ha^{-1})##
					Weedy rice	Barnyardgrass	Alexandergrass	Crabgrass	Goosegrass	Cyperus*	Broadleaves**	
1	Only for Clearfield® cultivars	Burndown	Glyphosate	1400								19.0
		PRE[1]	Imazethapyr + imazapic	56.25 + 18.75								20.4
		PAg[2]	Glyphosate	720								6.4
		POS[3]	Imazethapyr + imazapic	56.25 + 18.75								20.4
		POS[4]	Cyhalofop-butyl	360								59.8
2												=US$ 126
		Burndown	Glyphosate	1400								19.0
		PRE[1]	Imazapyr + imazapic	73.5 + 24.5								27.5
		PAg[2]	Glyphosate	720								6.4
		POS[3]	Cyhalofop-butyl	360								59.8
												=US$ 112.7

		Applic. time	Herbicide	Label dose (ge.a/i.a ha^{-1})	Main weeds and herbicide efficiency[#]							Estim. Cost (US$ ha^{-1})[##]
					Weedy rice	Barnyardgrass	Alexandergrass	Crabgrass	Goosegrass	Cyperus[*]	Broadleaves[**]	
3	For all rice cultivars	Burndown	Glyphosate	1400	green	green	green	green	green	green	green	19.0
		PRE[1]	Clomazone	360	red	green	green	green	green	red	red	21.8
		PAg[2]	Glyphosate	720	green	green	green	green	green	green	green	6.4
		POS[3]	Cyhalofop-butyl	360	red	green	green	green	green	red	red	59.8
		POS[4]	Cyhalofop-butyl	360	red	green	green	green	green	red	red	59.8
		POS[3]	Metsulfuron-methyl	2	red	red	red	red	red	red	green	2.7
		POS	Bentazon	960	red	red	red	red	red	green	red	14.1
												=US$ 183.6
4		Burndown	Glyphosate	1400	green	green	green	green	green	green	green	19.0
		PRE[1]	Penoxsulam	48	red	green	yellow	yellow	yellow	green	green	32.8
		PAg[2]	Glyphosate	720	green	green	green	green	green	green	green	6.4
		PRE[1]	Clomazone	360	red	green	green	green	green	red	red	21.8
		POS[4]	Cyhalofop-butyl	360	red	green	green	green	green	red	red	59.8
		POS[4]	Cyhalofop-butyl	360	red	green	green	green	green	red	red	59.8
												=US$ 199.6

		Applic. time	Herbicide	Label dose (ge.a/i.a ha^{-1})	Main weeds and herbicide efficiency[#]							Estim. Cost (US$ ha^{-1})[##]
					Weedy rice	Barnyardgrass	Alexandergrass	Crabgrass	Goosegrass	Cyperus[*]	Broadleaves[**]	
5	Conv. tillage system	PRE[1]	Clomazone	360								21.8
		PAg[2]	Glyphosate	720								6.4
		POS[3]	Cyhalofop-butyl	360								59.8
		POS[3]	Cyhalofop-butyl	360								59.8
												=US$ 147.8

[*]Ciperaceae (*C. difformis*, *C. esculentus*, *C. ferax*, *C. iria* and *C. brevifolia*).

[**]Broadleaves, as *Alternathera philoxeroides*, *Polygonum persicaria* and *Aeschynomene denticulata*.

[#]Herbicide efficiency: red = low control; yellow = intermediary control; green = high control.

[##]Prices surveyed in the local market on October 28, 2015 (1 US$ = 3.90 R$). Only herbicide costs.

[1]Preemergence (at the needle-point stage of rice) or at postemergence, closest to first postemergence application.

[2]At the needle-point stage of rice (no more than five days after seeding).

[3]Early postemergence (two to four leaves)—adjuvants can be required. Please consult the herbicide label.

[4]Late postemergence (more than five leaves). In some cases, even three applications can be needed to full grass control.

Table 4. Strategies for chemical weed control in sprinkler-irrigated rice, in a highly weed-infested field.

of the treatments, based on experimental data. The presented costs are the commercial prices paid by farmers for the treatments in south of Brazil, as in December 2015.

4. Final remarks

Weed occurrence can reduce grain yield in sprinkler-irrigated rice, and the reduction probably will be significant if the weeds are not efficiently controlled. Sprinkler irrigation is a convenient system for crop rotation, avoidance of drought effects, water savings in rice production and to obtain high yields from crops. These advantages, however, do not minimize the importance of weeds, which are still one of most important pests in fields irrigated by sprinklers. Cultural measures of weed management are needed to reduce the overall impact of weeds in all production systems. However, for the sprinkler-irrigated rice, the special condition provided by the frequent irrigation gives additional advantages to the weeds. Without any water restriction, weeds normally grow faster than in rainfed fields, and can attain high density, since the germination is stimulated by the high soil humidity. In this way, the chemical control has been the most important tool to reduce the impact of weeds in sprinkler-irrigated rice.

Rice conducted under sprinkler irrigation should, preferably, start without weeds growing together with the crop. This is important to avoid the initial competition between the weeds and the rice. Moreover, if the weeds are already established, the difficulties to control increase, and the control levels, consequently, are prejudiced. The simple increase in the herbicide doses not always increase weed control. This is especially valid for the fields with resistant weeds and for those situations where the farmer already applied the highest dose allowed for an herbicide.

For those reasons, it is important to associate several strategies for weed management, which can reduce weed density, attenuate the weed growth and improve the performance of chemical control. In fields with a large seed bank, for example, some techniques can be used to reduce seed viability, as the summer- or fall-tillage, which stimulates the weed seeds to germinate out of rice growing season; additionally, cover crops, no-tillage and crop rotation can be used to increase the amount of residues in and on the soil, which reduce seed viability.

Maintaining residues (mulching) in the soil surface, from crops or cover crops cultivated during winter, can reduce weeds in sprinkler-irrigated rice, cultivated in no-tillage in the succeeding summer. Some grass weeds, like *Urochloa plantaginea*, are very sensitive to this form of suppression. Apart from cultural measures, the use of preemergent herbicides, combined with nonselective weed control previously to rice emergence (the needle-point technique, using glyphosate or other nonselective herbicide) is an efficient option for chemical weed control in sprinkler-irrigated rice.

Besides these alternatives, the use of the Clearfield system, which uses rice cultivars resistant to imidazolinone herbicides, currently is one of the most powerful tools for weed control in irrigated rice in south Brazil. However, as the Clearfield system is based on ALS-inhibitor herbicides, some potential drawbacks related to weed resistance and carryover to nontolerant species cannot be ignored. For these reasons, farmers are advised to always monitor the fields regarding weeds not controlled (escapes) and strictly follow the doses indicated by the label.

Finally, the studies with sprinkler-irrigated rice emphasize the high potential of this system regarding productivity, water savings and diversification in the production system. Under the scope of pest management, the sprinkler-irrigated rice requires integrated strategies for a successful, effective weed control. Weeds are highly favored by this system, and the use of herbicides should be accompanied by preventive and cultural measures against the weeds. Moreover, it is important to monitor the fields to know which species of weeds are occurring, to choose the best alternative for chemical control. Finally, we advise farmers to always consult the label of herbicides before use, as well as follow the regional recommendations provided by official institutions.

Acknowledgements

The authors would like to thank Valmont Industries—Irrigation Division, Omaha-NE, USA, and Valmont Indústria e Comércio, Uberaba-MG, Brazil, for the research cooperation with Embrapa Clima Temperado, Pelotas-RS, Brazil, which made possible to install part of the experiments whose results are included into the present work.

Author details

André Andres[1*], Giovani Theisen[1], Gustavo Mack Teló[1], Germani Concenço[1], José Maria Barbat Parfitt[1], Leandro Galon[2] and Matheus Bastos Martins[3]

*Address all correspondence to: andre.andres@embrapa.br

1 Embrapa Temperate Agriculture, Pelotas, Brazil

2 Federal University of Fronteira Sul, Erechim, Brazil

3 Federal University of Pelotas, Pelotas, Brazil

References

[1] Magalhães Jr, A.M., Parfitt, J.M.B., Fagundes, P.R.R., Theisen, G., Nunes, C.D., Franco, D.F., Severo, A.C.M., Fonseca, G.M., Streck, E.A., Aguiar, G.A., Lopes, J.L., Oliveira, F.A., Silva, D.M., Silva, J.T. & Bretanha, G. (2012). Rice crop under sprinkler irrigation. In: Proceedings on the 6th International Crop Science Congress, Bento Gonçalves, RS.

[2] Stevens, G., Vories, E., Heiser, J. & Rhine, M. (2012). Experimentation on cultivation of rice irrigated with a center pivot system. In: Lee, T.S. (Ed.). Irrigation systems and practices in challenging environments. Rijeka, Croatia: InTech. p. 233–254.

[3] Theisen, G., Xavier, F.M., Bonow, J.F.L., Parfitt, J.M.B., Andres A. & Silva, J.J.C. (2013). Manejo integrado de plantas daninhas em sistema de produção de arroz irrigado por aspersão (Integrated weed management in a sprinkler-irrigated rice production system). In: Congresso Brasileiro de Arroz Irrigado (Brazilian Irrigated Rice Congress), 8. 2013, Santa Maria, RS. *Anais*... Santa Maria: SOSBAI, 2013. v. 1. p. 367–370.

[4] Merotto Jr., A., Lopes, S.I.G. & Kalsing, A. (2013). Latin American Seminar on Red Rice (2.: 2013 : Porto Alegre ,RS). Anais... Porto Alegre, RS: IRGA, UFRGS, 85 p. ISBN 978-85-66106-37-4. Available at: http://www.irga.rs.gov.br/upload/20141126093631 seminario_arroz_vermelho_2013___anais[1].pdf

[5] Sociedade Sul-Brasileira de Arroz Irrigado—Sosbai. (2014). Irrigated rice: technical recommendations for the South of Brazil. Pelotas: SOSBAI. 199 p. Available at: http://www.sosbai.com.br/recomendacoes.php. Accessed 03 Nov 2016.

[6] Andres, A. & Machado, S.L.O. (2004). Plantas daninhas em arroz irrigado. In: Gomes, A.S. & Magalhães Jr, A.M. (Eds.). Irrigated rice in southern Brazil. Brasília: Embrapa InformaçãoTecnológica. Pelotas:

[7] Freitas, G. D. (2004). Desempenho do arroz (*Oryza sativa* L.) cultivar BRS Pelota e controle de capim-arroz (*Echinochloa* spp.) submetidos a quatro épocas de entrada d'água após a aplicação de doses reduzidas de herbicidas (Performance of rice cv. BRS Pelota and control of barnyardgrass under four irrigation starting times and reduced doses of herbicides). 54p. Dissertation (Master in Plant Production). Universidade Federal de Pelotas, Pelotas, 2004.

[8] Andres, A., Theisen, G., Concenço, G. & Galon, L. (2013). Weed resistance to herbicides in rice fields in Southern Brazil. In: Price, A. (Ed.). *Herbicides—current research and case studies in use*. InTech. DOI: 10.5772/55947. Available from: http://www.intechopen.com/books/herbicides-current-research-and-case-studies-in-use/weed-resistance-to-herbicides-in-rice-fields-in-southern-brazil. Rijeka, Croatia: InTech. p. 3–25

Chapter 3

Weed Seedbank in Rice Fields

Mário Luiz Ribeiro Mesquita

Additional information is available at the end of the chapter

http://dx.doi.org/10.5772/66676

Abstract

The weed seedbank in the soil is the major source of weeds in rice fields. Therefore, information on ecological aspects of weeds occurring in rice, including their potential seed production, is crucial for weed management. The size of the weed seedbank in rice fields is highly variable depending on the climate, relief position, soil moisture content, depth of sampling, history of the areas and management practices used by farmers. As a survival strategy, colonization and persistence in the communities, most common weeds in rice fields produce huge number of seeds and vegetative propagules with physical and physiological dormancy mechanisms, insuring seed viability in the soil for long periods. A large proportion of weed seedbank remains generally on or close to the soil surface after seed rain. Sampling protocols involve the use soil cores at variable soil depths. Determination of the size of the weed seedbank can be made by seed direct extraction and germination methods. The latter is more precise with respect to enumeration of viable seeds in the soil. Weed management in rice fields should focus on methods suitable to decrease the weed population in the soil seedbank.

Keywords: competition, biological invasion, germination, weed management, allelopathy

1. Introduction

Many weeds grow in association with the rice crop and their distribution and occurrence intensity are determined by a complex of climate, soils and relief and management practices. Weeds interfere with rice growth and yield by means of competition for nutrients, water, light and space. Moreover, many weed species possess allelopathy mechanisms that hinder or even prevent the growth of other species associated with them, including rice, resulting in decreased yield by up to 96% [1].

Weeds are a major biological constraint for rice farmers. Many weed species that occur in rice fields can produce a huge number of small seeds and vegetative propagules as a strategy to survive stresses imposed by control methods [2–4]. After dispersal, seeds may remain on the soil surface or be buried by means of biotic and abiotic agents thus forming a seedbank which becomes the main source of weeds in rice cropping fields.

As a survival strategy, colonization and persistence in the communities, the weeds have developed a number of features, for example, seed dormancy, which enables the occurrence of discontinuous germination during the rice crop growing season in addition to ensuring the viability of the seeds in the soil for long periods.

The weed seedbank in the soil is a dynamic system with inputs and outputs. The inputs occur via seed rain as a result of efficient dispersion mechanisms and the outputs by means of germination, predation [5–7] and decay or seed death [8].

Various factors affect weed seed germination including variations in soil temperature and moisture [9–13] and physiological aspects of the seeds particularly seed dormancy [14]. When favorable environmental conditions occur and physiological constraints are overcome, seeds germinate; weeds grow and produce new propagules enriching the soil seedbank.

Research on identification and quantification of weed species germinated in the soil seedbank from rice fields were carried out by numerous authors [9, 15–20]. However, due to its ecological and economic importance, the status of the weed seedbank in rice cropping fields needs to be further investigated. Studies on weed seedbank ecology are crucial for improving weed control practices in rice fields.

Field and greenhouse studies are needed in order to understand the soil weed seedbank germination dynamics and its relationship with the weed flora on rice fields. These studies can contribute to predict infestations and could lead to improved management practices to decrease the negative effects of weed interference with rice crop growth and yield.

The goal of this chapter is to discuss general aspects of weed seedbank ecology including weeds associated with rice agroecosystems, types, sizes and major characteristics of the weed seedbank in rice fields, including seed dormancy, research methodology, factors affecting germination dynamics and some aspects of weed seedbank management in rice fields.

2. Major weeds in rice fields

2.1. South and Southeast Asia

Many weeds are associated with rice agroecosystems in different parts of the world. In South and Southeast Asia, 64 weeds were reported as the most important in upland rice [21]. These occur in 18 families; 37 are broadleaves, 20 are grasses and 7 are sedges. Twenty-seven of the cited weeds are primarily annuals, 20 are perennials and 17 are classified as annual or perennial [21]. Ninety weed species were reported competing with rice under aerobic systems [22]. In contrast [23] reported 47 weed species in the rice crop and [24] cited more than 1800 weed species reported in 15 South and Southeast Asian countries. *Cyperus iria* L., *Cyperus difformis* L., *Echinochloa colona* (L.) Link, *Ischaemum rugosum* Salisb. *Leptochloa chinensis* Nees, *Ludwigia hyssopifolia* (G. Don) Excel, *Oryza sativa* L., *Schoenoplectus juncoides* (Roxb.) Palla, *Sphenochlea zeylanica* Gaertn. are the 12 most troublesome weeds of rice in Asia [25].

2.2. Africa

One hundred and thirty weed species are reported to occur in rice-based cropping systems in Africa [26]. Major weed species of upland rice areas are *Rottboellia cochinchinensis* (Lour.) W. Clayton, *Digitaria horizontalis* Willd., *Ageratum conyzoides* L. and *Tridax procumbens* L., while *A. conyzoides* and *Panicum laxum* Sw. which were more cited in the hydromorphic areas and *Cyperus difformis* L., *Sphenoclea zeylanica* Gaertn., *Fimbristylis littoralis* Gaudich, *Oryza longistaminata* A. Chev. & Roehr., *Echinochloa colona* (L.) Link and *Echinochloa crus-pavonis* (Kunth) Schult. dominates the lowland rice fields. Poaceae (43%) and Cyperaceae (37%) are the most prevalent families in lowland rice while, in the uplands, weed species composition tends to be more diverse with Poaceae (36%) and Asteraceae (16%) most prevalent [26].

2.3. Latin America

In Latin America [27] reported 13,892 individuals belonging to 20 families, 40 genera and 60 species in the soil weed seedbank germination studies *in situ* and *ex situ* in which there were 11,530 individuals and 50 species *ex situ* and 2362 individuals and 34 species *in situ*. Total density was 3859 plants m^{-2} [27].

The families with the highest species richness were Cyperaceae with sixteen, Poaceae with ten and Fabaceae-Faboideae with six species each. These families contributed with 53.3% of total species. In contrast, ten families: Amaranthaceae, Euphorbiaceae, Lamiaceae, Loganiaceae, Marantaceae, Nyctaginaceae, Plantaginaceae, Portulacaceae, Solanaceae, Thelypteridaceae and Turneraceae had only one species each. These correspond to 50% of the total of all recorded families [27]. Similar results were observed by [15] who reported that that 86% of species present in seedbank from 22 rice fields in Camboja were Cyperaceae family. In Nepal, Ref. [28] also reported that 37% of the species present in the weed seedbank belonged to this family.

In the tropics, about 80% of seeds germinate until the 60th day of the study in the greenhouse. Germination peak is generally observed at 25 days after the beginning of the study which coincides with the period of the start of the rainy season in the region leading to an increase in weed germination and emergence in weed soil seedbank. Germination stabilization generally occurs at 115 days after start of study [16] (**Figure 1**).

Floristic diversity, based on Shannon Diversity Index, generally is greater *ex situ* study with $H' = 2.66$ nats ind^{-1}, against $H' = 2.53$ nats ind^{-1} *in situ*. The highest number of individuals and species found *ex situ* contribute for the greatest floristic diversity *ex situ* [16].

The most important species in the weed seedbank in Latin America based on the importance value were *Ludwigia octovalvis* (Jacq.) P. H. Raven, *Schoenoplectus juncoides* (Roxb.) Palla, *Lindernia crustacea* (L.) F. Muell, *Cyperus sphacelatus* Roth, *Cyperus iria* L., *Fimbristylis dichotoma* (L.) Vahl, *Boerhavia erecta* L., *Rhynchospora nervosa* (Vahl) Boeck, *Scleria lithosperma* (L.) Sw. and *Sida rhombifolia* L. [16]. In Latin America, species of the family Cyperaceae largely dominates the weed seedbank in the soil of rice fields [16]. Forrmation of a seedbank represents an important regeneration component for many species of this family [2].

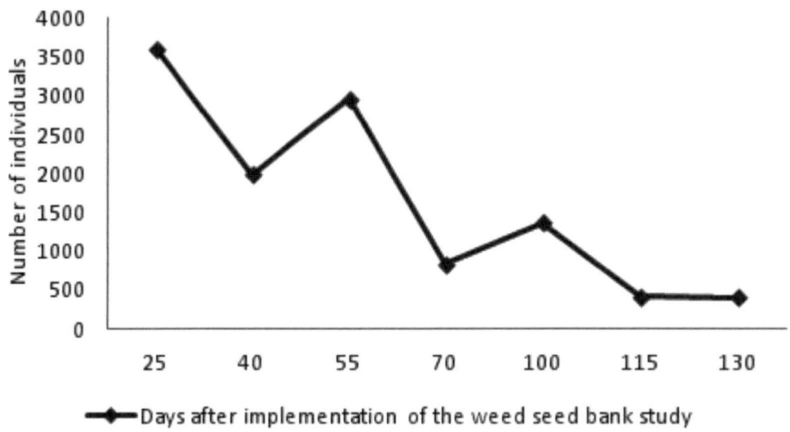

Figure 1. Germination curve of weed of the weed seedbank from a rice field, in Maranhão State, Northeast Brazil, Latin America.

The species dominance in weed seedbank in rice fields might be related not only to cultural practices and crop history but also to the reproductive capacity of the weed species. All species cited here are propagated exclusively by seeds, except for *F. dichotoma* and *S. lithosperma* (Cyperaceae), which also propagate asexually, by rhizomes [29].

The ability to produce a very high number of seeds is one of the main features developed by weeds that occur in rice fields. This is a strategy to escape the stress imposed by the control methods and ensure the species survival.

In the Philippines, for example, see [25], among the weed species occurring in paddy fields, one of the species *Ludwigia octovalvis* (L.) F. Muell (Onagraceae) is capable of producing 250,000 seeds, while *Echinochloa colona* (L.) Link and *Echinochloa crus-galli* (L.) P. Beauv both from the Poaceae family can produce 3100 and 2900 seeds per plant, respectively [5].

Schoenoplectus spp. (Cyperaceae) are able to produce on average 82,098 seeds.m^{-2} [2]. Other species of the same family, among which, *Fimbristylis miliaceae* (L.) Vahl, *Fimbristylis dichotoma* (L.) Vahl, can produce 10,000 and 6500 seeds per plant, respectively [29], while *Cyperus iria* (L.) can produce 5000 seeds per plant [30].

After dispersal, weed seeds are deposited on the soil forming the seedbank that becomes the main source of weeds in rice fields.

3. Weed seedbank types

Soil seedbanks vary according to the duration their seeds remain viable in the soil [30]. Weed scientists distinguished between transient seedbanks for species that have viable seeds present for less than 1 year, such as seeds from grasses, for example, and short-term persistent seedbanks for species with viable seeds that remain for at least 1 but less than 5 years and

http://dx.doi.org/10.5772/66676

long-term persistent, when seeds persist in the soil for at least 5 years [30]. Seeds of many weed species of the Malvaceae and Fabaceae families have long persistence in the soil because of their tegument impermeability to water and gases [14].

Seed persistence in the soil has been attributed to variation in fungal activity, soil fertility, particularly the presence of nitrates, oxygen supply, vegetation cover, burial depth via biotic and abiotic agents, seed density and predator pressure [31].

4. Characteristics of the weed seedbanks

4.1. Seed dormancy

Dormancy is the failure of the weed seeds to germinate under favorable environmental conditions. There are two types of seed dormancy. The first is known as primary or innate dormancy which occurs when seeds are dormant at the time of maturity and the second, as secondary dormancy which is when weed seeds can cycle in and out of a dormancy state due to variation on environmental conditions [14]. Seed dormancy in the soil is important because it maintains the weed seedbank over time and thus helps to ensure that for most weed species only a small proportion of buried weed seeds is recruited as seedlings from the soil seedbank in any given year [14].

The main dormancy mechanisms are physiological, by means of hormones, phytochromes and inhibitors; physical, due to impermeable seed coat to water and gases; and morphological, due to immature embryo [14].

In temperate climate regions, the weed seedbank declines 32% a year [32]. In contrast, in tropical regions, the weed seedbank is generally smaller and the decline tends to be faster because (a) there is a high seedling recruitment rate due to favorable climate conditions for seed germination, which persist for longer periods than in temperate regions; (b) high seed mortality due to attack of predators; (c) high relative humidity and higher temperatures, which favor biotic agents; (d) seedling mortality due to seed germination in short, hot dry periods that can occur during the rainy season; (e) a shorter duration or even the absence of seed dormancy in many weed species; and (f) low seed viability [33].

In post-dispersal weed seedbank studies carried out in rice fields in the Philippines, it was noted that in a period of only 14 days, the fire ants (*Solenopsis geminata*) were the main predators and responsible for the removal of 98%, 88% and 75% of *Digitaria ciliaris* (Retz.) Koeler, *Eleusine indica* (L.) and *Echinochloa colona* (L.) Link seeds, respectively, previously placed on soil surface [5].

Generally higher germination rates observed in the soil weed seedbank in rice fields in the first 60 days [27] is probably due to dormancy breaking because of greater sunlight exposition and temperature variation as observed by many authors [34–35]. This is corroborated by studies carried out in the Philippines where 50% of weed soil seedbank in rice fields germinated in first six weeks [36] and in rice field in Malaysia where it was noted that the highest germination peak occurred at 30 days [9].

4.2. Weed seedbank size in rice fields

The magnitude of weed seedbanks in rice fields is highly variable. Using the direct seed extraction method Ref. [17] found 260,000 seeds m^{-2} in Vietnã, Ref. [19] reported that the number of weed seeds in the soil ranged from 17,300 to 646,000 m^{-2} in New South Wales, Australia, Ref. [15] reported that in the top 5 cm of soil ranged from 52.1 to 167,000 seeds m^{-2} with overall mean of 8,500 seeds m^{-2} in Cambodian rice fields, Ref. [37] found from 116,812 to 294,761 seeds m^{-2} in China. In contrast, using the germination method Ref. [38] found from 1700 to 4000 seedlings m^{-2} in Northern Laos, Ref. [39] counted 878 seedlings m^{-2} and Ref. [18] found 4953 seedlings m^{-2} in weed seedbank in rice fields in Latin America.

Differences in the number of seeds or weed seedling density in the seedbank can be explained by several factors, including climate, relief position, soil moisture content, depth of sampling, history of the areas and management practices used by rice farmer [40].

4.3. Seed distribution in the soil profile

In cropping systems where there is no soil disturbance and no tillage, as is the case for subsistence farming, weed seeds tend to remain on the soil surface, where they are easier to control [42].

The seed location is an important feature because only those situated on or near the soil surface are able to germinate, which can lead to greater short-term germination flows accelerating the reduction of the seedbank. Moreover, the permanence of seeds at the soil surface favors predation [43].

Studies on the movement of weed seeds in a no-till soil have shown that after 1 year, the seeds reached deeper in sandy soils (10% > 6 mm) than in clayey soils (2% > 6 mm). It was also noted that the vertical movement is very small and is conditioned by soil texture, the cumulative rainfall and the seed size, weight and shape [43].

The smaller and lighter seed concentrate at the soil surface. With respect to the seed shape, those flattened are more difficult to penetrate the soil than spherical, discoidal or pyramidal [43].

5. Weed seedbank research methodology

5.1. Sampling

Weed scientists advocate the use of 5 cm diameter cores to sample weed seedbanks in the soil. They state that this size core is large enough to detect seeds, but small enough not to burden the researcher with too much soil [44]. The number of cores to be sampled and the depth to which soil cores should be taken depends upon the research objectives. If the research is to determine the seedbank size and composition or to relate seedbanks to aboveground weed flora, then seedbanks should be sampled at times that follow seed shed but precede seed germination [44].

5.2. Determination of the size and floristic composition of the weed seedbank

There are two methods to enumerate the number of seeds in the soil: Direct seed extraction and germination method

5.2.1. *Direct seed extraction*

In the direct seed extraction technique, seeds are separated from soil by washing or flotation. Initially the soil sample is placed on a screen with a mesh size smaller than the smallest expected seed. A mesh size of about 0.2 mm is enough to catch most small seeds [44].

The flotation method is often used after the soil sample has been washed. The objective is to separate seed from soil particles so that they will float in a solution made with water and potassium carbonate. After the seeds are separated using the direct seed extraction method, they must be identified. Identification is made under magnification using proper literature [44].

5.2.2. *Germination method*

The second technique for enumerating seeds in the soil seedbank is the germination method [44]. This technique is used to enumerate the density of nondormant seeds in the seedbank. Twenty cores are recommended from an experimental treatment [44]. The cores are mixed, composed, inserted in trays and placed in greenhouse. The most suitable soil depth in the trays should be within 2–3 cm with a maximum of 5 cm so that all seeds can germinate. Trays should be perforated in order to facilitate drainage. In case of sandy soils, water retention can be improved by lining the trays with vermiculite (**Figure 2**).

Figure 2. Germination method.

In recent years, research on seedbanks has focused more on the germination method instead of the direct seed extraction. The main reason for this is that the germination method is more accurate because it enables to estimate the actual weed seedbank size considering that all viable seeds will germinate even if it takes several months of work. Furthermore, the seedlings are easier to identify than the seeds.

6. Weed seedbank management in rice fields

The weed seedbank in rice fields is an indicator of weed community resulting from the present and past weed control practices and can provide valuable information for the development of ecologically friendly practices such as, for example, the reduction of herbicide application.

In the past few years, several authors have recommended that the weed management should integrate the different control methods in order to decrease weed population in the soil seedbank [45–48].

A reduction in the weed seedbank germination means minor problem with weeds and hence savings for rice farmers. Moreover, it can provide a healthier environment with less use of chemicals, creating the necessary conditions for the development of more efficient and environmentally acceptable weed management.

Therefore, it is important to limit the current contribution to the weed seedbank to reduce the population size and facilitate the use of future weed control practices.

6.1. Land preparation

Soil disturbance with tillage can promote weed seed germination by several mechanisms including exposition of weed seeds to light which releases seeds of some species from dormancy but can also bury some seeds that are on the soil surface [41]. Tillage prior to rice crop establishment may result in nitrogen mineralization which can promote some seed germination. On the other hand, off-season dry soil tillage at sufficient depth may help breaking and drying vegetative propagules including stolons, bulbs and subsoil rhizomes of perennial weeds. However, tillage may cause soil erosion and increase costs for the rice farmer [36]. Patterns of weed emergence as affected by tillage in upland and lowland rice soils have shown that 40–50% occurred within 6 weeks after tillage in both sites. A significant weed emergence was observed within 3 weeks in both soils but very little emergence occurred in lowland soil [36].

6.2. Mulching

Soil mulching reduces weed seed germination by 90% [49]. The reduction in seed germination in the weed seedbank occur because the mulch prevents the penetration of light or blocks certain spectrum of light wavelengths which are necessary for most of the weed seeds to germinate [50–52]. This is the case for the weed species that produce seeds that are photoblastic positive, that is, need light to germinate, such as *Amaranthus retroflexus* [53],

Eclipta alba [54], *Hyptis suaveolens* [55], *Digitaria* spp. [56], *Urtica dioica* [57], *Ageratum conyzoides* [23, 58], *Fimbristylis autumnalis* [23] and *Cyperus aggregatus* [59].

Moreover, the physical barrier formed by straw must contributes to the death of germinated seedlings from seeds located on the soil surface, whose reserves were not enough to overcome the mulch [60, 61] and provides cover for predators that feed on weed seeds. In addition, residues have a moderating effect on temperature fluctuations in the soil, which in turn can impact seed dormancy of many weed species.

In India, for example, see Ref. [62], the use wheat straw as mulch resulted in 54% reduction in weed density at 30 days after rice seeding. In Vietnam, the herbaceous legume *Tephrosia candida* (Roxb.) D.C. used as mulch caused a reduction in the weed growth and a significant increase in rice yield [63].

6.3. Herbicides

Herbicides are widely used in rice cropping systems all over the world and may be economically attractive in some cases as it requires less overall weeding times. In Africa, 26 herbicides as single application or mixtures are being used in upland and lowland rice [26]. They are effective in reducing weed populations and hence the number of seeds added to the soil seedbank. However, their use is sharply decreasing due to social and environmental concerns and major negative impacts on soil biology aside from promoting the appearance of herbicide resistance in 51 weeds in rice fields [64].

6.4. Interaction between weeds and ducks

In China, a form of organic rice farming called rice-duck farming (RDF) has proven to be very successful in controlling weeds and decreasing the weed bank size in rice fields [65]. Interaction between weeds and ducks after 9 years under RDF, resulted in a decline from 38 to 21 in the number of weed species and the density of both the weed seedbank and aboveground weed flora decreased by more than 90%. After 9 years of interaction between weeds and ducks, RDF resulted in a more uniform vertical distribution of the weed seedbank both quantitatively and qualitatively. The ecological indices point to a gradual change towards fewer species, lower density and lower diversity following continued RDF. The dominant species in the weed seedbank shifted [65].

7. Conclusions

In recent years, there is growing interest in the adoption of conservation practices in rice agricultural production. This involves reducing soil disturbance along with maintaining crop residues on the surface, reducing weed seed inputs and promoting seed depletion in the weed seedbank in the soil. The technology of no-till or minimum tillage and also the growing interest in the practice of organic agriculture and agroecology to develop more balanced rice production systems are current trends that converge to a healthy environmentally and economically sustainable agricultural model.

Acknowledgements

The author is funded by FAPEMA (Foundation for Research and Scientific and Technological Development of Maranhão) State, Brazil. We thank our main donor, for their past and present financial support.

Author details

Mário Luiz Ribeiro Mesquita

Address all correspondence to: mario-mesquita51@hotmail.com

Maranhão State University, Bacabal, Brazil

References

[1] Chauhan BS, Johnson DE. Row spacing and weed control timing affect yield of aerobic rice. Field Crops Research. 2011;**121**:226–231. http://dx.doi.org/10.1111/j.1365-3180.2010.00807.x

[2] Leck MA, Schütz W. Regeneration of Cyperaceae, with particular reference to seed ecology and seedbanks. Perspectives in Plant Ecology, Evolution and Systematics. 2005;**7**:95–133. http://dx.doi.org/10.1016/j. ppees.2005.05.001

[3] Libertino GV. Biology of *Boerhavia erecta* L. [thesis] Kabacan North Cotabato: University of Southern Mindanao Philippines; 1984.

[4] Munhoz CBR, Felfli JM. Phytosociology of the herbaceous and sub-shrub layer of a savannah in Federal District, Brazil. Acta Botanica Brasilica. 2006;**20**:671–685. http://dx.doi.org/10.1590/S0102-33062006000300017

[5] Chauhan BS, Migo T, Westerman PR, Johnson DE. Post dispersal of weed seeds in rice fields. Weed Research. 2010; **50**:553–560. http://dx.doi.org/10.1111/j.1365-3180.2010.00807.x

[6] Hesse E, Rees M, Müller-Schärer H. Seedbank persistence of clonal weeds in contrasting habitats: implications for control. Plant Ecology. 2007;**190**:233–243. http://dx.doi.org/10.1007/s11258-006-9203-7

[7] Rodriguez C. Garcia MA. Seed-bank dynamics of the tropical weed *Sida rhombifolia* (Malvaceae): incidence of seedling emergence, predators and pathogens. Seed Science Research. 2009;**9**:241–248. http://dx.doi.org/10.1017/S0960258509990146

[8] Mohler CL, Dykeman C, Nelson EB, Ditommaso A. Reduction in weed seedling emergence by pathogens following the incorporation of green crop residue. Weed Research. 2012;**52**:467–477. http://dx.doi.org/10.1111/j.1365-3180.2012.00940.x

[9] Begum M, Juraimi AS, Rastan SOBS, Amartalingam R, Man AB. Seedbank and seedling emergence characteristics of weeds in rice field soils of the muda granary area in north-west peninsular Malaysia. Biotropia. 2006;**13**:11–21. http://journal.biotrop.org/index.php/biotropia/article/view/215/184

[10] Hérault B, Hiernaux P. Soil seedbank vegetation dynamics in Sahelian fallows; the impact of past cropping and current grazing treatments. Journal of Tropical Ecology. 2004;**20**:683–69. http://dx.doi.org/10.1017/s0266467404001786

[11] Maia FC, Medeiros RB, Pillar VP, Focht T. Soil seedbank variation patterns according to environmental factors in a natural grassland. Revista Brasileira de Sementes. 2004;**26**: 126–137. http://dx.doi.org/10.1590/S0101-31222004000200018

[12] Vivian R, Gomes Junior FG, Chamma HMCP, Silva AA, Fagan EB, Ruiz ST. The effect of light and temperature on germination of *Alternathera tenella, Conyza bonariensis* and *Digitaria ciliaris*. Planta Daninha. 2008;**26**:507–513. http://dx.doi.org/10.1590/S0100-8358 14 2008000300005

[13] Batlla D, Benech-Arnold RL, Weed seed germination and the light environment: implications for weed management. Weed Biology and Management. 2014;**14**:77–87. http://dx.doi.org/10.1111/wbm.12039

[14] Baskin CC, Baskin JM, The natural history of soil seedbanks of arable land. Weed Science. 2006;**4**:549–557. http://dx.doi.org/10.1614/WS-05-034R.1.

[15] Kamoshita A, Ikeda H, Yamagishi J, Ouk M. Ecophysiological study on weed seedbank and weeds in Cambodian paddy fields with contrasting water availability. Weed Biology and Management. 2010;**10**:261–272. http://dx.doi.org/10.1111/j.1445-6664.2010.00393.x

[16] Mesquita MLR, Andrade LA, Pereira WE. Soil weed seedbank *in situ* and *ex situ* at a smallholder field in Maranhão State, northeastern Brazil. Ciência Agronomica. 2015;**11**:14–20. http://dx.doi.org/10.4025/actasciagron.v37i1.19360

[17] Hach CV, Chin DV, Nhiem NT, Mortimer M, Heonq KL, Nam NTH. Effect of tillage practices on weed infestation and soil seed naks in wet-seeded rice. In: Proceedings of the International Weed Science Congress; (WSSA 2000) 6–11 June 2000. Foz do Iguassu, Brazil: IWSS; 2000. pp. 51–52. http://www.iwss.info/download/iwsc-2000.pdf

[18] Silva MRM, Costa EA, Marques LJP, Corrêa MJP. Weed seedbank in upland rice fields in the Pré-Amazônia Maranhense Region, Brazil, Revista de Ciências Agrárias. 2014;**57**:351–357. http://dx.doi.org/10.4322/rca.1297

[19] McIntyre S. Seed reserves in temperate Australian rice fields following pasture rotation and continuous cropping. Journal of Appiled Ecology.1985;**22**: 875–884. DOI: 10.2307/2403236 Stable URL: http://www.jstor.org/stable/2403236

[20] de Rouw A, Casagrande M, Phaynaxay K, Soulileuth B, Saito K. Soil seedbanks in slash-and-burn rice fields of northern Laos. Weed Research. 2014;**54**:26–37. DOI: 10.1111/wre.12053

[21] Galinato MI, Moody K, Piggin CM. Upland rice weeds of South and Southeast Asia, Makati City: International Rice Research Institute; 1999. 156 p. http://books.irri.org/9712201309_content.pdf

[22] Jabran K, Chauhan BS. Weed management in aerobic rice systems. Crop Protection. 2015;**78**:151–163 http://dx.doi.org/10.1016/j.cropro.2015.09.005

[23] Caton BP, Mortimer M, Hill JE, Johnson DE. 2010. A practical field guide to weeds of rice in Asia. 2nd ed. Los Baños (Philippines): International Rice Research Institute; 2010. 118 p. http://books.irri.org/9789712202568_content.pdf

[24] Moody K.Weeds reported in rice in South and Southeast Asia. Manila: Philippines International Rice Research Institute; 1989. 442 p. http://pdf.usaid.gov/pdf_docs/PNABD500.pdf

[25] International Rice Research Institute. Rice knowledge bank. Makati: IRRI; 2010. http://keyserver.lucidcentral.org/key-server/data/0f080806-070a-460c-8709-0a0d0d0f0705/media/Html/F%20miliacea.htm

[26] Rodenburg J, Johnson DE. Weed management in rice-based cropping system in Africa. In: Sparks, D, editor: Advances in Agronomy, Vol. 103, Burlington: Academic Press; 2009, pp. 149–218. http://dx.doi.org/10.1016/S0065-2113(09)03004-1

[27] Mesquita MLR, Andrdae, LA, Pereira, WE. Floristic diversity in the soil weed seedbank in a rice growing area of Brazil: in situ and ex situ evaluation. Acta Botanica Brasilica. 2013;**27**:463–471. http://dx.doi.org/10.1590/S0102-33062013000300001

[28] Bhatt MD, Singh SP. Soil seedbanks dynamics of weed flora in upland and lowland rice paddy cultivation areas of far western Nepal. Scientific World. 2007;**5**:54–59.

[29] Lorenzi H. Invasive plants of Brazil: aquatic terrestrial, parasite and toxic. 4th ed. Nova Odessa – São Paulo: Instituto Plantarum, 2008. 640 p.

[30] Thompson K, Ceriani RM, Bakker JP, Bekker RM. Are seed dormancy and persistence in soil related? Seed Science Research. 2003;**13**:97–100 DOI: 10.1079/SSR2003128

[31] Saatkamp A, Poschlod P, Venable DL. The functional role of soil seedbanks in natural communities. In: Gallagher, RS editor. Seeds: The Ecology of Regeneration in Plant Communities, 3rd ed. CAB International; Wallingford, UK, 2014. pp. 263–295. http://www.eebweb.arizona.edu/faculty/venable/pdfs/Saatkamp_etal2014.pdf

[32] Roberts HA, Feast PM. Changes in the number of viable seeds in the soil under different regimes. Weed Research. 1973;**13**:298–303. DOI: 10.1111/j.1365-3180.1973.tb01278.x

[33] Garcia MA. Relationship between weed community and soil seedbank in a tropical agroecosystem. Agriculture, Ecosystem and Environment. 1995;**55**:139–146. http://dx.doi.org/10.1016/0167-8809(95)00604-Q

[34] Baskin, CC, Baskin JM. 1998. Seeds, ecology, biogeography and evolution of dormancy and germination. San Diego: Academic Press; 1998. 666 p. https://books.google.com.br/

books?hl=en&lr=&id=vXfNCgAAQBAJ&oi=fnd&pg=PP1&ots=-plfTbCXph&sig=_-2EJx6RgCQfGUEV_eWQlL_SPOI&redir_esc=y#v=onepage&q&f=false

[35] Benech-Arnold RL, Sanchez RA, Forcela F, Kruk BC, Ghersa CM. Environmental control of dormancy in weed seedbanks in soil. Field Crops Research. 2000;**67**:105–122. http://dx.doi.org/10.1016/S0378-4290(00)00087-3

[36] Zimdhal RL, Moody K, Lubigan RT. Castin EM. Patterns of weed emergence on tropical soils. Weed Science. 1988;**36**:603–608.

[37] Feng W, Pan G, Qiang S, Li R, Wei J. Influence of long-term different fertilization on soil weed seedbank diversity of a paddy soil under rice/rape rotation. Frontiers in Biology China. 2008;**3**:320–327. DOI: 10.1007/s11515-008-0056-4

[38] de Rowl A, Casagrande M, Phaynaxay K, Soulileuth B, Saito K. Soil seedbanks in slash-and-burn rice fields of northern Laos. Weed Research. 2014;**54**:26–37. DOI: 10.1111/wre.12053

[39] Mesquita MLR, Andrade LA, Pereira WE. Germination, floristic composition and phytosociology of the weed seedbank in rice intercropped with corn fields. Agraria Revista Brasileira de Ciências Agrárias. 2016;**11**:14–20. DOI:10.5039/agraria.v11i1a5359

[40] Maia FC, Medeiros RB, Pillar VP. Focht T. Soil seedbank variation patterns according to environmental factors in a natural grassland. Revista Brasileira de Sementes. 2004; **26**:126–137. http://dx.doi.org/10.1590/S0101-31222004000200018

[41] Chauhan BS, Johnson DE. Influence of tillage systems on weed seedling emergence pattern in rainfed rice. Soil & Tillage Research. 2009;**106**:15–21. http://dx.doi.org/10.1016/j.still.2009.10.004

[42] Grundy AC, Mead A, Burston S. Modelling the emergence response of weed seeds to burial depth: interactions with seed density, weight and shape. Journal of Applied Ecology. 2003;**40**:757–770. DOI: 10.1046/j.1365-2664.200300836.x

[43] Benvenuti S. Natural weed seed burial: effect of soil texture, rain and seed characteristics. Soil Science Research. 2007;**17**:211–219. DOI: 10.1017/S0960258507782752

[44] Forcella FT, Webster T, Cardina J. Protocols for weed seedbanks determination in agro-ecosystems. In: Adeendum 1 Labrada IR editor, Weed Management for Developing Countries, FAO – Rome, Plant Production and Protection Paper. 2003;**120**:3–18. http://www.fao.org/docrep/006/y5031e/y5031e00.htm

[45] Ghersa, CM, Benech-Arnold, RL, Satorre EH, Martinez-Ghersa MA. Advances in weed management strategies. Field Crops Research. 2000;**67**:95–104. http://dx.doi.org/10.1016/S0378-4290(00)00086-1

[46] Labrada, R. The need for improved weed management in rice. In: Proceedings of the 20th Session of the International Rice Comission (Bangkok, Thailand), 23–26 July 2002; Bangkpk, Thailand: FAO; 2002. http://www.fao.org/3/a-y4751e/y4751e0l.htm#bm21

[47] Bastiaans L, Paolini R, Baumann DT, Focus on ecological weed management: what is the hindering adoption? Weed Research. 2008;**48**:481–491. http://dx.doi.org/10.1111/j.1365-3180.2008.00662.x

[48] Saito K, Azoma K, Oikeh SO. Combined effects of *Stylosanthes guianensis* fallow and tillage management in upland rice. Soil and Tillage Research. 2010;**107**:57–63. http://dx.doi.org/10.1016/j.still.2010.03.001

[49] Teasdale JR. Principles and practices of using cover crops in weed management systems. In: Adeendum 1 Labrada IR editor, Weed Management for Developing Countries, FAO – Rome, Plant Production and Protection Paper.2003;120:Chapter 3. http://www.fao.org/docrep/006/y5031e/y5031e0d.htm#bm13.1

[50] Battla D, Benech-Arnold RL. Weed seed germination and the light environment: implications for weed management. Weed Biology and Management. 2014;**14**:77–87. http://dx.doi.org/10.1016/j.cropro.2005.07.014

[51] Steinmaus S, Elmore CL, Smith RJ, Donaldson D, Weber EA, Roncoroni JA, Miller PRM. Mulched cover crops as an alternative to conventional weed management systems in vineyards. Weed Research. 2008;**48**:273–281. http://dx.doi.org/10.1111/j.1365-3180.2008.00626.x

[52] Yamashita OM, Guimarães SC, Silva JL, Carvalho MAC, Camargo MF. Fatores ambientais sobre a germinação de *Emilia sonchifolia*. Planta Daninha. 2009;**27**:673–681. http://dx.doi.org/10.1590/S0100-83582009000400005

[53] Teasdale JR, Mohler CL. The quantitative relationship between weed emergence and the physical properties of mulches. Weed Science. 2000;**48**:385–392. http://www.jstor.org/stable/4046305

[54] Chauhan BS, Johnson DE. Seed germination ecology of purple-leaf button weed (*Borreria ocymoides*) and indian heliotrope (*Heliotropium indicum*): two common weeds of rain-fed rice. Weed Science. 2008;**56**:670–675. http://dx.doi.org/10.1614/WS-07-199.1

[55] Wulff R, Medina E. Germination of seeds of *Hyptis suaveolens* Poit. Plant Cell Physiology. 1971;**12**:567–579.

[56] Kobayashi H, Oyanagi A. *Digitaria ciliaris* seedbanks in tilled and untilled soybean fields. Weed Biology and Management. 2005;**5**:53–6. DOI: 10.1111/j.1445-6664.2005.00156.x

[57] Jankowska-Blaszczuk M, Daws MI. Impact of red : far red ratios on germination of temperate forest herbs in relation to shade tolerance, seed mass and persistence in the soil. Functional Ecology. 2007;**2**:1055–1062. DOI: 10.1111/j.1365-2435.2007.01328.x

[58] Sun P, Mantri N, Möller M, Jinbo Shen J. Shen Z, Jiang B. Chen C, Miao Q, Lu H. Influence of light and salt on the growth of alien invasive tropical weed *Ageratum conyzoides*. Australian Journal of Crop Science. 2012;**6**:739–748. http://www.cropj.com/lu_6_4_2012_739_748.pdf

[59] McIvor, JG, Reid DJ. Germination characteristics of tropical and subtropical rangeland species. The Rangeland Journal. 2011;**33**:195–208. http://dx.doi.org/10.1071/RJ10026.

[60] Gardarin A, Dürr C, Colbach N. Effects of seed depth and soil aggregates on the emergence of weeds with contrasting seed traits. Weed Research. 2010;50:91–101. DOI: 10.1111/j.1365-3180.2009.00757.x

[61] Gomes Junior FG, Christoffoleti PJ. Biologia e manejo de plantas daninhas em áreas de plantio direto. Planta Daninha. 2008;**26**:789–798. http://dx.doi.org/10.1590/S0100-83582008000400010

[62] Singh S, Ladha JK, Gupta RK, Bhusan L, Rao AN, Sivaprasad B, Singh PP. Evaluation of mulching, intercropping with Sesbania and herbicide use for weed management in dry-seeded rice (*Oryza sativa* L.) Crop Protection. 2007;**26**:518–524. DOI: 10.1016/j.cropro.2006.04.024

[63] Hoang Fagerstrom, MH, Niilsson SI, van Nordwijk M, Phien T, Olssom M, Hansson A, Svensson C. Does *Tephrosia candida* as fallow species, hedgerow or mulch improve nutrient cycling and prevent nutrient losses by erosion on slopes in northern Vietnam? Agriculture Ecosystems & Environment. 2002;**90**:291–304. http://dx.doi.org/10.1016/S0167-8809(01)00208-0

[64] Heap, I. The international survey of herbicide resistant weeds [Internet]. 2016 Available from www.weedscience.org [Accessed 2016-08-02].

[65] Li SS, Wei SH, Zuo RL, Wei JQ, Qiang S. Changes in the weed seedbank over 9 consecutive years of rice-duck farming. Crop Protection. 2012;**37**:42–50. http://dx.doi.org/10.1016/j.cropro.2012.03.001

Chapter 4

The Use of Rice in Brewing

Ombretta Marconi, Valeria Sileoni,
Dayana Ceccaroni and Giuseppe Perretti

Additional information is available at the end of the chapter

http://dx.doi.org/10.5772/66450

Abstract

Rice could be a useful raw material for the production of a gluten-free beer-like beverage. In today's beer brewing industry, rice is primarily used as an adjunct in combination with barley malt. But, recently, there is some information about rice malt for brewing an all-rice malt beer. The use of rice as an adjunct in brewing is described highlighting the quality attributes of the final beer. The rice grain quality attributes of different samples are reported in order to evaluate their attitude to malting and brewing and also considering their enzymatic activity. Then, the different brewing processes to produce all-rice malt beers will be described and the final gluten-free rice beers is evaluated and compared to a barley malt beer. Finally, the levels of major aroma-active components of an all-rice malt beer and the results of the sensory analysis assessing the beer-like character of the rice beverage are reported. The obtained beer samples show a content of volatile compounds comparable with a barley malt beer. The sensory profile of the rice malt beer is similar to a barley malt beer in aroma, taste and mouthfeel.

Keywords: rice, rice malt, gluten-free beer, adjunct

1. Introduction

Rice is a staple food for nearly 50% of the world's population. According to the Food and Agriculture Organization (FAO) of the United Nations, global paddy rice production in 2015 was of 738.2 million tons (490.3 million tons, milled basis) [1]. Rice does not contain gluten-like proteins, so it is particularly suitable for consumption by individuals with celiac disease [2]. Thus, rice could be a useful raw material for the production of a gluten-free beer-like beverage. Beer is an alcoholic beverage obtained from water, barley malt, hops and fermented by yeast but other cereals can be used as raw materials or adjunct. In today's beer brewing industry, rice is primarily used as an adjunct in combination with barley malt. As a brewing adjunct, rice has a very neutral flavor and aroma, and when properly converted in the brewhouse, it yields a light clean-tasting beer. Recently, there is a growing interest about the use of rice malt for brewing an all-rice malt beer. Malt is the product obtained from steeping, germination and drying of cereals, generally barley. The aim of malting is to develop enzymes needed for the brewing process. Some rice varieties showed good aptitude to be malted due to their good germinative energy and protein content. Rice malt beer can be produced by obtaining a gluten-free beverage comparable to conventional beer. The beverage represents a good alternative in the diet of individuals who suffer from celiac disease.

2. Rice varieties for brewing

Rice (*Oryza*), like barley, wheat and millet, belongs to the Poaceae or Graminae family. The two main species successfully cultivated are the African species *O. glaberrima* L. and the Asian species *O. sativa* L., of which 120,000 varieties are known. During its long history of cultivation, *O. sativa L.* Asian rice has undergone considerable differentiation, and thousands of cultivars have evolved as a response to the wide range of environmental condition into which it has been introduced. They fall into three groups, with different features:

(a) short-grained "japonica" or "sinica" forms, adapted to a relatively cool climate;
(b) long-grained "indica" and
(c) broad-grained "javanica" forms, which thrive under more tropical condition [3].

Short grain has the highest starch content, makes the stickiest rice, whereas long grain is lighter and tends to remain separate when cooked. The qualities of medium grain fall between the other two types [4]. These groups also show a different gelatinization temperature, namely the temperature at which the intermolecular bonds of starch molecules break down in the presence of water, that is, a key feature for brewing. In particular, the gelatinization temperature is about 65–68°C for short-grain rice and about 71–74°C for long-grain rice, which is also extremely viscous prior to liquefaction. For this reason, short-grain varieties are usually preferred [5]. In fact, the Californian short-grain varieties, such as Pearl, Mochi, Somi and Cahose, liquefy better than the medium-grain variety Nato [6]. Moreover, other quality parameters are important for the evaluation of suitability of rice variety for malting and brewing, such as thousand kernels weight, germinative energy, water sensitivity and total protein. Mayer et al. [7] analyzed 10 Italian rice varieties (8 short grain: Creso, Selenio, Kernak, Arborio, Vialone nano, Centauro, Crono and Balilla; 2 long grain: Sirio and Libero) for these parameters without finding significant differences between japonica and indica varieties, but finally only the varieties Centauro and Balilla were able to saccharify. In conclusion, not all the variety is suitable for brewing and the careful selection of the right varieties is important.

3. The rice grains quality attributes for brewing

The rice composition makes this cereal particularly suitable for human nutrition. The chemical composition of grains varies widely, depending on environment, soil and variety. The dry matter consists of about 70% starch, 5–8% protein, 0.2–2.2% oil and small amounts of inorganic substances. The chemical composition of rice, especially the high starch content, makes this cereal also perfectly suitable for brewing [5].

Usually, brewer's rice is a byproduct of the edible rice milling industry. Hulls are removed from paddy rice, and this hulled rice is then dry milled to remove the bran, aleurone layers and germ. The objective of rice milling is to completely remove these fractions with a minimal amount of damage to the starchy endosperm, resulting in whole kernels for domestic consumption. The broken pieces are considered esthetically undesirable for domestic use and sold to brewers at a low price. Rice is preferred by some brewers as adjuncts because of its lower oil content compared to corn grits. It has a very neutral aroma and flavor, and when properly converted in the brewhouse it results in a light, dry, clean-tasting and drinkable beer. The quality of brewer's rice can be judged by several factors: cleanliness, gelatinization temperature, mash viscosity, mash aroma, moisture, oil, ash and protein content. Rice grains contain more starch on a percentage dry weight basis than barley or wheat and they contain lower levels of fiber, lipid and protein, thus possessing some inherently useful properties for the brewer. The starch structure of rice is more granular than that of barley or wheat. Being small grained, rice is low yielding, in terms of brewer's extract [4].

Concerning proteins, their content range from 6.6% to 7.3% for brown rice, 6.2% to 6.9% for milled rice and 8.2 to 8.4% for basmati rice [8]. The average amount of proteins in rice 6–9% is lower than both barley (about 11.5%) and barley malt (about 10.5%). Quantitatively, the major proteins of barley are the prolamin storage proteins, which are endosperm specific. Uniquely, in rice, glutelin-type storage proteins, with a globulin-type amino acid sequence, are the major proteins. There is also some evidence that the endosperm storage proteins of cooked rice are very resistant to hydrolysis [9].

Moreover, just a little part of this nitrogen amount goes into solution during malting, and consequently the free amino nitrogen (FAN) needed for the yeast must be supplied by the malt. For this reason, the employment of high-yielding FAN malt may be important to balance this difference. Protein is most abundant in the subaleurone layers but is also present in aleurone cells [10].

Concerning lipids, their content is about 2.2%, a little bit higher than barley (1.8%). Cereal lipids are a chemically diverse group: neutral lipids, glycolipids and phospholipids. The ratio of these lipids classes does not differ between japonica and indica rice, but their distribution within the grain is not uniform and the endosperm lipids contain a higher proportion of polar lipids [11]. The lipids or oil content of rice is concentrated in the bran fraction, where it can contribute up to 20% by mass (dry basis), specifically as lipids bodies or spherosomes about 0.1–1 μm size in the aleurone layer and bran. The crude oil content in brown rice is about 2.9%, of which 51% was found in the germ, 32% in the polish and only 17% in the endosperm [8]. A high lipid content can cause increased yeast growth and reduced ester formation during fermentation, reduced foam stability, flavor problem and gelatinization difficulties. Brewer's rice should therefore contain less than 1.5% lipid. In such concentration, lipids do not affect beer quality unless they become rancid. On the other hand, rice grain polishing and repeated washing can decrease the fat content. This results in a decrease of the fat-derived metabolites γ-nonalactone and 1-hexanol. Both are specific flavor components, which become perceptible only after fermentation [12].

Concerning starch, the granules are evenly distributed in amyloplasts packed in the endosperm cells [10]. The two outermost cell layers (the subaleurone layer) are rich in protein and lipid and have smaller amyloplasts and compound starch granules than the inner endosperm. Rice endosperm cell walls seem to be different from barley, although arabinoxylans and β-glucans account for the major proportion (~47 to 49%) of endosperm cell wall, there are also substantial proportions of cellulose (~23 to 28%) and pectic substances containing polygalacturonides (~27%) and variable amounts of glucomannans. It does not, however, seem that these pectic substances have an adverse effect on wort filtration when rice is used as an adjunct [9]. Rice starch granules are the smallest produced by plants, with an average size of 3–8 μm and a polygonal but irregular shape. Compound granules having diameters up to 150 μm form clusters containing 20 and 60 individual granules and fill most of the central space within the endosperm cells. However, in waxy varieties, which are essentially 100% amylopectin (0% amylose), the endosperm is opaque because of air spaces between the starch granules [8].

The starch granules are accumulations of numerous starch molecules that can be fractionated into essentially linear chain amylose and the highly branched amylopectin. The main variation in composition of rice starch is caused by the relative proportions of these two fractions in the starch granules. Amylose content varies greatly between varieties, from a low of 0–2% in waxy rice (milled rice and dry mass basis) to a high of greater than 25% in non-waxy rice. The starch content, and the relative proportions of amylose and amylopectin, together with the chain length distribution and the frequency and spacing of branch points within the amylopectin molecule, has a profound influence on the physicochemical properties of starch, such as the gelatinization temperature, which is mainly important for brewing [13].

3.1. Gelatinization temperature of rice

Gelatinization describes the irreversible collapse (disruption) of molecular order within a starch granule when heated in excess water. This feature is particularly important for brewing because if the starch granules swell and lose their structure, then they become susceptible to rapid enzyme attack during mashing. Rice has a relatively broad gelatinization temperature range (65–85°C) but, even considering this great variation, the gelatinization temperature of rice is generally higher than the barley malt (64–67°C), probably because of the smaller starch granules. Consequently, rice need to be gelatinized or pre-cooked before brewing, otherwise malt enzymes will be rapidly inactivated at the elevated temperatures, which is needed to gelatinize the rice starch granules. In fact, for brewing, it is necessary that starch gelatinization happens in a temperature range where the amylolytic enzymes are still active, otherwise they cannot degrade the starch to fermentable sugars and dextrins. Starches with higher gelatinization temperature require longer cooking time than those with lower gelatinization temperature. Gelatinization is affected by several factors including water content of the gel, amylose content, degree of crystallinity in the amylopectin fraction and amylopectin chain length. Other factors that influence gelatinization include placement and content of starch granule-associated protein and lipids. As already illustrated, the short-grain varieties have a lower gelatinization temperature than the long-grain varieties. Thus, the careful selection of varieties that liquefy well is important [5–14]. The selection of suitable grades is also important, rice liquefies more easily the fine the particle grist is and particles less than 2 mm are considered adequate [6]. Moreover, even if the gelatinization temperature is quite high for unmalted cereals, it can be lowered up to 20°C in the presence of enzymes. It depends on the malting conditions. So the high gelatinization temperature of rice (65–85°C) should also be lowered after malting. But there is no knowledge about the real optimal temperature range of the amylolytic enzymes of rice malt, which can differ from that of barley malt. Moreover the starch granules are already attacked by enzymes during germination, and therefore the starch can be also hydrolyzed below the gelatinization temperature [7].

4. Malting

Malting is a process involving steeping, germination and drying of cereal seed to obtain the malt. Generally, malt production from barley is the first step in beer production but it is possible to obtain malt from other cereals such as rye, sorghum, wheat, quinoa, amaranth or rice [15]. The aims of malting are to produce enzymes in the kernel and modify its chemical constituents. In stored kernel, the enzymes that are important for the malting process are inactive. The first step in malting is the steeping where the kernel absorbs the water, until about 40%, for germination. As a result, the enzyme synthesis is induced and the germination begins. The germination takes place in large tanks or chamber where the temperature and the humidity are controlled. During germination, the formation of enzymes is one of the main requirements of malting because these enzymes are absolutely essential, subsequently, for the breakdown processes that occur during mashing in the brewhouse transforming insoluble substances in the soluble form. The final step is the kilning at high temperature to dry the malt. The temperature of the kilning depends on the type of malt produced and it ranges from about 82°C for a pilsner malt to about 220°C for chocolate malt. During the kilning, the water content is lowered from 40% to about 5%, the germination is stopped but the enzymes are preserved and the color and flavor compounds are formed.

Rice does not contain gluten-like proteins, so it is particularly suitable for consumption by individuals with celiac disease [7, 16]. Thus, rice could be a useful raw material for the production of a gluten-free beer-like beverage in a traditional way, saccharifying completely the wort without the addition of exogenous enzymes. Since rice and barley are physiologically and chemically different, the malting process commonly used for barley is not suitable for rice (**Table 1**) [9].

There are many enzymes involved in the germination process but among these enzymes the amylases have an important role. They breakdown the starch into fermentable sugars. In barley malt, an important attribute for the activity of starch-degrading enzymes is the diastatic power, which measures the combined activity of α- and β-amylases, and for this reason it is always taken into account for the evaluation of malt. Rice malt, however, shows lower diastatic power values, and for this it seems less suitable for brewing [17]. But it contains other amylolytic enzymes that can act synergistically with α- and β-amylases [9, 18, 19]. One of these amylolytic enzymes is limit dextrinase, a debranching enzyme, which

Enzyme	Rice malt	Barley malt
α-Amylase	120 IU/mg protein	206 IU/mg protein
	28–42 DU	44 DU
		365 IU/g
β-Amylase	23–175 IU/g	1017 IU/g
		234 IU/mg protein
α-Glucosidase	0.22–0.30 IU	1.8 IU
Limit dextrinase	2.2–6.0 EU/g	0.2–0.4 EU/g
Endo-β-(1,3)(1,4)-glucanase	0.0–0.1 U/g	100–135 U/g

IU (international unit), amount of enzyme that releases 1 μmol of p-nitrophenol from the substrate per minute at the defined pH and temperature. DU (dextrinizing unit), quantity of α-amylase that dextrinizes soluble starch in the presence of an excess of β-amylase at the rate of 1 g/h at 30°C. EU (enzyme activity unit), amount of enzyme that releases 1 μmol of glucose reducing sugar equivalent per minute at 40°C and pH 5.0 or 5.5.

Table 1. Activity of enzymes important in lager brewing in barley and rice malts.

catalyzes the hydrolysis of the α-1,6-glucosidic linkages in starch and related oligosaccharides and exhibits activity levels in rice several fold higher than in barley malt [20]. Another enzyme, α-glucosidase, shows high activity in rice malt and an optimum temperature of 55°C, whereas in barley malt its activity is low and the optimum temperature is 35–40°C. Above 50°C in barley malt, the enzyme is inactivated very rapidly [21]. Evidently, the same enzyme can have different features in different cereals. With the notable exception of β-amylase, it appears that rice malt has all enzymes required to produce a well-fermentable wort. Nevertheless some authors [9, 22] suggested that supplementation with exogenous cell wall-degrading enzymes, namely, β-1,3–1,4 -d-glucanase, would benefit brewing because rice malt contains low amounts of these enzymes compared to barley malt. Degradation of cell walls in the rice malt endosperm is an important step during the malting and mashing process, because it exposes starch granules to amylolytic enzymes. This behavior is different from rice barley malt. The cell walls of the endosperm of rice, unlikely barley, contain mainly pectin and xyloglucan. So the addition of (1−3,1−4)-β-D-glucanase, which were found in rice malt only in a negligible amount, is not necessary. Pectin might simply be washed out at higher temperatures [23].

Some authors have studied the optimization of the malting condition for rice improving the steeping, the germination and the kilning steps (**Table 2**). Steeping is a critical stage in malting because the enzymes production in grains depends on the steeping period. The steeping time and temperature influence the water uptake and the warmer the steeping water the faster the water is taken up, and vice versa [27]. Some authors performed a short steeping because increasing the steeping period directly increases the steeping losses and the mold development. The optimum steeping period for rice grains should be a balance between time, temperature and losses [17]. However in Ref. [7], it was decided to perform a long an unusual steeping period to increase the water uptake and solubilize any inhibitor for

Grain	Steeping	Germination	Kilning	Rice malt characteristics
Ofada rice [24]	24–48 h	6–7 days	48–60°C	Moisture 11.01%; cold water extract 1.001
Rice [25]	Wet-steep: 20°C for 42 h (20+22); air rest: 4 h	5 days at 20°C	50°C for 24 h	Hot water extract 352 Ldeg/kg; total soluble nitrogen 0.46%; free amino nitrogen 133 mg/l
Italian rice varieties [16]	Wet-steep: 18–20°C for 39 h (24+15); 8 h air rest: 9 h (8+1)	7 days at 20°C	From 45°C to 58°C within 8 h, 4 h at 58°C, from 58°C to 63°C within 4 h, 6 h at 63°C	Final moisture 4.0–5.2%; extract 64.3–73.5% d.m.; saccharification time > 60 min, Kolbach index 15.1–25.7%, viscosity 1.53–1.94 mPas
Jasmine 85 rice [17]	28°C for 48 h	12 days	Not reported	Steeping degree about 36%; diastase activity 668 U/dry malt
6 *Oryza sativa* L. Indica cultivars [26]	30°C for 24 h	4–5 days at 30°C	24 h at 50°C	Steeping degree about 30%; high enzyme activity and satisfactory modification of the rice endosperm
Non-waxy and waxy Black rice [20]	Wet-steep: 25°C for 5 h; air rest: at 95% RH until 44% moisture	8 days at 30°C	24 h at 50°C	Extract 59.3–62% d.m.; Kolbach index 20–23%; viscosity 1.33–1.37 mPas
10 paddy rice varieties of *Oryza sativa* L. japonica and indica [7]	Wet step: 25–23°C 40 h (8×5); air rest: 25–23°C 32 h (8×4)	24 h at 22°C, 24 h at 21°C, 72 h at 20°C	Withering: 12 h at 45°C, 12 h at 50°C, 13.5 h at 55°C. Curing: 6 h, 70°C	Steeping degree about 42%; final moisture <5%; diastatic power 42 WK; fine extract >70% d.m.; saccharification time 50 min, Kolbach index 28.5–40%, viscosity 1.48 mPas

RH: air relative humidity %.

Table 2. Comparative studies on the optimization of the rice malting process.

germination. The longer steeping time was compensated for the lower water temperature of 23–25°C. In this condition no mold was developed and a steeping degree of about 42% was reached, which is necessary for good rice modification. The germination step has a duration of about 4–8 days and it was performed at 20°C or 30°C. In Ref. [20], it was reported that the optimal germination conditions for black rice varieties were at a temperature of 30°C for 8 days obtaining an extract content of 60% (w/w) and a good enzymatic activity related to limit dextrinase and α-glucosidase, which compensate the low activity of β-amylase. Other authors obtained good rice malts performed the germination for 5–7 days at 20°C. In fact the germination temperature affects the extract level and some authors [25] demonstrated that they obtained the highest level of extract at the lowest germination temperature (20°C). As reported in literature, good extract values for barley malt are greater than 82%

and poor values lower than 79% but highest rice malt extract is generally around 70% [7, 16]. Furthermore the European Brewing Congress (EBC) official analytical method to determine the extract is set for barley malt but it was shown that the amount of the endogenous rice malt enzymes is high enough to saccharify the rice malt starch. Anyway, the optimal temperature and pH conditions of rice malt enzymes are different from those of barley malt. So the method to obtain the conventional Congress mash must be slightly changed. With different temperature rests and the addition of $CaCl_2$ and lactic acid, a complete saccharification of the rice malt starch is achieved [7]. The conditions of this new standardized mashing program are 30 min at 45°C, 30 min at 64°C and 30 min at 74°C. The increase of the temperature of the final rest is due to the fact that the temperature at 70°C did not permit the complete gelatinization. The kilning step is important to stop the germination and have a moisture content of about 5%. The kilning step influences the aroma and the color of the rice malt. Generally, the kilning temperature for rice malt is not higher than 70°C, as reported in **Table 1**, to preserve the endogenous rice enzymes. Therefore the color of the rice malts is around 2 EBC units, derived from absorbance at 430 nm. The rice malts are paler than pilsner malt (2.5–4.5 EBC units).

Table 3 reports the quality attributes of rice malt [7]. The fine extracts showed very good results considering the fact that the hull accounts for up to 20% of the kernel's mass and contributes less to the extract. The soluble nitrogen content was low if compared to barley malt. The FAN values were surprisingly high compared to barley malt wort in relation to the soluble nitrogen content. Normally in a good brewing malt, the FAN content is around 22%

Attributes	
Moisture	4.3
Fine extract (% d.m.)	71.7–72.7
Saccharification time (min)	50
pH	5.76–5.96
Color (EBC-U)	1.7–2
Total nitrogen (g/(100 g of d.m.))	1.34–1.43
Soluble nitrogen (mg/l at 8.6°P)	450–506
Kolbach index (%)	29.5–27.4
Viscosity (mPa S at 8.6°P)	1.40–1.48
Free amino nitrogen (FAN) (mg/l at 8.6°P)	148–166
Fermentability (%)	71.2–73.2
Diastatic power (WK)	42

°P, Plato degree; d.m., dry matter; WK, Windish Kolbach.

Table 3. Quality attributes of a rice malt.

http://dx.doi.org/10.5772/66450

of the soluble nitrogen. This could be an overestimation that depends on the method for FAN determination, which is a color reaction with ninhydrin.

5. Brewing

Brewing process consists of mashing, the mix of grinded malt with water to obtain conversion of starch in sugar (wort); lautering, filtration of wort through the grain bed for the clarification and recover of sugar; boiling of the wort for the inactivation of enzymes, concentration and formation of color and flavor and fermentation of the wort by the yeast to produce alcohol. The resulting beers can vary depending on raw materials, temperatures and times used for the production of the worts, type of yeast and temperatures of the fermentation. In most conventional breweries, the malted grains are primarily barley and wheat. Other grains are used as adjuncts to provide a carbohydrate source for alcohol production, to add body or foaming characters to the final product and also to attenuate the negative aspects of malt that may be more evident in lighter style beers. There are breeding efforts to produce dual food and brewing varieties of rice, but worldwide, it is malting barley that breeders focus on to produce a cereal specifically adapted for use in brewing [14].

5.1. The rice as adjunct

Rice is one of the most important cereals used as adjunct in brewing. There are several advantages of using adjunct in brewing. It has lower cost compared to malt, in case of suboptimal malting facilities and malting conditions, adjunct material can be used to supplement the sugar content of barley and wheat malt [28]. A lower production tax in some countries, for example, in Kenya and Japan, beer made from high percentage of adjunct is less taxed than beer made from malted grain [29, 30]. An improvement in the quality and a characterization of the product can also be obtained using adjunct. Rice indeed has a decisive impact on the flavor, color and colloidal stability of an American pale lager. As brewing adjunct, rice has a very neutral flavor and aroma, and when properly converted in the brewhouse it yields to a light clean-tasting beer. Rice for brewing is a by-product of the edible rice milling industry, any kernels that may get fractured during the milling process (~30%) are considered undesirable are therefore sold to the brewing industry at a cheaper price. The main rice adjunct types currently available to the international brewing industry are: gritz, flaked, extrusion cooked and flour/starch. The physicochemical properties of that adjunct will dictate its addition rates to a grist recipe, its time of addition and how it will be processed. The quality of rice can be judged by several factors. A high lipid content can cause reduced foam stability, flavor problems and gelatinization difficulties. Rice should therefore contain less than 1.5% lipid. Rice supplies little free amino nitrogen, and this deficiency should be balanced. Extract yield differences depend greatly on the rice cultivar. It is important that rice is finely milled before brewing, otherwise gelatinization problems will occur. Not all varieties of rice are acceptable brewing varieties. As already stated (see Section 3.1.), short-grain rice is preferred because medium- and long-grain varieties can lead to viscosity problems [31]. The rice starch swells greatly during the gelatinization, and this can lead to the sticking of the pasta to the hot surfaces of the plant and cause burns. It is therefore necessary to find a compromise so that the malt α-amylases are able to liquefy the viscous paste and make it less viscous. The α-amylases are, however, rapidly inactivated at temperatures of about 80°C and are no longer able to liquefy the starch. As described, is it clear that the use of unmalted rice in brewing is not very easy. There are different solution for the use of rice in mashing [5, 14, 27]:

(1) An adjunct cooker is preferably used. This is a closed kettle in which the rice mash is gelatinized under pressure (i.e. at a temperature above 100°C).

(2) The rice grist is mashed with 10–20% of the malt mash and held at 78°C. Thereby almost all the starch is gelatinized and liquefied.

(3) There are rice varieties that do not gelatinize below 80°C. The rice mash must therefore, to be certain, be heated to 85–90°C, gelatinized at that temperature and then cooled again to 70–75°C in order to be saccharified in a shorter time by the addition of a malt mash.

(4) Another possibility is to heat the rice mash with 10–20% of the malt mash slowly to above 80°C to liquefy the gelatinizing starch by the still active α-amylase in the malt.

(5) A very reliable method is to add commercial, heat-resistant bacterial α-amylase, which is still active at temperatures above 80°C and consequently liquefies the rice mash.

(6) A double-mashing system was developed in North America, to deal with grist containing large proportions of rice. The adjunct mash containing a small proportion of enzyme-rich malt or bacterial amylase is mashed with different temperature steps and is brought to 85±100°C standing until the starch is gelatinized. Meanwhile, the malt mash has been mashed in at 35°C and then mixed together.

(7) Flaked rice has the advantage of being pre-gelatinized.

5.2. The all rice malt beer

5.2.1. Wort production

Once the malt grain is milled, it is added to a large vessel called the mash tun and mixed with hot water to form the mash. The heat from the water activates the enzymes in the malt. These enzymes then convert the starch of the grains into sugars. There are several different types of enzymes within malt, which work at specific temperatures (see Section 4). The mashing process can be carried out by infusion or decoction procedure. In the mashing decoction method, a part of the mash is separated and brought to high temperatures to gelatinize the starch and re-added to the main vessel, thus increasing the temperature of the mash. This process can be repeated several times (one, two or three mash procedure). This procedure could be useful when brewing rice malt because of its high gelatinization temperature. Some authors performed a three mash decoction procedure to obtain an all rice malt beer [32]. Grinded rice malt and water are mixed and mashed at 50°C for 30 min. The mash is decanted and the supernatant rich in enzymes is separated. The bottom part of the mash is heated to 88°C to completely gelatinize the starch and then collected with supernatant reaching thus 62°C. Two further decoction steps are then performed increasing the temperature of the mash to 67°C and 70°C. Finally, the mash temperature is raised to 74°C. Otherwise, a one mash decoction procedure can be used to obtain an all rice malt beer [33]. Rice malt and water are mixed at 35–37°C for 10 min, one-third of the mash is transferred to a pot and then heated to the boiling point and re-added. Then, the mash is rested to different temperatures to allow different enzymes work properly; 10 min at 50–53°C for proteolytic enzymes, 15 min at 63–65°C for amylolytic enzymes and 60min at 71–73°C, 20 min at 75°C and 5 min at 77°C. Concerning infusion process is simpler than decoction because the entire mash is always kept together. The total mash is heated with rests being used at temperatures determined by the enzyme properties [27]. Nevertheless having rice malt different enzymatic pool from barley malt, an adapted mashing step temperatures in the infusion method need to be used. Some authors reported the infusion mashing program for rice malt in a pilot scale [33]. Rice malt and water were mashed at about 40°C. The mash was allowed to rest at 50°C for 10 min, then at 62°C for 50 min and finally at 72°C for 90 min. During the last rest, the conversion of starch in fermentable sugars was tested by the iodine test and it resulted incomplete. At the end of the process, the mash was heated at 77°C. The pH of the mash was adjusted to a value of 5.5–5.6. Otherwise rice malt can be mashed in a pilot scale brewery by the following procedure [34]. The pH is adjusted with lactic acid to 5.3, and the brewing water is added with CaCl. The infusion mash program is the following: 30 min at 45°C, 45 min at 65°C, 60 min at 74°C and 10 min at 78°C. The used pH value enhances the activity of the most part of hydrolytic enzymes in the starch and cell wall degradation except for α-amylase whose optimum pH is 5.6–5.8. Whereas limit dextrinase (optimum pH value 5.4–5.5) in rice malt is 10-fold higher than in barley malt, the adequate temperature rests at 65°C for limit dextrinase and β-amylase and 74°C for α-amylase, lead to complete saccharification 1 h after reaching 74°C. After the mashing step, the lautering process is carried out in a filtering system (mash tun) for the separation of the spent grains from the wort. The wort passes through a "bed" of spent grains to allow the recovery of the sugar and the clarification. It is a critical step because a slowdown of the wort flow, or even a stop can occur if the bed is not suitable. Barley husks play an important role in the lautering bed allowing the correct flow. Also rice husks are suitable for this process. Lautering problems occur with rice malt if the saccharification result is incomplete and the starch forms a sticky paste [33]. On the contrary, no lautering problems are reported when complete saccharification occurred [34]. Boiling of the wort is a step for the concentration of the wort to the desired sugar content and the formation of aroma and can be 60–90 min time long.

5.2.2. *Wort quality*

Typical quality attributes for rice malt worts are shown in **Table 4**. pH values are quite similar to worts produced from barley malt (5.2–5.6). The lowest color of the boiled wort is 3.5 EBC units and the highest is 9.6 EBC units. A very clear color of the wort is a consequence of the low color of the malt. Rice malt kilning temperature is usually lower than in barley malt. Thus the rice malt is paler. Also the composition of the grain influence the color, in particular the low rice malt nitrogen content lead to low soluble nitrogen in the wort. Indeed, less Maillard products are produced that are responsible for the wort color and are produced by the interaction of sugar nitrogen and heating. Anyway a caramelization of wort sugars can occur because of direct-fired vessel used in some boiling system causing highest color in the wort.

	Decoction [33]	Infusion [33]	[34]
Original gravity (°P)	10.34–12.20	11.85–12.30	12.65–12.78
pH	5.20–5.98	5.12–5.91	5.20–5.41
Color (EBC)	3.5–9.9	3.7–9.6	5.4–6.9
Brewhouse yield (%)	42.4–63.3	49.2–65.4	64–66
Saccharification (min)	Incomplete	Incomplete	Complete

Table 4. Comparison of the quality attributes of the different rice malt worts obtained by different production process and rice varieties.

The brewhouse yield indicates what percentage of the grist charge is available as extract content in the cast wort. The brewhouse yield is therefore an important internal brewery measure of the efficiency of the brewhouse operations. This percentage (brewhouse yield) is usually between 75% and 80%. The brewhouse yields obtained for rice malt beer are lower than those usually obtained with barley malt. This can be due to an incomplete saccharification, which is sometimes further reduced by a more difficult filtration process. However, it is to be noted that also when the starch is completely converted a bad filtration process adversely affects the brewhouse yield. Complete saccharification, ascertained with the iodine test, is achieved by the optimized mashing program with the starting temperature of 45°C for 30 min and the correction of the pH to 5.3 that improved protein degradation. This is in fact a prerequisite for complete saccharification of the starch.

Sugar profile of rice malt wort is the result of the enzymatic pool that worked during mash steps. Successful fermentation also depends on the wort sugar composition. **Table 5** shows the different sugar profiles of the rice malt worts compared to barley malt wort.

	Decoction [33]	Infusion [33]	[34]	Barley malt wort [35]
Fructose			1.0–1.1	2.1±0.3
Glucose	17–20	16–23	33.8–36.9	13.8±2.1
Sucrose			1.1–1.6	3.0±0.7
Maltose	27–40	34–45	22.6–38.3	73.5±5.4
Maltotriose	9–12	9–10	15.5–15.8	19.6±2.2
Maltotetraose	4–6	4–6	12.3	5.0±1.2
Maltopentaose			4.3–5.9	1.6±0.2
Maltohexaose			12.7–13.7	1.8±0.3
Maltoheptaose			0.5	1.5±0.2

Table 5. Sugar profile of different rice malt worts (g/L) from different production process and rice varieties.

All worts contained an acceptable sugar composition and besides glucose and maltose, maltotriose and maltotetraose were also present but in lower amounts. In the barley malt wort, the most abundant fermentable sugar is maltose. By using an infusion program that lead to a complete saccharification, the obtained rice malt worts show a maltose content in the same range of glucose and there are higher levels of maltotetraose, maltopentaose and maltohexaose. This is an indication of the different amylolytic enzymes working in the rice malts. Moreover, the sum of the fermentation onset stage sugars glucose, fructose and sucrose are between 42% and 49% of the total content of the fermentable sugars. This value should not exceed 25%, because maltose and maltotriose utilization could be delayed causing a slow primary fermentation [34]. On the contrary, some authors find maltose as main sugar, ranging between 27 and 45 g/L, followed by glucose in the range 16–23 g/L. Anyway, with the mashing programs used in these studies, both decoction and infusion method, a complete saccharification was not achieved [33].

5.2.3. *Wort fermentation*

In the fermentation of the wort, the fermentable sugars are converted into alcohol and carbon dioxide by the yeast (*Saccharomyces cerevisiae* or *Saccharomyces pastorianus*). Fermentation can be conducted at high (18–24°C) or low temperatures (8–13°C), giving top fermented or bottom fermented beer, respectively. Rice beer worts are fermented at low temperatures.

5.2.4. *Beer quality*

Typical quality attributes for rice malt beer are shown in **Table 6**. The sugars in the wort are not totally converted by the yeast, depending on the wort suitability in terms of sugars and nitrogen composition. The attenuation limit is the highest apparent degree of attenuation, which can be reached by fermentation of all fermentable materials in the extract. It is predetermined by the action of starch-degrading enzymes in the brewhouse. In the rice malt beer, attenuation limit values calculated according to original and final gravity is acceptable, even if in barley malt wort, higher values can be found (80–84%) [27]. In some beers because of problems encountered during fermentation, the attenuation results low (e.g. 50–60%). Problems in the fermentation can occur because of incomplete saccharification and also because of the low content of nitrogen. The rice beer has an alcoholic content of about 3.5–5.1% vol., that is, a usual value for a barley malt beer. The beers have a color very similar to the worts from

	Decoction			Infusion
	[32]	[33]	[33]	[34]
Attenuation (%)	79–81	50–73	59–70	69–76
Alcohol wt/wt (%)-% vol	3.9–4.2	3.7–4.5	3.6–4.5	4.6–5.1
Original gravity (°P)	10.8–11.9	10.3–12.20	11.85–12.30	12.5
pH	4.3–4.4	4.23–4.81	4.35–4.84	4.21–4.24
Color (EBC)	7.2–8.0	3.3–9.6	3.5–9.41	4.3–5.0
Foam stability (s)	30–40.5	not detected	not detected	157–169

Table 6. Quality attributes of rice malt beers from different production process and rice varieties.

which they are produced. The pH is comparable to that of a barley malt beer. Foam as color is a hedonistic aspect of the beer that is very appreciated by the consumer in most of the case. It is formed by the interaction of medium molecular weight proteins and carbon dioxide. Rice malt wort in most case have a sub optimal content of nitrogen that lead to a low foam stability. In a barley beer, good values of foam are higher than 200 s.

Flavor of beer is characterized of typical volatile compound profile affected principally by the yeast metabolism. The low amount of free amino nitrogen present in rice malt wort promotes an overproduction of fermentation by-products like higher alcohols. Indeed, higher alcohols are present in great amounts in the rice malt beers (**Table 7**) but they do not exceed

	Rice malt beer	**Literature value of barley malt bottom-fermented beer**	**Threshold limit in barley malt beer****
Higher alcohols			
1-propanol (mg/L)	16.2–17.6	5–20*	800
2-Methyl-1-propanol (mg/L)	34.1–37.4	5–20*	200
3-Methyl-1-butanol (mg/L)	58.8–60.1	30–70*	65
2-Methyl-1-butanol (mg/L)	26.4–28.8	8–30*	70
2-Phenylethanol (mg/L)	23.0–30.4	8–40*	125
2-Furanmethanol (mg/L)	0.5–0.6	–	3
Esters			
Ethyl acetate (mg/L)	9.9–23.0	10–40*	33
Ethyl butanoate (mg/L)	0.1–12.0	0.05–0.15*	0.4
3-Methylbutyl-1-ethanoate (mg/L)	0.1–0.6	0.5–3*	(1.2) 1.6
Ethyl hexanoate (mg/L)	0.1–0.3	0.05–0.3*	0.23
Aldehydes			
Ethanal (mg/L)	19.1–41.9	2–10*	10
2-Methyl-1-butanal (μg/L)	7.6–12.3	60**	1250
3-Methyl-1-butanal (μg/L)	24.2–46.0	20**	600
Hexanal (μg/L)	19.1–21.8	4.5**	300
Furfural (μg/L)	41.7–66.6	40**	150,000
Methional (μg/L)	15.4–25.1	–	250
Phenylacetaldehyde (μg/L)	15.7–57.8	45**	1600
Trans-2-nonenal (μg/L)	not detected	–	0.3
Sulfur compounds			
Dimethylsulfide (μg/L)	65.5–72.9	<100**	100

Ref. [36].
Ref. [27].

Table 7. Volatile compounds in rice malt beer.

the threshold limit. The aroma-active esters, other fermentation by-products that are highly desired in beer because of their fruity, were present in small amounts [37]. Aldehydes are generally off-flavors in beer, and they are produced by oxidation of the corresponding alcohols or are derived from fatty acids and lipids present in the malt and formed during the various stages along the malting and brewing process [38]. Anyway no particular off-flavor was revealed even though ethanal exceeded the perception threshold limit of 10–25 mg/L. The concentrations of volatile compounds in the rice malt beers are in the range of a barley malt beer [27]. The dimethyl sulfide content is an off-flavor derived from malt whose content is below the threshold limit in rice malt beer.

The sensory analysis conducted on the beers shows a pale yellow color and a white coarse foam which rapidly collapsed. Rice malt beer has a relatively flat character. From a sensory test, Vanillin flavor is revealed. The sensory profile of the rice malt beer was similar to a barley malt beer in aroma, taste and mouthfeel, but more flat.

Nowadays, the advancement in research about the use of rice as raw material to produce malt and rice malt beer shows encouraging results. The rice is a suitable grain despite the believed low enzymes content for use in brewing. Rice malt beer can be produced in a traditional way leading to well-fermented beers with no off-flavor comparable to barley malt bottom-fermented beer.

Author details

Ombretta Marconi[1*], Valeria Sileoni[1], Dayana Ceccaroni[1] and Giuseppe Perretti[2]

*Address all correspondence to: ombretta.marconi@unipg.it

1 Department of Agricultural, Food and Environmental, University of Perugia, Perugia, Italy

2 Italian Brewing Research Centre, University of Perugia, Perugia, Italy

References

[1] Food and Agriculture Organization of the United Nations (FAO) "Rice Market Monitor". http://www.fao.org/economic/RMM, volume XIX, issue no. 1, April 2016.

[2] de Lourdes Moreno M, Comino I, Sousa C, Alternative Grains as Potential Raw Material for Gluten Free Food Development in the Diet of Celiac and Gluten Sensitive Patients. Austin Journal of Nutrition and Food Sciences. 2014;**2(3)**: 1016–1025. DOI:10.3390/nu5104250

[3] Meussdoerffer F, Zarnkow M. Starchy Raw Materials. In: Esslinger HM editor. Handbook of Brewing – Process, Technology, Marketing. 1st ed. Wiley-VCH Verlag GmbH and Co. KGaA, Weinheim, Germany, 2009. pp. 43–84. DOI:10.1002/9783527623488.ch2

[4] Hornsey I. Barley and Malt. In: Hornsey I editor. Brewing. 2nd ed. The Royal Society of Chemistry; Thomas Graham House, Science Park, Milton Road, Cambridge CB4 0WF, UK, 2013. pp. 25–65. ISBN:978-1-84973-602-2

[5] Briggs DE. Grain and Pulses. In: Briggs DE editor. Malts and Malting. 1st ed. Blackie Academic and Professional, 2-6 Boundary Row, London SE18HN, UK, 1998. pp. 35–78. ISBN:0 412 29800 7

[6] Stewart G. Adjuncts. In: Bamforth CW editor. Brewing Materials and Processes – A Practical Approach to Beer Excellence. 1st ed. Academic Press 125 London Wall, London EC2Y 5AS, UK, 2016. pp. 27–46. DOI:10.1016/B978-0-12-799954-8.00002-2

[7] Mayer H, Marconi O, Regnicoli GF, Perretti G, Fantozzi P. Production of a Saccharifying Rice Malt for Brewing Using Different Rice Varieties and Malting Parameters. Journal of Agricultural and Food Chemistry. 2014; **62**: 5369–5377. DOI:10.1021/jf501462a

[8] Zhou Z, Robards K, Helliwell S, Blanchard C. Composition and Functional Properties of Rice. International Journal of Food Science and Technology. 2002; **37**: 849–868. DOI:10.1046/j.1365-2621.2002.00625.x

[9] Taylor JRN, Dlamini BC, Kruger J. Anniversary Review: The Science of the Tropical Cereals Sorghum, Maize and Rice in Relation to Lager Beer Brewing. Journal of the Institute of Brewing. 2013; **119**: 1–14. DOI:10.1002/jib.68

[10] Azhakanandam K, Power JB, Lowe KC. Qualitative Assessment of Aromatic Indica Rice *Oryza sativa* L. Proteins, Lipids and Starch in Grain from Somatic Embryo- and Seed-Derived Plants. Journal of Plant Physiology. 2000; **156**: 783–789. DOI:10.1016/S0176-1617(00)80248-5

[11] Mano Y, Kawaminami K, Kojima M, Ohnishi M, Ito S. Comparative Composition of Brown Rice Lipids (Lipid Fractions) of Indica and Japonica Rices. Bioscience, Biotechnology, and Biochemistry. 1999; **63**: 619–626. DOI:10.1271/bbb.63.619

[12] Resurreccion AP, Juliano BO, Tanaka Y. Nutrient Content and Distribution in Milling Fractions of Rice Grain. Journal of the Science of Food and Agriculture. 1979; **30**: 475–481. DOI:10.1002/jsfa.2740300506

[13] Jane J, Chen YY, Lee LF, McPherson AE, Wong KS, Radosavljevic M, Kasemsuwan T. Effects of Amylopectin Branch Chain Length and Amylose Content on the Gelatinization and Pasting Properties of Starch. Cereal Chemistry. 1999; **76**: 629–637. DOI:10.1094/CCHEM.1999.76.5.629

[14] Goode DL, Arendt EK. Developments in the Supply of Adjunct Materials for Brewing. In: Bamforth CW editor. Brewing New Technologies. Woodhead Publishing Ltd; 2006, pp. 30–31, 36–37. DOI:10.1533/9781845691738.30

[15] De Meo B, Freeman G, Marconi O, Booer C, Perretti G, Fantozzi P. Behaviour of Malted Cereals and Pseudo-cereals for Gluten-free Beer production. Journal of the Institute of Brewing. 2011; **117** (**4**): 541–546. DOI:10.1002/j.2050-0416.2011.tb00502.x

[16] Ceppi ELM, Brenna OV. Experimental Studies to Obtain Rice Malt. Journal of Agricultural and Food Chemistry.2010; **58**: 7701–7707. DOI:10.1021/jf904534q

[17] Owusu-Mensah E, Oduro I, Sarfo KJ. Steeping: A Way of Improving the Malting of Rice Grain. Journal of Food Biochemistry. 2011; **35**: 80–91. DOI:10.1111/j.1745-4514.2010.00367.x

[18] Nakai H, Ito T, Hayashi M, Kamiya K, Yamamoto T, Matsubara K, Kim YM, Jintanart W, Okuyama M, Mori H, Chiba S, Sano Y, Kimura A. Multiple Forms of Alpha-glucosidase in Rice Seeds (*Oryza sativa* L., var Nipponbare). Biochemistry. 2007; **89**: 49–62. DOI:10.1016/j.biochi.2006.09.014

[19] Yamasaki Y, Nakashima, S, Konno H. Pullulanase from Rice Endosperm. Acta Biochimica Polonica. 2008; **3**: 507–510.

[20] Usansa U, Burberg F, Geiger E, Black W, Chokchai W, Arendt EK, Kreisz S, Boonkerd N, Teamuroong N, Zarnkow M. Optimization of Malting Conditions for Two Black Rice Varieties, Black Non-waxy Rice and Black Waxy Rice (*Oryza sativa* L. indica). Journal of the Institute of Brewing. 2011; **117**: 39–46. DOI:10.1002/j.2050-0416.2011.tb00441.x

[21] Keßler M, Zarnkow M, Kreisz S, Back W. Gelatinisation Properties of Different Cereals and Pseudocereals. Monatsschrift fur Brauwissenschaft. 2005; **58**: 82–88.

[22] Dziedzoave NT, Graffham AJ, Westb A, Komlaga G. Comparative Assessment of Amylolytic and Cellulolytic Enzyme Activity of Malts Prepared from Tropical Cereals. Food Control. 2010; **21**: 1349–1353. DOI:10.1016/j.foodcont.2010.04.008

[23] Bunzel M, Steinhart H. allaststoffe aus Pflanzenzellwänden. Ernährungs-Umschau. 2003; **50** (**12**): 469–475.

[24] Adebowale AA, Sanni SA, Karim OR, Ojoawo JA. Malting haracteristics of Ofada Rice: Chemical End Sensory Qualities of Malt from Ofada Rice Grains. International Food Research Journal. 2010; **17**: 83–88.

[25] Agu R, Yukihiro C, Goodfellow V, MacKinlay J, Brosnan J, Bringhurst T, Jack F, Macdonald Harrison B, Pearson S, Bryce JH. Effect of Germination Temperatures on Proteolysis of the Gluten-free Rice and Buckwheat During Malting and Mashing. Journal of Agricultural and Food Chemistry. 2012; **60**: 10147–10154. DOI:10.1021/jf3028039

[26] Usansa U, Sampong N, Wanapu C, Boonkerd N, Teaumroong N. The Influences of Steeping Duration and Temperature on the α- and β-amylase Activities of Six Thai Rice Malt Cultivars (Oryza sativa L. indica). Journal of the Institute of Brewing. 2009; **115**: 140–147. DOI:10.1002/j.2050-0416.2009.tb00359.x

[27] Kunze W. Technology Brewing and Malting. 3rd ed. Berlin: Versuchs und Lehranstalt fùr Brauerei in Berlin; 2004. p. 949. ISBN:978-3-921690-77-2

[28] Grujicâ O, Application of Unconventional Raw Materials and Procedures in Wort Production. Journal of the Institute of Brewing. 1999; **105**: 275–278. DOI:10.1002/j.2050-0416. 1999.tb00520.x

[29] Cege P, Shah S, Kubai E, Kenyan Beer Brewed with Unmalted Barley. Ferment. 1999; **12**: 41–45.

[30] Shimizu C, Ohno M, Araki S, Furusho S, Watari J, Takashio M. Effect of Reduction of Carbonyl Compounds by Yeast on Flavor Stability of Happoshu. Journal of the American Society of Brewing Chemists. 2002; **60**: 122–129. DOI:10.1094/ASBCJ-60-0122

[31] Teng J, Stubits M, Lin E. The Importance of Rice Variety Selection for Optimum Brewhouse Operation, Proceedings of the 19th European Brewery Convention Congress, London. Oxford: IRL Press. 1983, pp 47–54.

[32] Okafor N Iwouno J. Malting and Brewing Qualities of Some Nigerian Rice (Oryza sativa L.) Varieties and Some Thoughts on the Assessment of Malts from Tropical Cereals. World Journal of Microbiology Biotechnology. 1990 **6**: 187. DOI:10.1007/BF01200940

[33] Ceppi ELM, Brenna OV. Brewing with Rice Malt – A Gluten-free Alternative. Journal of the Institute of Brewing, 2010; **116**: 275–279. DOI:10.1002/j.2050-0416.2010.tb00431.x

[34] Mayer H, Ceccaroni D, Marconi O, Sileoni V, Perretti G, Fantozzi P. Development of an All Rice Malt Beer: A Gluten Free Alternative. LWT – Food Science and Technology. 2016; **67**: 67–73. DOI:10.1016/j.lwt.2015.11.037

[35] Floridi S, Miniati E, Montanari L, Fantozzi P. Carbohydrate Determination in Wort and Beer by HPLC-ELSD. Monatsschrift Brauwissenschaft. 2001; **9/10**: 209–215.

[36] Mitteleuropeaische Brautechnische Analysenkommission. Collection of Brewing Analysis Methods: Wort, Beer, Beer-based Beverages. Freising-Weihenstephan: MEBAK e. V. 2013: p. 247.

[37] Rossi S, Sileoni V, Perretti G, Marconi O. Characterization of the Volatile Profiles of Beer using Headspace Solid-phase Microextraction and Gas Chromatography–Mass Spectrometry. Journal of the Science of Food and Agriculture 2014; **94**: 919–928 DOI:10.1002/jsfa.6336

[38] Baert JJ, De Clippeleer J, Hughes PS, De Cooman L, Aerts G. On the Origin of Free and Bound Staling Aldehydes in Beer. Journal of Agricultural and Food Chemistry. 2012; **60**: 11449–11472. DOI:10.1021/jf303670z

Section 2

Breeding for Quality

Chapter 5

Breeding Rice for Improved Grain Quality

Maxwell Darko Asante

Additional information is available at the end of the chapter

http://dx.doi.org/10.5772/66684

Abstract

Rice grain quality improvement has become very crucial for most breeding programs around the world. Grain quality is a complex trait which comprises milling, appearance (grain size and chalkiness), cooking and eating (starch properties including apparent amylose content (AAC), gelatinization temperature (GT), gel consistency and paste viscosity measured using rapid visco analyzer measured using rapid visco analyzer (RVA) as well as nutritional quality. Many genes/quantitative trait loci (QTLs) for the various quality traits have been identified/cloned. This has enabled the development of functional markers to facilitate the selection for this complex trait. Functional markers, especially those targeting mutations in the *BADH2*, *waxy*, *alk* and *GS3* genes, are highly associated with aroma, *AAC/RVA*, *GT* and grain size, respectively; and thus effective for marker-assisted breeding. Different alleles can be combined through gene pyramiding to improve rice grain quality for various consumers. To be able to meet future needs, rice breeders must exploit modern marker technologies such as genomic selection (GS) to take care of the effects of both major and minor genes for grain quality as well as high yield, abiotic and biotic stress tolerance.

Keywords: rice, grain quality, molecular markers, aroma, *waxy* gene, *alk* gene, *GS3* gene

1. Introduction

Rice is the most important source of calories for at least 50% of the world's population. Consequently, many countries around the world have strategies to achieve self-sufficiency in rice production by expanding the area under cultivation and or increasing yield per unit area. However, for rice, grain quality is as important as yield. This is because unlike other cereals, which are usually processed as food or feed (for animals), rice is mainly eaten as whole cooked grains by humans.

Breeding for consumer-preferred grain qualities have thus become a major goal for breeding programs around the world. To be able to breed for specific consumer preferences, grain quality must be clearly defined and the genes underlying their control deciphered. The rice grain is basically composed of the lemma and palea which form the hull, the bran, embryo and the endosperm (white rice) (**Figure 1**). Scientists have classified the grain quality of rice as milling, appearance, cooking and eating and nutritional aspects.

The milling quality of rice determines the yield and appearance of the rice after the milling process. It is thus sometimes classified as under appearance quality. The first step in milling involves the removal of the lemma and palea to obtain de-hulled rice called brown rice.

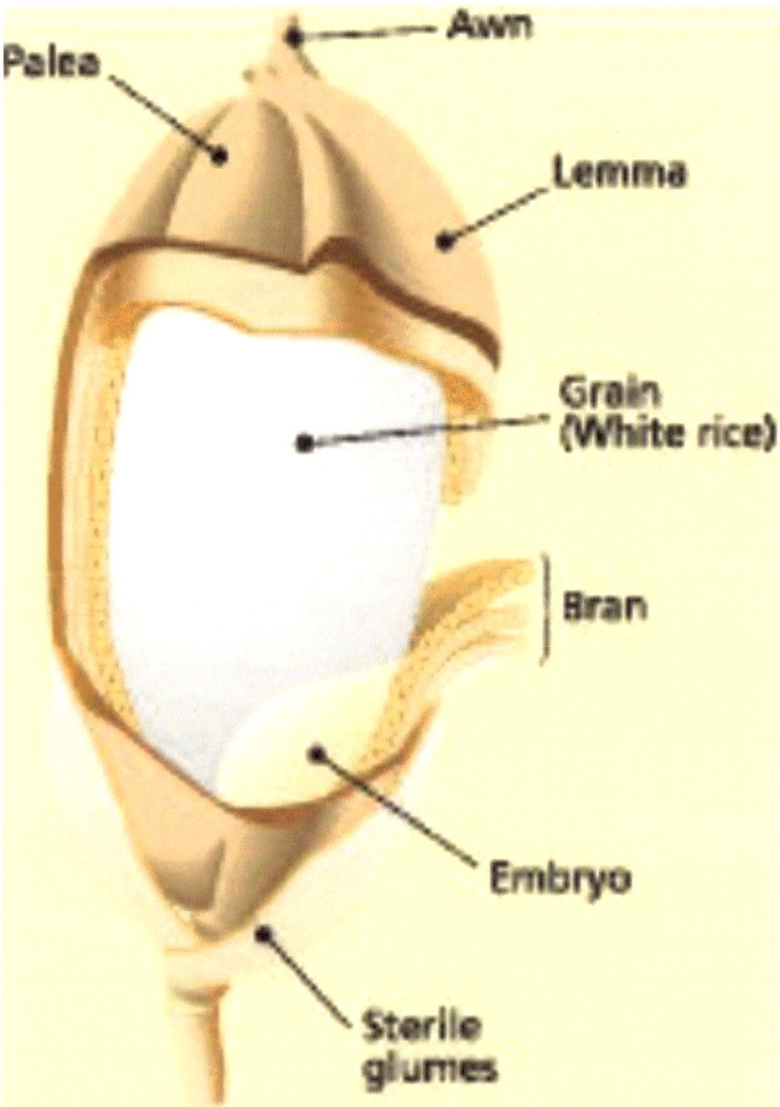

Figure 1. The rice grain. Source: IRRI.

Brown rice has become very important because it has good nutritional value obtained from the presence of the bran. The bran which consists of the aleurone, pericarp and embryo is removed to obtain milled rice. Milling quality thus comprises of brown, milled and head rice recovery (HRY). Brown rice recovery is the percentage of brown rice obtained after de-hulling a sample of paddy. Brown rice absorbs water poorly and does not cook as rapidly as milled rice. Milling recovery is the percentage of milled rice (including brokens) obtained from a sample of paddy. HRY is the percentage of head rice (excluding brokens) obtained from a sample of paddy. It normally includes broken kernels that are 75–80% of the whole kernel. High HRY is one of the most important criteria for measuring milled rice quality. Broken grain has normally only half of the value of head rice. HRY is influenced by genotype (i.e. the potential HRY), production factors and harvesting, drying and milling process. HRY and the degree of milling greatly influence the way the rice appears and thus consumer preference for rice on the market.

Appearance quality is how the rice appears after milling and it is associated with grain length, width, length-width ratio (shape) and translucency/chalkiness of the endosperm. Generally, most markets prefer translucent rice as opposed to chalky ones. Preferences for grain size and shape vary across different countries and cultures. Short grain varieties are preferred in Japan, Korea, Northern China and Sri Lanka while Southern China, India, Pakistan, Thailand, the USA, as well as most African countries prefer long and slender grain rice. Appearance quality has a direct influence on marketability and success of commercial varieties.

Cooking and eating quality is the easiness of cooking as well as texture, springiness, stickiness and chewiness of cooked rice. These characteristics are controlled by starch physicochemical properties comprising of apparent amylose content (AAC), gelatinization temperature (GT), gel consistency (GC) and paste viscosity properties. Starch makes up about 90% of the rice grain.

Rice starch is composed of two classes of glucose polymers—amylopectin and amylose. Amylose is a lightly branched linear molecule with a degree of polymerization (DP) of 1000–5000 glucose units and amylopectin has a much larger polymer unit containing frequent α-1,6 branching linkages [1, 2].

Amylose content is usually referred to as apparent amylose content (AAC) in literature because the iodine-based assay used for measuring it often detects long-chain amylopectin in addition to the "true" amylose [3]. AAC is the most important element influencing the cooking, eating and processing characteristics of rice [4]. The AAC of rice is known to play a crucial role in determining its cooked texture. It is directly related to water absorption, volume expansion, fluffiness and separability of cooked grains and inversely related to cohesiveness, tenderness and glossiness [5]. Typically, cereal grains contains 20–30% amylose with the remainder (70–80%) being amylopectin [6]. Rice can be classified based on its amylose content: waxy rice (0–2% amylose), very low amylose (3–9%), low amylose (10–18%), intermediate amylose (19–23%) and high amylose (>23%) [7, 8]. Synthesis of amylose is catalyzed by the granule-bound starch synthase (GBSS) protein which is encoded by the *waxy* gene *(Wx)* [9].

There are three alleles of the *waxy* gene—*Wx*, Wx^a and Wx^b which exist in waxy (sticky) rice, indica and japonica sub-species, respectively [10]. Rice plants with the Wx^a allele accumulate more GBSS protein in the endosperm during grain filling than plants with the Wx^b allele; hence indica rice generally has a higher content of AAC than japonica. The splicing pattern of the first intron of the *waxy* gene is reported to be highly correlated to the level of AAC [11]. The authors reported that transcripts in which the intron is completely spliced out would produce high AAC, ones with introns completely un-spliced would produce grains with no AAC (sticky rice) and those with partially spliced introns would produce intermediate AAC.

Gelatinization temperature (GT) is the range of temperature wherein at least 90% of starch granules swell irreversibly in hot water with loss of crystallinity and birefringence [12]. GT ranges from 55 to 85°C and determines the cooking time of rice [13, 14]. According to Tian et al. [72], "GT is a physical-chemical property which directly reflects the cooking quality of rice grain in terms of energy and time needed for cooking". Cooking time for rice can be reduced by up to four minutes by lowering the GT of the grain [14]. GT of rice can be classified as low (55–69°C), intermediate (70–74°C) or high (75–79°C) [15]. GT also plays an important role in water uptake, volume expansion and kernel elongation of cooked rice [16]. For example, high GT rice elongates less and is likely to be undercooked when standard cooking procedure is applied but its texture could be softened through overcooking. Although rice grain quality preferences differ around the world, varieties with intermediate GT are generally preferred [15]. GT is predominantly determined by amylopectin structure [17]. Three classes of enzymes including starch synthase, starch branching and starch de-branching enzymes have been implicated in the synthesis of amylopectin [9, 18, 19]. The gene, Starch synthase II (*SSIIa*) has been found to be the major determinant of GT [1, 17].

There two types of rice amylopectin—L- or S-types [19]. For L-type amylopectin the number of short α-1,4-glucan chains with degree of polymerization (DP) of ≤10 was less than 20% with DP ≤24. The amylopectin fine structure of japonica has more DP 7–11 chains and fewer $12 \leq DP \geq 24$ chains than indica [17]. Amylopectin side chains of $DP \geq 10$ form double helices; the length of these double helices determines gelatinization temperatures of starches [20]. Starch granules with amylopectin containing longer A and B1 chains would, therefore, be more resistant to gelatinization [17]. Consequently, japonica starch granules have lower GT than indica starch granules [21]. GT is highly correlated with alkali spreading value (ASV), which reflects the disintegration of milled rice in dilute KOH [8, 21].

Amylose content is the primary measure of the texture of cooked rice, but it fails to describe precisely the texture of certain types of rice [2]. Gel consistency (GC) is, therefore, used to complement AAC. Rice with the same amylose content can be classified as hard gel consistency (26–40 mm); medium gel consistency (41–60 mm); or soft gel consistency (61–100 mm) [22]. The length of gel flow is inversely proportional to GC; therefore, the long gel length corresponds to soft GC and the short gel length means hard GC. Amylopectin, rather than AAC, has been reported to be the major determinant of GC [23]. GC is reported to be affected by milling (lipid content), protein content, aging of milled rice (fat oxidation) and rice flour particle size (efficiency of dispersion) [24].

GC measurements are reported to have poor repeatability, and some laboratories have replaced them with starch paste viscosity parameters [8]. Starch viscosity curves are useful for breeding because the shape of the curve is unique to each class of rice [25]. Their use for breeding purposes was however limited because the traditional equipment used for measurement, Brabender Viscoamylograph (Brabender OHG, Duisberg, Germany), has many disadvantages. It requires a large amount of rice flour (40–50 g) and a long period (80–94 min) is required to run a sample [8, 25]. It has now largely been replaced by Rapid Visco Analyser (RVA, Newport Scientific Pty Ltd., Warriewood, Australia). RVA is popular because it is easy to operate, gives rapid results and requires a small sample (3 g) to run [25]. The rice starch viscosity profile, tested on the Rapid Visco Analyser, is usually called RVA profile. The RVA profile is generated by subjecting rice flour to a "heat-hold-cool-hold" temperature cycle [25]. This cycle (RVA) mimics the process of cooking and monitors the changes to slurry of rice flour and water, during the test [8]. The primary RVA parameters include peak viscosity, PV (first peak viscosity after gelatinization); trough or hot paste viscosity, HPV (paste viscosity at the end of the 95°C holding period) and final or cool paste viscosity, CPV (paste viscosity at the end of the test) [26]. Secondary parameters derived from the primary ones include breakdown (BD = PV − HPV); setback (SB = CPV − PV); consistency (CS = CPV − HPV); set back ratio (SBR = CPV/HPV) and stability (ST = HPV/PV) [26–28]. Other parameters include peak time (time required to reach peak viscosity), and pasting temperature (temperature of initial viscosity increase) [29].

Breeding for improved grain is complex because many of the quality traits are phenotyped using subjective and or expensive biochemical methods. Consequently, the scientific community has map/clone many quantitative trait locus (QTLs)/genes for various quality traits and developed molecular marker to facilitate selection for specific grain quality types.

In this chapter, the major breakthroughs in studying the genetics of grain quality will be highlighted and its usefulness in grain quality improvement as well as future trends in rice breeding is discussed.

2. Genetic analyses of grain quality in rice

2.1. Genetics of milling and appearance quality

2.1.1. Genetics of milling quality

Milling quality comprises of brown, milled and head rice recovery. It is a complex trait and its genetics has not been fully deciphered. A good review of studies carried out on milling quality has been done by Bao [30]. At least 20, 19 and 34 QTLs have been identified for brown, milled and head rice recovery, respectively [30].

2.1.2. Genetics of appearance quality

Many QTLs associated with grain size and length have been identified, including one QTL on chromosome 3 with a major effect on grain length/size [31–36]. The QTL on chromosome 3 was mapped to a region of 93.8 kb in length [31, 32]. Wan et al. [40] established that a major QTL controlling grain length, *qGL-3a*, was a single Mendelian gene—long grain was controlled by a recessive gene, *gl-3*. They further mapped the *gl-3* gene to a region of 87.5 kb and suggested that the *gl-3* gene could be the same as a grain size gene mapped for rice grain weight [31]. This gene, subsequently referred to as *GS3*, was cloned through comparative sequence analysis [37] and position cloning confirmed by transformation [38]. The researchers found that all the varieties with large grains had a nonsense mutation, in the second exon of the *GS3* gene which caused a 178-aa truncation in the C-terminus of the putative protein [37–39]. They found that the *GS3* locus was a major QTL for grain length and weight, and a minor QTL for grain width and thickness. Slender shape grain (large length:width ratio) was also reported to be controlled by the *GS3* locus (*gl-3* gene) [38, 40].

A seed width (*SW5*) QTL on chromosome 5 [41, 42] and a grain width and weight (*GW2*) QTL on chromosome 2 [43] have also been cloned. In all these cases, the authors reported that genotypes with the recessive allele(s) have longer, wider and/or heavier seeds than genotypes with the wild type allele and concluded that, in each case, the genes affect cell division.

More recently, a gene on chromosome 3, that controls grain length (*GL3*) and encodes protein phosphatase with Kelch-like repeat domain (*OsPPKL1*) has been identified and cloned [44–46]. The novel *qGL3-1* allele increased in a new variety [46]. Two other genes for grain width chromosomes 5 [47] and 8 [48] have been cloned.

The cloning of these genes provides a basis for marker-aided selection and QTL pyramiding in breeding for appearance-quality. Transgressive segregation for grain dimensions has also been reported by several authors [49–51]. This could make gain from selection very feasible.

2.2. Genetics of cooking and eating quality

2.2.1. Genetics of amylose content

Conventional genetic studies have revealed that AAC is controlled by one major gene with several modifiers [52–54]. High amylose content was reported to be dominant over low and intermediate amylose content [52, 53]. Other studies concluded that AAC was under the control of two complementary genes [54]. Using diallel analysis, the gene action for AAC was found to be mainly additive although dominance effects were also involved [55]. The authors found no evidence of maternal effects in their study. However, He et al. [60] said that the "inheritance of grain quality is more complicated than that of other agronomic traits in cereals due to epistasis, maternal and cytoplasmic effects, and the triploid nature of endosperm". The complex nature of inheritance of amylose content is supported by various reports [56, 57].

The advent of molecular marker technology is helping scientists to better understand complex quantitative traits [58]. Grain quality traits, including AAC, have been extensively studied using molecular marker–based quantitative trait locus (QTL) [49, 59–62]. AAC is reported to be mainly controlled by the *waxy* gene locus (*Wx*) on chromosome 6, which encodes the granule-bound starch synthase (GBSS) [9, 11, 60]. Minor QTLs for AAC have also been detected on chromosomes 1, 3, 4, 7, 8 and 11 [49, 62–64]. Using association analysis, the *Wx* gene was reported to act additively with five minor genes—*AGPlar, PUL, SSI, SSII-3* and *SSIII-2*—to affect AAC [65]. In at least three studies, QTLs for AAC were not associated with the *Wx* locus [32, 66, 67].This is presumably because the parents shared the same *Wx* allele, making it easier to detect other loci associated with the trait [32].

2.2.2. Genetics of gelatinization temperature

Monogenic [54], digenic [68] and polygenic inheritance [69] for gelatinization temperature or ASV, have been reported. Using an 8 × 8 diallel cross, the inheritance of GT was found to fit into additive-dominance with dominance effects being predominant [70]. Crosses in which one parent has high GT have been found to produce many high GT F2 individuals [2, 71]. High GT individuals in an F2 population are essentially homozygous and do not segregate in subsequent generations. F2 individuals with low and intermediate GT will continue to segregate into all three classes of GT for several generations.

Many molecular marker–based QTL analyses have been done for GT [49, 51, 59, 60, 67]. GT is reported to be mainly under the control of the *alk* locus on chromosome 6 that encodes soluble starch synthase II *(SSIIa)* [1, 17]. The *SSIIa* locus was reported to explain 25.5% of the variation in GT [1]. The authors found that, gene interactions as well as the *Wx* gene (4.7%) were important contributors to the variation in GT. The contribution of the *Wx* locus to the control of GT has also been reported by other researchers [29, 58, 62]. However, Tian et al. [72] reported that the control of GT was independent of the *Wx* locus. Other minor QTLs or modifier genes have been found to contribute to variations in GT [49, 58, 63]. Four minor genes including the *Wx* gene—*Wx, SBE3, ISA* and *SSIV-2*—have been reported to act additively with *SSII-3* (also referred to as *SSIIa* by other authors), the major for GT, to affect the trait [65].

2.2.3. Genetics of gel consistency

Some conventional genetic analyses have reported that GC is under the control of one major gene and several minor genes [73–75]. The difference between hard and soft gel, hard and medium as well as medium and soft was reported to be under monogenic control [73]. Hard gel was dominant over medium and soft gel; medium gel was dominant over soft. The authors concluded that it was possible to select for GC in early segregating generations. However, multigenic control of GC with genes acting additively has also been reported [76]. Using an 8 × 8 diallel cross, GC was found to fit into an additive-dominance model of inheritance with dominance being predominant [70]. Transgressive segregation has been observed for GC [63, 70, 77]. Using a genetic model for studying quantitative traits of triploid endosperm, cooking quality traits were found to be controlled mainly by genetic effects—seed, maternal and cytoplasmic. Cytoplasmic effects were the main components of GC [78]. Genotype by environment interactions as well as epistatic effects has also been found to play important roles in the control of GC [58, 78].

Many QTLs for GC have been mapped using molecular markers [1, 29, 61–63]. Some researchers reported that a single locus in the *Wx* region and some modifier genes control GC [62, 63]. He et al. [1] also reported that the *Wx* locus is the major determinant of GC but it explained only 38.9% of the phenotypic variation. Using association analysis, the *Wx* was confirmed as the major gene controlling GC [65]. A major QTL for GC at the *Wx* locus has been cloned [79].

In addition to the *Wx* gene, GC is reported to be affected by three other minor genes in the starch biosynthesis pathway—*AGPiso, SBE3* and *ISA* [65]. Two QTLs with minor effects on GC had earlier been reported [60, 67] and the *alk* locus has also been found to make minor contributions to GC [29].

2.2.4. Genetics of paste viscosity

The genetics of rice paste viscosity has not been studied as widely as AAC, GC and GT. RVA was reported to be controlled by a single locus [80]. The authors' finding was based on both $F_{2:3}$ segregation and diallel analyses. Subsequently, QTL mapping has been used to locate the chromosomal positions of genes controlling various RVA parameters in rice [29, 67, 81, 82].

In general, the major QTLs for most RVA parameters related to eating quality including HPV, CPV, BD, SB and CS were found at the *waxy* locus [29, 81–83]. In at least three studies, QTLs for PV were not found at the *Wx* locus [29, 67, 81]. Recently, a major QTL for PV was detected at the *Wx* locus [82].

QTLs related to the cooking process such as pasting temperature and peak time have generally been found at the *alk* locus on chromosome 6 [29, 82]. Minor QTLs for the various RVA parameters have been found on all 12 chromosomes of rice [29, 81–84].

2.2.5. Association between RVA parameters and AAC

RVA parameters including HPV, CPV, SB and CS have been found to be positively correlated with AAC while BD is negatively correlated to AAC [28, 29, 85]. AAC has been reported to be negatively [13, 86], positively [87] and not significantly [28, 29, 85] correlated with PV. It has been suggested that the SNPs in Ex 10 of the *waxy* gene has a confounding effect that influences the relationship between PV and AAC [88]. The authors found that the effect of AAC on PV would be different depending on the amount of TAC and GAT *waxy* gene haplotypes in the germplasm used by the various researchers. The significant correlations between AAC and most RVA parameters is not surprising because both traits are mainly controlled by the *waxy* gene [81, 89]. An SSR marker in the *Wx* gene (RM 190) explains a large portion of the variation in AAC and most RVA parameters [86, 90, 91]. The RVA parameters and AAC are also highly associated with three SNPs—intron 1 (G→T), exon 6 (A→C) and exon 10 (C→T) substitutions—in the *Wx* gene [86, 91, 92]. Using the three SNPs together, four *waxy* SNP haplotypes were found in a collection of world germplasm [86]. These *waxy* SNP haplotypes include TAC, GCC, GAC and GAT for low, intermediate, high and high AAC plus high RVA, respectively [86, 93].

2.2.6. Relationships between AAC, GC and GT

Since the rice grain is basically composed of starch (approximately 90%), genes involved in starch biosynthesis are naturally expected to affect cooking and eating qualities. Starch biosynthesis is a complex system composed of multiple subunits or isoforms of four classes of enzymes: ADP-glucose pyrophosphorylase (*AGP*), starch synthase *(SS),* starch branching enzyme (*SBE*) and starch debranching enzyme (DBE) [94, 95]. The effect of 18 genes involved in different steps of starch synthesis on AAC, GC and GT was investigated through association analysis [65]. These genes include: *AGP*, ADP-glucose pyrophosphorylase; *AGPlar, AGP* large subunit; *AGPiso, AGP* large subunit isoform; *AGPsma, AGP* small subunit; *GBSS*, granule-bound starch synthase; *SS (SS-I, SS-II-1, SS-II-2, SS-II-3, SS-III-1, SS-III-2, SS-IV-1* and *SS-IV-2*) soluble starch synthase; *SBE (SBE1, SBE3* and *SBE4*) starch branching enzyme; *ISA*, isoamylase; *PUL*, pullulanase; *ISA* and *PUL* belong to starch debranching enzyme (DBE). The authors found that genes related to starch synthesis cooperate with each other to form a fine regulating network that controls the eating and cooking quality of rice.

Tian et al. [65] confirmed earlier reports that *Wx* gene affects AAC, GC and GT ("three in one" function of the *Wx* locus)[29, 58, 62] and found strong evidence that the *Wx* gene does not only have a major effect on AAC, but also regulates GC as a major gene and GT as a minor one. Some authors have proposed that the *Alk* locus (*SSIIa*) only affects GT and GC (a "two in one function") [29, 58]. However, Tian et al. [65] found that the *Alk* locus (referred to as *SSII-3* in their report) plays an essential role not only in controlling GT and GC but also AAC. The authors also showed that some other starch synthesis related genes affect additively AAC, GC and GT as minor genes resulting in the fine complex network controlling cooking and eating qualities of rice grains.

AAC, GC and GT were found to be highly correlated: AAC is negatively correlated with GC (−0.91) and GT value (−0.46), whereas GC is positively correlated with the GT value (0.50) [65]. The negative correlation between AAC and GT was due to the natural occurrence of different haplotype combinations of *Wx* and *SSII-3* in rice germplasm. The *Wx* gene has three haplotypes: *Wx-III* is the wild type allele results in high AAC rice, *Wx-II* is associated with medium level of AAC and *Wx-I* has a loss-of-function mutation that produces waxy rice varieties [11]. Since AAC is negatively correlated with GC, varieties with *Wx-I* show high GC values, those with *Wx-II* show medium GC values, and those with *Wx-III* have low GC values [65]. The *SSII-3* has two allelic states: *SSII-3-I*, which leads to varieties that have higher GT values and *SSII-3-II* which leads to phenotypes with lower GT values.

The effect of either *Wx* or *SSII-3* on AAC and GT values was found to fall into a consistent pattern: *SSII-3-I* contributed to higher AAC under the same *Wx* background, whereas *SSII-3-II* led to lower AAC. *Wx* also combined with *SSII-3* to influence GT. For varieties that had the *SSII-3-I* background, *Wx-I* caused lower AAC and GT. GT value were however increased by *Wx-II* and *Wx-III*. They found five natural haplotype combinations between AAC and GT in their panel—*Wx-I/SSII-3-I, Wx-II/SSII-3-I, Wx-II/SSII-3-II, Wx-III/SSII-3-I* and *Wx-III/SSII-3-II*. Varieties belonging to *Wx-II/SSII-3-I* had medium AAC and high GT values and varieties belonging to *Wx-III/SSII-3-II* had high AAC and low GT values, these two haplotype combination form 71% of the germplasm. At the same time, varieties with *Wx-III/SSII-3-I* haplotype had high AAC and high GT values, and those with *Wx-II/SSII-3-II* had low AAC and low GT values. The *Wx-II/SSII-3-II* haplotype combination would be most useful for rice breeding in many parts of the world especially if it is combined with long-grain and fragrant/aroma alleles.

2.2.7. Inheritance of aroma

Monogenic, digenic and trigenic control of aroma has been reported by various authors [96–100]. However, some authors believed that aroma was quantitatively inherited [15]. The lack of agreement among researchers appears to be related to the different aromatic varieties and methods used in evaluating aroma [101].

However, the use of molecular markers to study the inheritance of fragrance appeared to favor monogenic recessive inheritance of fragrance [102–107]. A gene associated with fragrance was originally mapped by Ahn et al. [102] to rice chromosome 8, where it was associated with the RFLP marker, RG28. Bradbury et al. [105] identified a gene encoding betaine aldehyde dehydrogenase 2 (*BADH2*) as the likely cause of aroma in Basmati and Jasmine styled rices.

An eight base pair deletion (8-bp) and three SNPs in exon 7 of the *BADH2* gene distinguished fragrant from non-fragrant rices in that study. These polymorphisms served as the basis for developing an allele-specific marker for fragrant (Bradbury et al. [104]). A new fragrance allele with sequence identical to that of the *BADH2* allele in exon 7, but with a 7-bp deletion in exon 2 was identified as the cause of fragrance in some varieties [107]. Based on this information, the authors developed functional markers which can distinguish non-fragrant from fragrant rice and differentiate fragrance caused by 8-bp deletion on exon 7 from that caused by the 7-bp deletion on exon 2 of chromosome 8. In addition, 8 new alleles were discovered at the *BADH2* locus, all of which conferred fragrance in 24 accessions that did not carry any of the previously identified alleles [108, 109]. Another molecular marker study reported that three genes, located on chromosomes 3, 4 and 8, caused fragrance in Pusa 1121 [51]. The authors identified a BADH1 gene in the aroma QTL on chromosome 4 and also mapped the QTL on chromosome 8 to the *BADH2* region. The *BADH2* gene is known to code for 2-acetyl-1-pyrroline, or 2AP [110, 111]. The accumulation of 2AP has been explained by the absence of *BADH2* activity leading to increased levels of 4-aminobutyraldehyde/Δ^1-pyrroline, the immediate precursor of 2AP [112]. However, in another study, it was concluded that *BADH2* had no direct role in the synthesis of 2AP [113]. The authors found that Δ^1-pyrroline-5-carboxylate, usually the immediate precursor of proline, synthesized from glutamate, reacts directly with methylglyoxal to form 2AP. Fitzgerald et al. [14] declared that "the genetic and biochemical stories of 2AP synthesis are yet to be fully written".

2.3. Marker-assisted breeding for grain quality in rice

Breeding for improved grain is complex because many of the quality traits are phenotyped using subjective and or expensive biochemical methods. Consequently, the scientific community has map/clone many QTLs/genes for various quality traits and developed molecular markers to facilitate selection for specific grain quality types.

Most fragrant rices including Jasmine and Basmati types have the 8-bp deletion on exon 7 of the *BADH2* gene and an allele-specific marker has been developed for selecting rice with this mutation [104]. This marker is being used widely for selecting for aroma. It is a co-dominant marker (**Figure 2**) and thus very useful for marker-assisted backcrossing for recessive trait such aroma because selection of lines carrying the aroma gene can done in the heterozygote state without progeny testing.

Other researchers have also developed markers for the 8-bp deletion in exon 7 of chromosome 8 [51, 115]. Functional markers have also been developed for other alleles in the *BADH2* gene including a 7-bp deletion in exon 2 [107] and a 3-bp insertion in exon 13 found in aromatic rice varieties from Myanmar [116].

Functional markers for a *waxy* gene SSR called RM 190, and *waxy* SNPS on intron (In1), exon 6 (Ex6) and exon 10 (Ex10) are used to select for AAC and RVA around the world [93]. The haplotype across these three SNPs in the *waxy* gene (*waxy* SNP haplotypes) have been found to be more efficient in selecting for AAC and RVA than the RM 190 [86, 88]. Across the *waxy* SNP haplotypes (In1-Ex6-Ex10) TAC, GCC, GAC and GAT is highly associated with low AAC, intermediate AAC, high AAC and high AAC accompanied with high RVA paste viscosity.

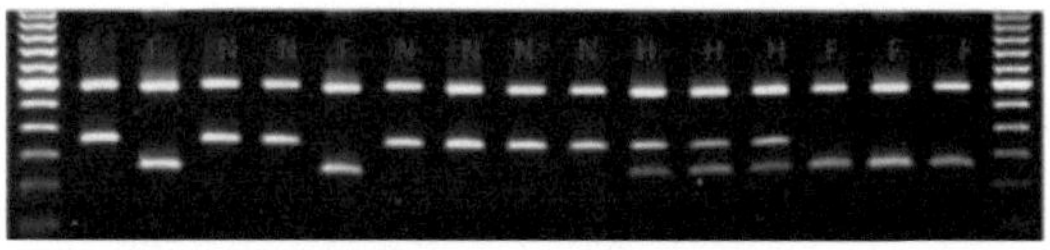

Figure 2. Agarose gel showing allele-specific marker for the 8-bp deletion in the *BADH2* gene. Lanes 1 and 17 = 100 bp DNA ladder. Fragrant individuals (F), non-fragrant individuals (N) and heterozygote types (H) [114].

The *alk* gene has been cloned [117] and validated as being the major gene for GT through genetic transformation [118]. Two SNPs (GC/TT and G/A) in the *alk* gene was found to be highly associated with GT [118]. These two functional SNPs have been used to developed DNA markers for selection of GT [118, 119].

Functional markers have also been developed for grain size [39, 120, 121]. These markers are very highly associated with the C-A SNP mutation in exon 2 of the *GS3* gene which is responsible for 80–90% of the variation in kernel length.

These validated markers will facilitate marker-assisted breeding for grain quality in rice. Various alleles of these important genes can be pyramided together to obtain the different consumer preferences for grain quality across countries and regions.

2.4. Future trends in rice breeding and grain quality improvement

By 2050, rice production must double in order to keep pace with population growth. Population growth will come with income growth so consumers will demand even higher quality rice. In addition to this challenge, new biotic and abiotic stresses are merging to due to climate change. Consequently, Rice breeders have to consider a huge number of simple and quantitative traits in combination when developing new lines while, at the same time, maintaining and improving grain quality.

Even though, MAS has been successfully used to improve some biotic, abiotic and quality traits in rice it is based on large effect QTLs/genes and does not take care of epistatic and genetic background effects. Most traits of interest to rice breeders are not controlled only by a few large-effect genes, but by a combination of many genes of small effect and/or major genes.

Genomic selection (GS) has been projected as alternative to conventional MAS. GS has huge potential to enhance breeding efficiency by increasing gain per selection per unit time [122]. GS breeding allows breeders to select the most desirable parents for the next generation using genome-wide DNA marker data. These parents are selected based on the relationship between the genome-wide markers and phenotypes of the individuals undergoing selection. The major advantage of GS over MAS is that genotyping is not restricted to selected markers that target genes with large effects, but rather all available marker data are used to predict breeding value. This helps to prevent loss of information. Genes with small effect can be tracked and selected for using information on all the marker data. GS would become more effective tool for increasing the efficiency of rice breeding as the costs of genotyping continue to decline [122].

2.5. Conclusion

Since rice is eaten mainly by humans as whole grain in cooked form, its grain quality is extremely important. The quality of the rice grain can be classified as milling, appearance, cooking and eating as well as nutritional quality. Different consumers around the world demand very specific measurements and combinations of the various aspects of rice grain quality. Breeding for these specific consumer demands can be challenging because grain quality is phenotyped using subjective, biochemical analyses that can be very expensive. Marker-assisted selection is thus a very good option for breeding for grain quality. The sequencing of the rice genome over a decade ago has made it possible for researchers to identify genes for the various grain quality traits. Functional molecular markers have been developed that are highly efficient in selecting for grain size, aroma, AAC, GT and paste viscosity parameters. These markers are increasingly being used for breeding for consumer-preferred grain qualities around the world. Modern genome-wide marker technologies which will take care of genes with small effect and allow breeders to simultaneously select for grain quality, yield and stress tolerance are recommended for future rice breeding work.

Acknowledgements

The Alliance for a Green Revolution in Africa (AGRA) funded the author's research on rice grain quality.

Author details

Maxwell Darko Asante

Address all correspondence to: mdasante@gmail.com

Council for Scientific Industrial Research, Crops Research Institute, Kumasi, Ghana

References

[1] He Y, Han Y, Jiang L, Xu C, Lu J, Xu M. Functional analysis of starch-synthesis genes in determining rice eating and cooking qualities. Molecular Breeding. 2006;18(4):277–90.

[2] Juliano BO. Rice chemistry and quality. Philippine Rice Research Institute, Manila; 2003.

[3] Takeda Y, Hizukuri S, Juliano BO. Structures of rice amylopectins with low and high affinities for iodine. Carbohydrate Research. 1987;168(1):79–88.

[4] Juliano BO. Amylose analysis in rice—a review. In Proceedings of the workshop on chemical aspects of rice grain quality, International Rice Research Institute, Los Banos, Laguna, Philippines. 1979;251–60.

[5] Juliano BO. A simplified assay for milled-rice amylose. Cereal Science Today. 1971;16: 334–40,60.

[6] Preiss J. Biology and molecular biology of starch synthesis and its regulation. Oxford Surveys of Plant Molecular and Cell Biology. 1991;7:59–114.

[7] Fitzgerald M. Starch. In Champagne ET, editor. Rice: Chemistry and Technology. AACC: St Paul, USA; 2004. 109–41.

[8] Bergman C, Bhattcharya K, Ohtsubo K. Rice end-use quality analysis. In Champagne E, editor. Rice Chemistry and Technology. AACC: St Paul; 2004. 415–72.

[9] Smith AM, Denyer K, Martin C. The synthesis of the starch granule. Annual Review of Plant Physiology and Plant Molecular Biology. 1997;48(1):67–87.

[10] Sano Y, Katsumata M, Okuno K. Genetic studies of speciation in cultivated rice. 5. Inter- and intraspecific differentiation in the *waxy* gene expression of rice. Euphytica. 1986;35(1):1–9.

[11] Wang Z-Y, Zheng F-Q, Shen G-Z, Gao J-P, Snustad DP, Li M-G, et al. The amylose content in rice endosperm is related to the post-transcriptional regulation of the *waxy* gene. The Plant Journal. 1995;7(4):613–22.

[12] R.K. Singh ,U.S. Singh G.S. Khush. Grain quality evaluation procedures. In Aromatic rices. Science Publishers, Inc.: Oxford; 2000. 15–28.

[13] Tan Y, Corke H. Factor analysis of physicochemical properties of 63 rice varieties. Journal of the Science of Food and Agriculture. 2002;82(7):745–52.

[14] Fitzgerald MA, McCouch SR, Hall RD. Not just a grain of rice: The quest for quality. Trends in Plant Science. 2009;14(3):133–9.

[15] Khush GS, Paule CM, DeLaCruz MN. Rice quality evaluation and improvement at IRRI. In Proceedings of the workshop on chemical aspects of rice grain quality, International Rice Research Institute, Los Banos, Laguna, Philippines. 1979;21–31.

[16] Tomar JB, Nanda JS. Genetics and association studies of kernel shape in rice. Indian Journal of Genetics and Plant Breeding. 1985;45(2):278–83.

[17] Umemoto T, Yano M, Satoh H, Shomura A, Nakamura Y. Mapping of a gene responsible for the difference in amylopectin structure between japonica-type and indica-type rice varieties. Theoretical and Applied Genetics. 2002;104(1):1–8. Epub 2003/02/13.

[18] Myers AM, Morell MK, James MG, Ball SG. Recent progress toward understanding biosynthesis of the amylopectin crystal. Plant Physiology. 2000;122(4):989–98.

[19] Nakamura Y, Sakurai A, Inaba Y, Kimura K, Iwasawa N, Nagamine T. The fine structure of amylopectin in endosperm from Asian cultivated rice can be largely classified into two classes. Starch-Starke. 2002;54(3–4):117–31.

[20] Safford R, Jobling SA, Sidebottom CM, Westcott RJ, Cooke D, Tober KJ, et al. Consequences of antisense RNA inhibition of starch branching enzyme activity on properties of potato starch. Carbohydrate Polymers. 1998;35(3–4):155–68.

[21] Little RR, Hilder GB, Dawson EH. Differential effect of dilute alkali on 25 varieties of milled white rice. Cereal Chemistry. 1958;35:111–26

[22] Cagampang GB, Perez CM, Juliano BO. A gel consistency test for eating quality of rice. Journal of the Science of Food and Agriculture. 1973;24(12):1589–94.

[23] Juliano BO, Perdon AA. Gel and molecular properties of nonwaxy rice starch. Starch-Starke. 1975;27(4):115–20.

[24] Perez CM. Gel consistency and viscosity of rice. In Proceeding of the workshop on chemical aspects of rice grain quality, International Rice Research Institute, Los Banos, Laguna, Philippines. 1979; 303–11.

[25] Juliano BO. Rice quality screening with the rapid visco analyser. In Walker CE, Hazelton JL, editor. Applications of the rapid visco analyser. Newport Scientific: Sydney; 1996. pp. 19–24.

[26] Bao JS, Xia YW. Genetic control of paste viscosity characteristics in indica rice (*Oryza sativa* L.). Theoretical and Applied Genetics. 1999;98(6):1120–4.

[27] Collado LS, Corke H. Properties of starch noodles as affected by sweet potato genotype. Cereal Chemistry Journal. 1997;74(2):182–7.

[28] Bao J, Shen S, Sun M, Corke H. Analysis of genotypic diversity in the starch physicochemical properties of nonwaxy rice: apparent amylose content, pasting viscosity and gel texture. Starch—Starke. 2006;58(6):259–67.

[29] Wang L, Liu W, Xu Y, He Y, Luo L, Xing Y, et al. Genetic basis of 17 traits and viscosity parameters characterizing the eating and cooking quality of rice grain. Theoretical and Applied Genetics. 2007;115(4):463–76.

[30] Jinsong Bao (2014). Genes and QTLs for Rice Grain Quality Improvement, Rice - Germplasm, Genetics and Improvement, Wengui Yan (Ed.), InTech, DOI: 10.5772/56621.

[31] Li J, Thomson M, McCouch SR. Fine mapping of a grain-weight quantitative trait locus in the pericentromeric region of rice chromosome 3. Genetics. 2004;168(4):2187–95. Epub 2004/12/22.

[32] Li J, Xiao J, Grandillo S, Jiang L, Wan Y, Deng Q, et al. Qtl detection for rice grain quality traits using an interspecific backcross population derived from cultivated Asian (*O. sativa* L.) and African (*O. glaberrima* S.) rice. Genome/National Research Council Canada. 2004;47(4):697–704. Epub 2004/07/31.

[33] Redoña ED, Mackill DJ. Quantitative trait locus analysis for rice panicle and grain characteristics. Theoretical and Applied Genetics. 1998;96(6):957–63.

[34] Tan YF, Xing YZ, Li JX, Yu SB, Xu CG, Zhang Q. Genetic bases of appearance quality of rice grains in shanyou 63, an elite rice hybrid. Theoretical and Applied Genetics. 2000;101(5–6):823–9.

[35] Thomson MJ, Tai TH, McClung AM, Lai XH, Hinga ME, Lobos KB, et al. Mapping quantitative trait loci for yield, yield components and morphological traits in an advanced backcross population between oryza rufipogon and the oryza sativa cultivar jefferson. Theoretical and Applied Genetics. 2003;107(3):479–93. Epub 2003/05/09.

[36] Bai X, Luo L, Yan W, Kovi MR, Zhan W, Xing Y. Genetic dissection of rice grain shape using a recombinant inbred line population derived from two contrasting parents and fine mapping a pleiotropic quantitative trait locus qgl7. BMC Genetics. 2010;11(1):16. Epub 2010/02/27.

[37] Fan C, Xing Y, Mao H, Lu T, Han B, Xu C, et al. *Gs3,* a major qtl for grain length and weight and minor qtl for grain width and thickness in rice, encodes a putative transmembrane protein. Theoretical and Applied Genetics. 2006;112(6):1164–71. Epub 2006/02/03.

[38] Takano-Kai N, Jiang H, Kubo T, Sweeney M, Matsumoto T, Kanamori H, et al. Evolutionary history of *gs3*, a gene conferring grain length in rice. Genetics. 2009;182(4):1323–34. Epub 2009/06/10.

[39] Fan C, Yu S, Wang C, Xing Y. A causal c-a mutation in the second exon of *gs3* highly associated with rice grain length and validated as a functional marker. Theoretical and Applied Genetics. 2009;118(3):465–72. Epub 2008/11/21.

[40] Wan X, Wan J, Jiang L, Wang J, Zhai H, Weng J, et al. Qtl analysis for rice grain length and fine mapping of an identified qtl with stable and major effects. Theoretical and Applied Genetics. 2006;112(7):1258–70.

[41] Shomura A, Izawa T, Ebana K, Ebitani T, Kanegae H, Konishi S, et al. Deletion in a gene associated with grain size increased yields during rice domestication. Nature Genetics. 2008;40(8):1023–8.

[42] Weng J, Gu S, Wan X, Gao H, Guo T, Su N, et al. Isolation and initial characterization of gw5, a major qtl associated with rice grain width and weight. Cell Research. 2008;18(12):1199–209. Epub 2008/11/19.

[43] Song XJ, Huang W, Shi M, Zhu MZ, Lin HX. A qtl for rice grain width and weight encodes a previously unknown ring-type e3 ubiquitin ligase. Nature Genetics. 2007;39(5):623–30. Epub 2007/04/10.

[44] Hu Z, He H, Zhang S, Sun F, Xin X, Wang W, et al. A kelch motif-containing serine/threonine protein phosphatase determines the large grain qtl trait in rice. Journal of Integrative Plant Biology. 2012;54(12):979–90.

[45] Zhang X, Wang J, Huang J, Lan H, Wang C, Yin C, et al. Rare allele of *osppkl1* associated with grain length causes extra-large grain and a significant yield increase in rice. Proceedings of the National Academy of Sciences. 2012;109(52):21534–9.

[46] Qi P, Lin Y-S, Song X-J, Shen J-B, Huang W, Shan J-X, et al. The novel quantitative trait locus *gl3.1* controls rice grain size and yield by regulating cyclin-t1;3. Cell Research. 2012;22(12):1666–80.

[47] Li Y, Fan C, Xing Y, Jiang Y, Luo L, Sun L, et al. Natural variation in gs5 plays an important role in regulating grain size and yield in rice. Nature Genetics. 2011;43(12):1266–9.

[48] Wang S, Wu K, Yuan Q, Liu X, Liu Z, Lin X, et al. Control of grain size, shape and quality by osspl16 in rice. Nature Genetics. 2012;44(8):950–4.

[49] Aluko G, Martinez C, Tohme J, Castano C, Bergman C, Oard JH. Qtl mapping of grain quality traits from the interspecific cross *Oryza sativa* x *O. glaberrima*. Theoretical and Applied Genetics. 2004;109(3):630–9. Epub 2004/04/24.

[50] Asante MD, Dartey PKA, Akromah R, Ofori J. Genetic analysis of grain size and shape in two rice crosses. Journal of Ghana Science Association. 2007;9(1):20–7.

[51] Amarawathi Y, Singh R, Singh AK, Singh VP, Mohapatra T, Sharma TR, et al. Mapping of quantitative trait loci for basmati quality traits in rice (*Oryza sativa* l.). Molecular Breeding. 2008;21(1):49–65.

[52] Kumar I, Khush GS. Genetic analysis of different amylose levels in rice. Crop Science. 1987;27(6):1167–72.

[53] Kumar I, Khush GS. Inheritance of amylose content in rice (*Oryza sativa* L.). Euphytica. 1988;38(3):261–9.

[54] McKenzie KS, Rutger JN. Genetic analysis of amylose content, alkali spreading score, and grain dimensions in rice. Crop Science. 1983;23:306–11.

[55] Kuo Y-C, Webb BD, Stansel JW. Griffing and hayman diallel analysis of variance for eating and processing quality parameters of miled rice. Journal of Agriculture Research China. 1997;46(1):15–31.

[56] Pooni HS, Kumar I, Khush GS. Genetical control of amylose content in a diallel set of rice crosses. Heredity. 1993;71(6):603–13.

[57] Bollich CN, Webb BD. Inheritance of amylose in two hybrid populations of rice. Cereal Chemistry. 1973;50:631–6.

[58] Fan CC, Yu XQ, Xing YZ, Xu CG, Luo LJ, Zhang Q. The main effects, epistatic effects and environmental interactions of qtls on the cooking and eating quality of rice in a doubled-haploid line population. Theoretical and Applied Genetics. 2005;110(8):1445–52.

[59] Govindaraj P, Vinod K, Arumugachamy S, Maheswaran M. Analysing genetic control of cooked grain traits and gelatinization temperature in a double haploid population of rice by quantitative trait loci mapping. Euphytica. 2009;166(2):165–76.

[60] He P, Li SG, Qian Q, Ma YQ, Li JZ, Wang WM, et al. Genetic analysis of rice grain quality. Theoretical and Applied Genetics. 1999;98(3–4):502–8.

[61] Septiningsih EM, Trijatmiko KR, Moeljopawiro S, McCouch SR. Identification of quantitative trait loci for grain quality in an advanced backcross population derived from the *Oryza sativa* variety ir64 and the wild relative *O. rufipogon*. Theoretical and Applied Genetics. 2003;107(8):1433–41. Epub 2003/09/27.

[62] Tan YF, Li JX, Yu SB, Xing YZ, Xu CG, Zhang Q. The three important traits for cooking and eating quality of rice grains are controlled by a single locus in an elite rice hybrid, shanyou 63. Theoretical and Applied Genetics. 1999;99(3–4):642–8.

[63] Lanceras JC, Huang Z-L, Naivikul O, Vanavichit A, Ruanjaichon V, Tragoonrung S. Mapping of genes for cooking and eating qualities in Thai jasmine rice (kdml105). DNA Research. 2000;7(2):93–101.

[64] Zheng X, Wu JG, Lou XY, Xu HM, Shi CH. The qtl analysis on maternal and endosperm genome and their environmental interactions for characters of cooking quality in rice (*Oryza sativa* l.). Theoretical and Applied Genetics. 2008;116(3):335–42. Epub 2007/11/09.

[65] Tian Z, Qian Q, Liu Q, Yan M, Liu X, Yan C, et al. Allelic diversities in rice starch biosynthesis lead to a diverse array of rice eating and cooking qualities. Proceedings of the National Academy of Sciences. 2009;106(51):21760–5.

[66] Cho Y-G, Kang H-J, Lee Y-T, Jong S-K, Eun M-Y, McCouch SR. Identification of quantitative trait loci for physical and chemical properties of rice grain. Plant Biotechnology Reports. 2010;4(1):61–73.

[67] Bao JS, Wu YR, Hu B, Wu P, Cui HR, Shu QY. Qtl for rice grain quality based on a dh population derived from parents with similar apparent amylose content. Euphytica. 2002;128(3):317–24.

[68] Stansel JW. Influence of heredity and environment on endosperm characteristics of rice (*Oryza sativa* L.). Microfilms. Purdue University: Ann Arbor, Michigan; 1965, Dissertation Abstract, 27:48B.

[69] Singh NB, Singh HG, Singh P. Heterosis and combining ability for quality components in rice. Indian Journal of Genetics and Plant Breeding. 1977;37(2):347–52.

[70] Leng Y, Hong D-L. Grain quality and genetic analysis derived from different ecological types in japonica rice (*Oryza sativa*) (abstract). Rice Science 2004;2:165–70.

[71] Jennings PR, Coffman WR, Kauffman HE. Rice improvement: International Rice Research Institute. Los Banos: Philippines; 1979.

[72] Tian R, Jiang G-H, Shen L-H, Wang L-Q, He Y-Q. Mapping quantitative trait loci underlying the cooking and eating quality of rice using a dh population. Molecular Breeding. 2005;15(2):117–24.

[73] Tang SX, Khush G, Juliano BO. Genetics of gel consistency in rice (*Oryza sativa* L.). Journal of Genetics. 1991;70:69–78.

[74] Tang SX, Zhang YK, Yu HY. Genetics of gel consistency in the crosses between indica and japonica rice. Scientia Agricultura Sinica. 1996;29:51–5.

[75] Tang SX, Khush GS, Juliano BO. Diallel analysis of gel consistency in rice (*Oryza sativa* L.). SABRAO Journal. 1989;21:135–42.

[76] Zaman FU, Siddiq EA, Phasod AB. Genetical analysis of gel consistency in rice (*Oryza sativa* L.). Indian Journal of Genetics and Plant Breeding. 1985; 45:111–8.

[77] Harrington SE, Bligh HFJ, Park WD, Jones CA, McCouch SR. Linkage mapping of starch branching enzyme iii in rice (*Oryza sativa* l.) and prediction of location of orthologous genes in other grasses. Theoretical and Applied Genetics. 1997;94(5):564–8.

[78] Shi CH, Zhu J, Zang RC, Chen GL. Genetic and heterosis analysis for cooking quality traits of indica rice in different environments. Theoretical and Applied Genetics. 1997;95(1–2):294–300.

[79] Su Y, Rao Y, Hu S, Yang Y, Gao Z, Zhang G, et al. Map-based cloning proves qgc-6, a major qtl for gel consistency of japonica/indica cross, responds by waxy in rice (*Oryza sativa* L.). Theoretical and Applied Genetics. 2011;123(5):859–67. Epub 2011/06/24.

[80] Gravois K, Webb B. Inheritance of long grain rice amylograph viscosity characteristics. Euphytica. 1997;97(1):25–9.

[81] Bao JS, Zheng XW, Xia YW, He P, Shu QY, Lu X, et al. Qtl mapping for the paste viscosity characteristics in rice (*Oryza sativa* L.). Theoretical and Applied Genetics. 2000;100(2):280–4.

[82] Zheng L, Zhang W, Liu S, Chen L, Liu X, Chen X, et al. Genetic relationship between grain chalkiness, protein content, and paste viscosity properties in a backcross inbred population of rice. Journal of Cereal Science. 2012;56(2):153–60.

[83] Liu X, Wan X, Ma X, Wan J. Dissecting the genetic basis for the effect of rice chalkiness, amylose content, protein content, and rapid viscosity analyzer profile characteristics on the eating quality of cooked rice using the chromosome segment substitution line population across eight environments. Genome/National Research Council Canada. 2011;54(1):64–80.

[84] Bao J, He P, Xia Y, Chen Y, Zhu L. Starch RVA profile parameters of rice are mainly controlled bywx gene. Chinese Science Bulletin. 1999;44(22):2047–51.

[85] Bao J, Kong X, Xie J, Xu L. Analysis of genotypic and environmental effects on rice starch. 1. Apparent amylose content, pasting viscosity, and gel texture. Journal of Agricultural and Food Chemistry. 2004;52(19):6010–6.

[86] Chen M-H, Bergman CJ, Pinson SRM, Fjellstrom RG. Waxy gene haplotypes: associations with pasting properties in an international rice germplasm collection. Journal of Cereal Science. 2008;48(3):781–8.

[87] Singh N, Kaur L, Sandhu KS, Kaur J, Nishinari K. Relationships between physicochemical, morphological, thermal, rheological properties of rice starches. Food Hydrocolloids. 2006;20(4):532–42.

[88] Asante MD, Offei SK, Gracen V, Adu-Dapaah H, Danquah EY, Bryant R, et al. Starch physicochemical properties of rice accessions and their association with molecular markers. Starch—Starke. 2013;65:1022–8.

[89] Larkin P, McClung A, Ayres N, Park W. The effect of the *waxy* locus (granule bound starch synthase) on pasting curve characteristics in specialty rices (*Oryza sativa* L.). Euphytica. 2003;131(2):243–53.

[90] Bergman CJ, Delgado JT, McClung AM, Fjellstrom RG. An improved method for using a microsatellite in the rice *waxy* gene to determine amylose class. Cereal Chemistry. 2001;78(3):257–60.

[91] Chen M-H, Bergman C, Pinson S, Fjellstrom R. waxy gene haplotypes: Associations with apparent amylose content and the effect by the environment in an international rice germplasm collection. Journal of Cereal Science. 2008;47(3):536–45.

[92] Larkin PD, Park WD. Association of waxy gene single nucleotide polymorphisms with starch characteristics in rice (*Oryza sativa* L.). Molecular Breeding. 2003;12(4):335–9.

[93] Chen M-H, Fjellstrom RG, Christensen EF, Bergman CJ. Development of three allele-specific codominant rice waxy gene pcr markers suitable for marker-assisted selection of amylose content and paste viscosity. Molecular Breeding. 2010;26(3):513–23.

[94] James MG, Denyer K, Myers AM. Starch synthesis in the cereal endosperm. Current Opinion in Plant Biology. 2003;6(3):215–22.

[95] Nakamura Y. Towards a better understanding of the metabolic system for amylopectin biosynthesis in plants: rice endosperm as a model tissue. Plant and Cell Physiology. 2002;43(7):718–25.

[96] Reddy PR, Sathyanarayanaiah K. Inheritance of aroma in rice. Indian Journal of Genetics and Plant Breeding. 1980;(40):327–9.

[97] Sood BC, Siddiq EA. A rapid technique for scent determination in rice. Indian Journal of Genetics and Plant Breeding. 1978;38:268–71.

[98] Berner DK, Hoff BJ. Inheritance of scent in American long grain rice. Crop Science. 1986;26(5):876–8.

[99] Pinson SRM. Inheritance of aroma in six rice cultivars. Crop Science. 1994;34(5):1151.

[100] Dong Y, Tsuzuki E, Terao H. Trisomic genetic analysis of aroma in three Japanese native rice varieties (*Oryza sativa* L.). Euphytica. 2001;117(3):191–6.

[101] Tsuzuki E, Shimokawa E. Inheritance of aroma in rice. Euphytica. 1990;46(2):157–9.

[102] Ahn SN, Bollich CN, Tanksley SD. RFLP tagging of a gene for aroma in rice. Theoretical and Applied Genetics. 1992;84(7):825–8.

[103] Bourgis F, Guyot R, Gherbi H, Tailliez E, Amabile I, Salse J, et al. Characterization of the major fragance gene from an aromatic japonica rice and analysis of its diversity in Asian cultivated rice. Theoretical and Applied Genetics. 2008;117(3):353–68. Epub 2008/05/21.

[104] Bradbury L, Henry R, Jin Q, Reinke RF, Waters DLE. A perfect marker for fragrance genotyping in rice. Molecular Breeding. 2005;16(4):279–83.

[105] Bradbury LMT, Fitzgerald TL, Henry RJ, Jin Q, Waters DLE. The gene for fragrance in rice. Plant Biotechnology Journal. 2005;3(3):363–70.

[106] Chen S, Wu J, Yang Y, Shi W, Xu M. The fgr gene responsible for rice fragrance was restricted within 69kb. Plant Science. 2006;171(4):505–14.

[107] Shi W, Yang Y, Chen S, Xu M. Discovery of a new fragrance allele and the development of functional markers for the breeding of fragrant rice varieties. Molecular Breeding. 2008;22(2):185–92.

[108] Fitzgerald MA, Sackville Hamilton NR, Calingacion MN, Verhoeven HA, Butardo VM. Is there a second fragrance gene in rice? Plant Biotechnology Journal. 2008;6(4):416–23. Epub 2008/03/12.

[109] Kovach MJ, Calingacion MN, Fitzgerald MA, McCouch SR. The origin and evolution of fragrance in rice (*Oryza sativa* L.). Proceedings National Academy of Science. 2009;106(34):14444–9. Epub 2009/08/27.

[110] Buttery RG, Ling LC, Juliano BO, Turnbaugh JG. Cooked rice aroma and 2-acetyl-1-pyrroline. Journal of Agricultural and Food Chemistry. 1983;31(4):823–6.

[111] Lorieux M, Petrov M, Huang N, Guiderdoni E, Ghesquière A. Aroma in rice: genetic analysis of a quantitative trait. Theoretical and Applied Genetics. 1996;93(7):1145–51.

[112] Bradbury L, Gillies S, Brushett D, Waters D, Henry R. Inactivation of an aminoaldehyde dehydrogenase is responsible for fragrance in rice. Plant Molecular Biology. 2008;68(4):439–49.

[113] Huang TC, Teng CS, Chang JL, Chuang HS, Ho CT, Wu ML. Biosynthetic mechanism of 2-acetyl-1-pyrroline and its relationship with delta1-pyrroline-5-carboxylic acid and methylglyoxal in aromatic rice (*Oryza sativa* L.) callus. Journal Agriculture and Food Chemistry. 2008;56(16):7399–404. Epub 2008/08/06.

[114] Asante MD, Kovach MJ, Huang L, Harrington S, Dartey PK, Akromah R, et al. The genetic origin of fragrance in nerica1. Molecular Breeding. 2010;26(3):419–24.

[115] Sakthivel K, Shobha Rani N, Pandey MK, Sivaranjani AKP, Neeraja CN, Balachandran SM, et al. Development of a simple functional marker for fragrance in rice and its validation in Indian basmati and non-basmati fragrant rice varieties. Molecular Breeding. 2009;24(2):185–90.

[116] Myint K, Arikit S, Wanchana S, Yoshihashi T, Choowongkomon K, Vanavichit A. A pcr-based marker for a locus conferring the aroma in Myanmar rice (*Oryza sativa* l.). TAG Theoretical and Applied Genetics. 2012;125(5):887–96.

[117] Gao ZY, Zheng DL, Cui X, Zhou YH, Yan MX, Huang DN, et al. Map-based cloning of the *alk* gene, which controls the gelatinization temperature of rice. Science China (Series C). 2003;46:661–8.

[118] Gao Z, Zeng D, Cheng F, Tian Z, Guo L, Su Y, et al. *Alk*, the key gene for gelatinization temperature, is a modifier gene for gel consistency in rice. Journal of Integrative Plant Biology. 2011;53(9):756–65. Epub 2011/06/30.

[119] Bao JS, Corke H, Sun M. Nucleotide diversity in starch synthase iia and validation of single nucleotide polymorphisms in relation to starch gelatinization temperature and other physicochemical properties in rice (*Oryza sativa* L.). Theoretical and Applied Genetics. 2006;113(7):1171–83. Epub 2006/07/20.

[120] Wang C, Chen S, Yu S. Functional markers developed from multiple loci in *gs3* for fine marker-assisted selection of grain length in rice. Theoretical and Applied Genetics. 2011;122(5):905–13. Epub 2010/11/26.

[121] Ramkumar G, Sivaranjani A, Pandey M, Sakthivel K, Shobha Rani N, Sudarshan I, et al. Development of a 2010 PCR-based SNP marker system for effective selection of kernel length and kernel elongation in rice. Molecular Breeding. 26(4):735–40.

[122] Spindel J, Begum H, Akdemir D, Virk P, Collard B, Redoña E, et al. Genomic selection and association mapping in rice (*oryza sativa*): effect of trait genetic architecture, training population composition, marker number and statistical model on accuracy of rice genomic selection in elite, tropical rice breeding lines. PLoS Genetics. 2015;11(2):e1004982.

Chapter 6

Evaluation of Palatability of Cooked Rice

Ken'ichi Ohtsubo and Sumiko Nakamura

Additional information is available at the end of the chapter

http://dx.doi.org/10.5772/66398

Abstract

Quality evaluations of rice in Japan are performed by sensory testing and physicochemical measurements. The former is a basic method that requires large amounts of samples and several panelists. The latter is an indirect method that estimates the eating quality based on the chemical composition, cooking quality, gelatinization properties, and physical properties of cooked rice. Satake Co Ltd. developed a taste analyzer in the 1980s that is equipped with a palatability estimation formula that was based on the combination of near-infrared spectroscopy (NIR) and physicochemical measurements related with sensory test. A novel method to evaluate the quality of the cooked rice is necessary to breed high-quality rice cultivars and to select the suitable rice for each consumer and each purpose. We try to develop the novel method to evaluate the rice quality using various kinds of apparatus, such as Tensipresser, RVA, NIR, and spectrophotometer. Simple, rapid, and accurate method to evaluate the quality of rice grains is very valuable. We evaluated 16 Japanese and Chinese rice cultivars in terms of their physicochemical properties. Based on these quality evaluations, we concluded that Chinese rice cultivars are characterized by a high protein and that the grain texture after cooking has higher hardness and lower stickiness than Japanese ones reflecting the difference in consumers' preference. The relationship between the palatability of rice and agronomical condition to preserve the bio-diversity for Crested Ibis was investigated. Furthermore, the quality of rice grown in Sado Island, Japan, was assayed using rice grains grown in mountainous areas and in the field areas as samples.

Keywords: palatability, physicochemical measurements, cooked rice

1. Introduction

Rice, wheat, and maize are the most important crops in the world. Rice consumers need sufficient quantities and high-quality rice grains.

The rice quality depends on various aspects such as consumer safety, nutrition, appearance, price level, and palatability. Because rice is eaten every day, it is indispensable that it should be safe to eat and highly nutritious. Furthermore, because rice provides income for farmers and is prepared, milled, and sold by wholesalers and retailers, the rice yield and price are extremely important to the farmers. Consumers have been asking for more palatable rice because they have recently become more affluent. According to the Food Agency, 45.1% consumers purchase rice grains based on their palatability, 16.3% select it on the basis of its brand name, and 4.8% choose it on the basis of the location of rice production, implying that more than 66% Japanese consumers prefer palatable rice grains to cheaper, nonpalatable rice.

Quality evaluations of rice in Japan are performed by sensory testing and physicochemical measurements. The former is a basic method that requires large amounts of samples and several panelists. The latter is an indirect method that estimates the eating quality based on the chemical compositions, cooking quality, gelatinization properties, and physical properties of cooked rice.

A rapid, simple, and more accurate method is required to evaluate rice. Therefore, if a novel method that used near-infrared spectroscopy could be developed, producers and consumers would have a rapid and nondestructive testing method which assessed the quality of the rice grains.

2. Sensory test

Sensory testing is the fundamental method that evaluates the eating quality of rice grains. The members of a trained panel taste cooked rice samples and give scores on the basis of appearance, flavor, taste, hardness, stickiness, and overall sensory evaluation. The test results can be expressed as numerical values and treated statistically; there are disadvantages to using this test. The results differ depending on the preference of panel members, and the results cannot be directly compared if the time or country differs.

3. Physicochemical evaluations

Physicochemical evaluation [1] is only indirect evaluation of rice palatability, and its results are common all over the world if we use the same method and apparatus. Physicochemical measurements include component analysis, such as protein, amylose, and fibers, pasting properties of starch, texture measurements of the cooked rice grains, etc. For the sake of estimating the total palatability, statistical treatment is adopted using the results of each measurement as the variables. Multiple regression analysis, principal component analysis, and PLS analysis are very useful to clarify the characteristics of the rice samples. Recently, spectrophotometric analyses, such as NIR or visible-light spectrometer, are utilized to estimate the chemical components and to develop the estimation formulae for palatability. "Taste Analyzer" is an example of the NIR system as a nondestructive estimation apparatus for palatability of rice grains. Examples of sensory test and physicochemical evaluation of rice palatability are shown in **Table 1**.

3.1. Amylose content

Amylose content is the most important factor that determines rice quality because starch shares approximately 85% (w/w, dry basis) of the milled rice grains. Because Japanese con-

Sensory test	Physicochemical test
Appearance	Gloss meter, "Mido" meter
Aroma	GC, GCMS, Electric nose
Taste	HPLC, Enzyme kit, Taste sensor, Taste analyzer
Texture	Texturometer, Texture analyzer, Tensipresser
Overall evaluation	Multivariate analysis

Table 1. Evaluation of rice palatability.

sumers prefer soft and sticky cooked rice, low amylose rice is the dominant rice sold in the Japanese rice market. Amylose content is generally measured by the colorimetric method of Juliano [2].

3.2. Protein content

Protein content is another extremely important factor that determines rice quality. Cooked rice with high protein content tends to be hard and nonsticky. Protein content is measured by the Kieldahl or the combustion method.

3.3. Moisture content

Moisture content affects the eating quality of cooked rice. Low-moisture rice grains absorb water faster than high-moisture grains. In cooked rice, the final moisture content from the low-moisture rice grains is higher than that from the high-moisture rice grains. Moisture content is generally measured by the oven dry method.

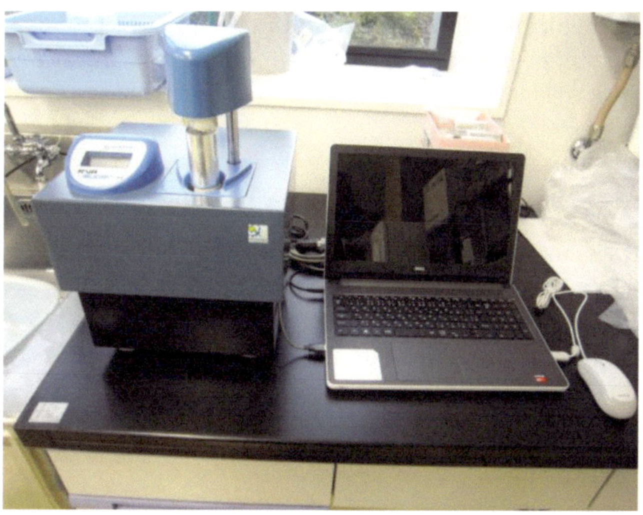

Figure 1. Rapid visco analyzer.

3.4. Gelatinization properties

The maximum viscosity, final viscosity, and "break-down" viscosity (maximum viscosity-minimum viscosity) are measured in an Amylograph. These viscosity measurements were considered good indices for the eating quality of rice. The Rapid-Visco-Analyzer (RVA) was developed in Australia and was recently introduced in Japan (**Figure 1**) [3]. Time (13–19 min) and samples (3.5 g) can be saved using an RVA.

3.5. Fat acidity

Fat acidity is a good index to measure the quality deterioration of rice grains during storage. In Japan, newly harvested rice is preferred to aged rice because consumers like soft and sticky cooked rice. Ohtsubo proposed a new colorimetric method to accurately measure fat acidity [4].

3.6. Cooking quality test

A cooking quality test was developed by the researchers of USDA in the 1950s [5]. An expanded volume and water uptake ratios are measured after cooking rice samples in excess amounts of water. Dissolved amylose is measured by color generation by combining iodine and amylose.

3.7. Physical properties of cooked rice grains

Physical properties of cooked rice grains, such as hardness and stickiness, are measured by a Texturometer, Tensipresser, Texture analyzer, or Rheolograph-micro. Low-compression and high-compression test using a Tensipresser [6] is shown in **Figure 2**.

3.8. Palatability estimation formula

Chikubu et al. developed a new equation to estimate the palatability of rice grains using a multiple regression analysis based on physicochemical measurements, such as an amylograph viscosity profile, protein assay, and iodine blue value measurements [7]. The formula showed a high correlation between the results of a sensory test ($R = 0.84$) of the rice samples harvested next year.

3.9. Application of near-infrared spectroscopy to evaluate rice quality

Near-infrared spectroscopy is a rapid, nondestructive, and extremely promising technique to evaluate qualities of the rice grains. In Japan, NIR spectroscopy was initially used to determine the protein and moisture content of the rice flours. Satake Co. Ltd subsequently combined NIR and palatability estimation formula and developed a new system called "Taste Analyzer."

3.9.1. Protein content

Iwamoto et al. reported that NIR was equally sensitive to both rice and wheat proteins. A standard error of estimation (SEP) was decreased for calibrations using a derivative absorbance.

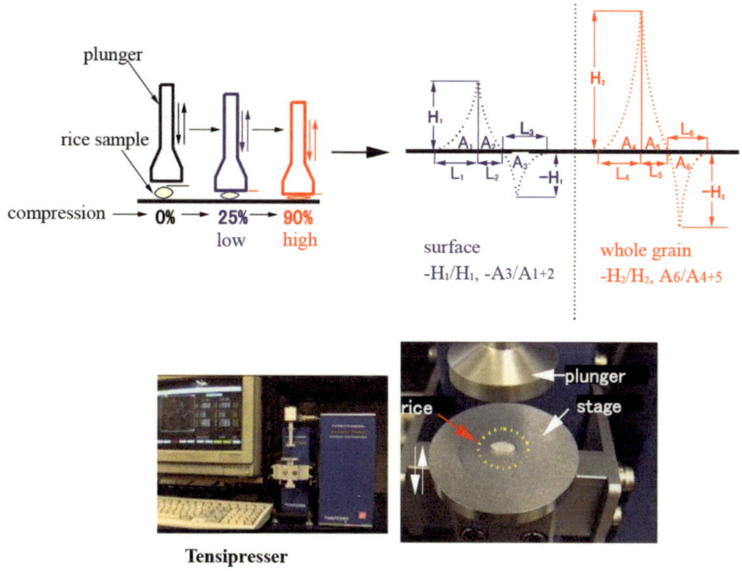

Figure 2. Tensipresser used for texture measurement of cooked rice grains.

However, the derivative procedure may possibly make an increment bias in cases that predict "unknown" samples [8]. Inatsu utilized NIR spectroscopy to select palatable rice cultivars by measuring the protein content of rice because low protein rice is preferred in Japan because of their palatability [9].

3.9.2. Estimation of rice amylose contents by NIR [10]

Japanese rice breeders have tried to develop palatable rice cultivars by selecting low amylose rice; however, the amylose content among Japanese rice grains are not significantly different (15–20%). NIR spectroscopy can easily compare protein and moisture; however, it does not easily differentiate amylose from amylopectin. Villareal and Delwiche had already developed excellent calibrations to detect amylose in various rice samples through NIR spectroscopy [11, 12]. Japanese consumers request high-performance calibration for the measurement of narrow range amylose of Japanese rice cultivars. Apparent amylose content (AAC) analysis is a rapid and simple technique that uses NIT spectroscopy. It was developed based on the near-infrared transmission spectra and the reference value of AAC determined by the iodine colorimetric method. The wide AAC range (0–35.3%) PLS model was inadequate for the accurate prediction of the extremely narrow range of the AAC (13.2–20.7%) of Japanese milled rice samples. The statistics performed on the 11-factor PLS model for the extremely narrow range of AAC analyses was 0.78, 0.74, and 2.01 in SECv, r^2, and RPD, respectively. The performance of this model was 1.25 and 0.49 for SEP and r^2, respectively, on the validation set. The previous models developed a wide range of AAC samples, which were difficult to apply to the narrow range of AAC rice. The present AAC analysis technique is based on NIT spectroscopy, which can discriminate between the extremely narrow differences in AAC of Japanese milled rice in the same subfamily.

3.9.3. Eating quality evaluation system by NIR

Satake Co. Ltd. developed a taste analyzer in the 1980s that combined the palatability estimation formula that was based on the physicochemical measurements and NIR. The taste analyzer principle is based on the multiple regression analysis that uses NIR data (protein, moisture, amylose, and fat acidity) against the sensory test results. A previous study by the Japanese Association of Milling Companies evaluated three different types of palatability evaluation system using NIR. The results showed that the taste scores correlated significantly with the results of the sensory test (r = 0.54–0.63), but the scores were affected by the moisture contents and milling yield [13].

There were eight Japanese companies that developed the characteristic systems to evaluate the palatability of rice grains based on the spectroscopic technique. More than 1000 agricultural cooperatives, wholesalers, and retailers introduced these NIR systems to evaluate the quality of their rice grains (**Figure 3**). In 1996, the Japanese Food Agency conducted a survey regarding the palatability evaluation system, and the results showed that 67% users were satisfied with the performance of the system. The protein and moisture data was extremely reliable, but some users requested an improvement in the taste score and amylose accuracy.

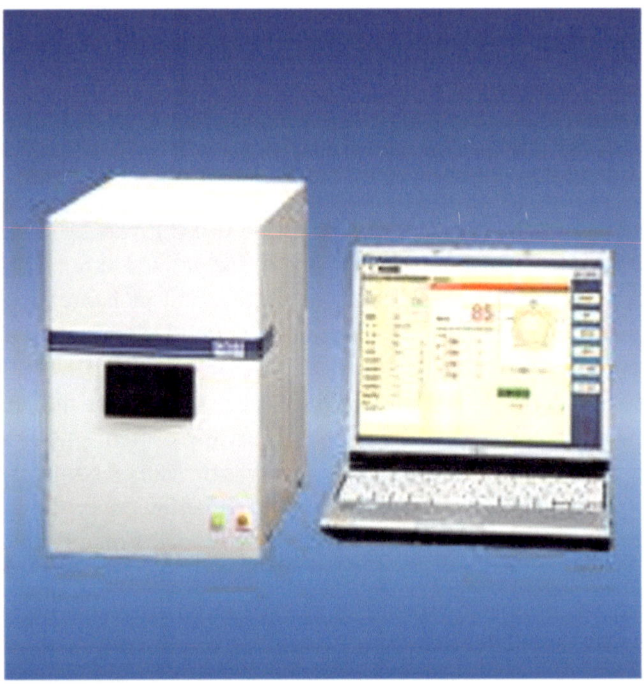

Figure 3. Taste analyzer based upon NIR technology.

4. Proposal as novel physicochemical measurement, iodine colorimetric analysis, for rice palatability evaluation [14]

4.1. Introduction

As described in Section 3, physicochemical test is very important because it is time-saving and labor-saving objective test for rice palatability. Main component of rice grains is starch, therefore, amylose content affects eating quality [15]. Low-amylose rice generally becomes soft and sticky after cooking, whereas high-amylose rice becomes hard with fluffy separated grains [16].

The most widely used method for amylose determination is a colorimetric assay where iodine binds with amylose to produce a blue-purple color, which is measured spectrophotometrically at a single wavelength (620 nm) [2].

We here characterized the starch of various rice cultivars; evaluated the relationship between their iodine absorption curve and apparent amylose contents, amylopectin fractions, and resistant starch. We improved the iodine colorimetric method, and developed the novel estimation formulae against the amylose contents, resistant starch, or a certain fractions of definite chain-length amylopectin. This novel method would lead to an easy and low-cost spectroscopic method for analyses of starch characteristics.

4.2. Contents

We used Japonica rice, Iindica rice, Indica-Japonica hybrid rice, *ae* mutant rice, and waxy rice produced in Japan and China.

We polished brown rice to a milling yield of 90–91%, using an experimental friction-type rice milling machine (Yamamotoseisakusyo, Co. Ltd., Tendoh, Japan). White rice flour was produced with a cyclone mill (SFC-S1; Udy, Fort Collins, CO, USA). Starch granules were prepared from polished rice flour using the cold alkali method.

The iodine absorption spectrum of the rice starch was measured using a Shimadzu UV-1800 spectrophotometer. The AAC of rice starch was measured by the iodine colorimetric method of Juliano [2]. The iodine absorption spectrum was scanned from 200 to 900 nm. The absorbance was measured at 620 nm, at λ_{max} (wavelength at the peak between 400 and 900 nm), and absorbance at λ_{max} ($A\lambda_{max}$) (**Figure 4**).

The area of F1 (from B to 400 nm), area of F2 (from 400 nm to λ_{max}), and area of F3 (from λ_{max} to 900 nm) were calculated.

We developed a formula for apparent amylose content based on iodine colorimetric analysis. Estimation formula is as follows:

$$\mathrm{AAC} = 73.31 \times A\lambda_{max} + 0.11 \times \lambda_{max} - 73.02 \qquad (1)$$

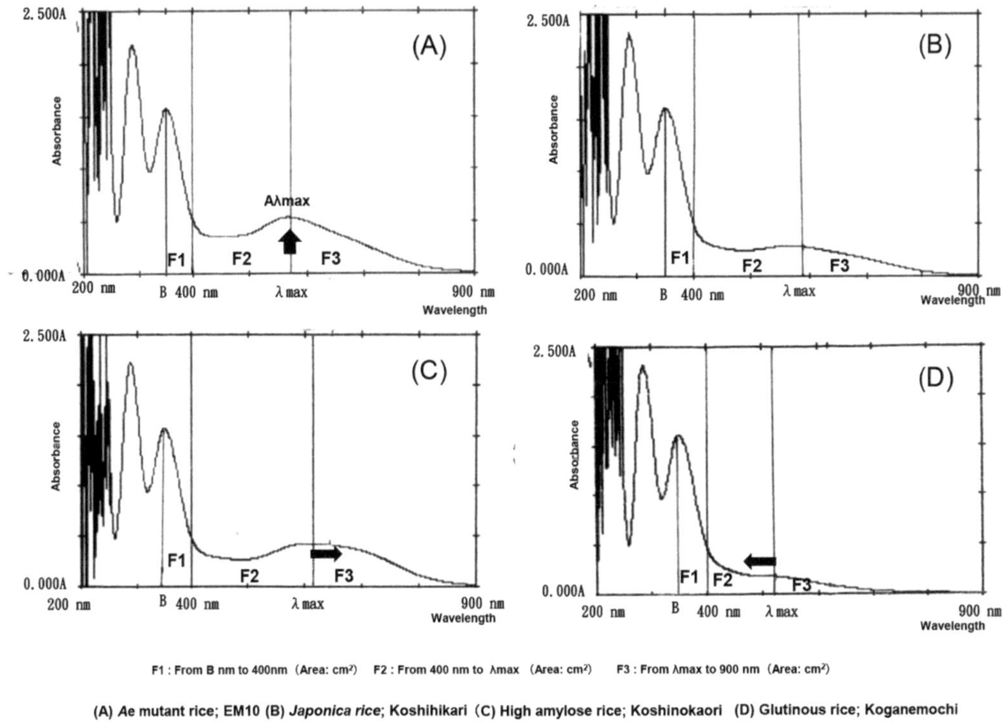

Figure 4. Analysis of iodine absorption curve of various starch samples.

Chain length distribution (CD) was carried out by the method of Hanashiro et al. using HPAEC-PAD of isoamylase-debranched materials of starch:

$$Fb3 = 44.69 \times A\,\lambda_{max} - 0.77 \quad (2)$$

RS in the starch of rice flour was measured according to the AOAC method using a RS assay kit (Megazyme, Wicklow, Ireland) in which each sample (100 mg) was digested by pancreatin and amyloglucosidase at 37°C for 6 h, followed by the measurement of glucose at 510 nm:

$$RS = 21.31 \times A\,\lambda_{\max} - 0.030 \times \lambda_{\max} + 12.25 \quad (3)$$

Fraction of DP10 (short chain glucans of amylopectin) and fraction of DP22 (medium chain glucans of amylopectin) were estimated based on F2 area and $A\lambda_{\max}$ based on the iodine colorimetric analyses, respectively, as shown in **Figure 5**.

4.3. Conclusion

We found that the iodine absorption curve differed between the Indica rice, Japonica rice, Japonica-Indica hybrid rice, *ae* mutant rice cultivars, and glutinous rice cultivars. We propose a novel index, "new $\lambda_{\max}$" based on iodine absorption curve, which shows higher correlation with amylose, resistant starch, or CD of amylopectin than ordinary $\lambda_{\max}$.

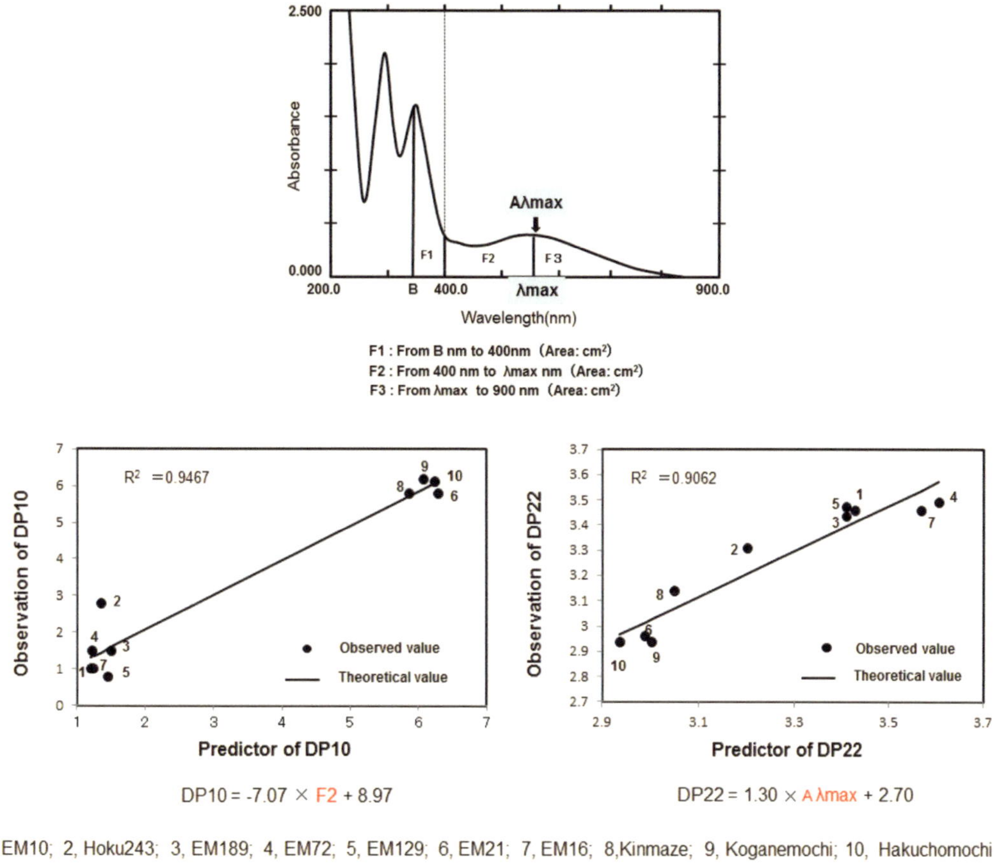

Figure 5. Estimation of fraction DP10 and fraction DP22 based on iodine colorimetric analyses.

We developed the novel estimation formulae for AAC, RS contents, amylopectin chain lengths (Fa; DP ≦ 12, Fb3; DP≦ 37), and fraction of DP10 (short chain glucans of amylopectin) and fraction of DP22 (medium chain glucans of amylopectin) on the basis of the iodine absorption curve. These formulae would lead to an easy and low-cost spectroscopic method for the starch characteristics.

5. Examples of physicochemical measurements of rice palatability

5.1. Comparison of physicochemical properties between Japanese and Chinese rice cultivars

As described in Section 3, in the case of physicochemical evaluation, we can get the common results for the rice samples produced in different countries if we use the same method. As an example, we report the results of physicochemical measurements of rice palatability using rice samples produced in Japan and China.

5.1.1. Introduction

China is the largest rice-producing country, accounting for 32% of the global production from 20% of the global rice-growing area. China produces Indica subspecies mainly in the southern region and Japonica subspecies mainly in the northern region (Heilongjiang, Liaoning), and eastern or southern region (Jilin, Jiangsu, Zhejiang, and Yunnan), whereas the other three countries, India, Indonesia, and Bangladesh, primarily grow Indica rice. In China, consumers are now choosing Japonica rice based on its shape and color as well as its texture and taste. Zhang et al. performed a sensory test of Chinese Japonica rice cultivars [17], in which a Chinese panel mainly determined the overall eating quality based on the stickiness and hardness. In the present section, we conducted physicochemical evaluations of some Chinese and Japanese Japonica rice cultivars using traditional and novel indicators based on the iodine absorption curve and RVA.

5.1.2. Materials and methods

We used high-quality premium Japanese Japonica rice (Koshihikari, Tsuyahime, Yumepirika, Sagabiyori, and Kinumusume) and Chinese Japonica rice cultivars (Kenjing 5, Shendao 529, Jinyuan 45, Changyou 5, Lianjing 7, Longjing 31, Nanjing 9108, Jinongda 878, Shennong 265, Daohuaxiang, and Jinchuan 1) as rice samples.

5.1.2.1. Measurement of the moisture content of rice flour

The moisture content of the milled rice grains was measured using an oven-drying method by drying 2 g flour samples for 1 h at 135°C.

5.1.2.2. Protein content

Nitrogen was determined using a nitrogen analyzer based on the combustion method. The protein content was obtained by multiplying a nitrogen-protein conversion factor of 5.95.

5.1.2.3. Measurement of iodine absorption spectra

The iodine absorption spectrum of milled rice was measured using a Shimadzu UV-1800 spectrophotometer. The apparent amylose content (AAC) of milled rice was estimated using the iodine colorimetric method (as described by Juliano [2]). The iodine absorption spectrum [14] was analyzed from 200 to 900 nm using a square cell of which inner dimension was 1 cm × 1 cm.

5.1.2.4. Measurement of pasting properties of rice flours

The pasting properties of starch rice flours were analyzed using an RVA (model Super 4; Newport Scientific Pty Ltd, Warriewood, Australia). A programmed heating and cooling condition was followed as described by Toyoshima et al. [3]. Novel indices, such as the setback/consistency (SB/Con) and maximum viscosity/minimum viscosity (Max/Fin) ratios, have very strong correlations with the proportion of intermediate and long chains of amylopectin: Fb_{1+2+3} (DP≧ 13) [14].

5.1.2.5. Physical properties of boiled rice grains

The boiled rice samples were kept in the vessel at 25°C for 2 h and then subjected to the measurements. The hardness and stickiness of the boiled rice grains were measured using a Tensipresser (My Boy System, Taketomo Electric Co., Tokyo, Japan) using "low compression (25%) and high compression (90%) test" [6]. The average of each parameter was calculated by measuring 20 individual grains. As a staling test of the cooked rice, the cooked samples were stored at 10°C for 16 h and measured again with a Tensipresser according to the previously described method [6].

5.1.2.6. Sodium dodecyl sulfate-polyacrylamide gel electrophoresis (SDS-PAGE)

Proteins were extracted from milled rice flour samples (0.5 g) by shaking with 2 mL of buffer A (50 mM Tris-HCl, pH 6.8, 2% SDS, 5% 2-mercaptoethanol) at 37°C for 30 min and then centrifuged for 5 min at 3000 × *g*. The supernatant (1 mL) was diluted with an equal volume of sample buffer (0.125 M Tris-HCl, pH 6.8, 10% 2-mercaptoethanol, 4% SDS, 10% sucrose, 0.004% bromophenol blue) and mixed well, followed by heating for 2 min at 100°C. In total 10 μg of extracted protein was charged into each lane. SDS-PAGE was conducted with a 12% polyacrylamide gel according to the modified method described by Laemmli [18]. The values were calculated based on the intensities of various bands on the gel after SDS-PAGE analysis using the ATTO densitograph software library (CS Analyzer ver 3.0).

5.1.3. Results and discussion

5.1.3.1. Main chemical components

Protein is a very important component of milled rice, and it affects the physical properties of cooked rice grains. Higher protein content makes the rice grains harder and less sticky. The protein content of milled Japonica rice is around 6.5%. The protein contents of Chinese rice cultivars (6.8–9.0%; mean = 7.8%) were significantly higher than those of Japanese rice varieties (6.2–6.8%; mean = 6.5%).

AAC has been used as a good parameter for estimating the cooking or eating quality of rice grains, and the iodine colorimetric method for AAC measurement at 620 nm was developed by Juliano [2]. AAC comprises a large amount of amylose and a small amount of SLC in amylopectin. In general, low amylose rice becomes soft and sticky after cooking, whereas high amylose rice becomes hard and separated. The AAC values for Chinese rice cultivars (6.6–17.2%; mean = 14.3%) were higher than those of Japanese rice cultivars (9.7–14.6%; mean = 12.7%). In our previous study [14], we showed that the iodine absorption curve differed among various samples of rice cultivars, and we developed a novel formula for estimating AAC. ($AAC = 73.31 \times A\lambda_{max} + 0.11 \times \lambda_{max} - 73.02$). It was shown that a multiple coefficient of determination of 0.996 was obtained when we used this formula to estimate AAC for Chinese rice cultivars.

The molecular structures of many starches, including the molecular sizes of amylose and the amylopectin branch chain lengths, have been reported previously [19–22]. The high molecular weight amyloses tend to have a longer wavelength for λ_{max}. We found that the glutinous rice cultivars had very low λ_{max} values, and Indica rice, Japonica-Indica hybrid rice cultivars, and a high-amylose Japonica rice cultivar had higher λ_{max} values. The λ_{max} values of Japonica rice cultivars were intermediate [14].

The "new λ_{max}" values are assumed to be related to the SLC content of amylopectin [14]. It was shown that the "new λ_{max}" values of the Chinese rice cultivars (0.11–0.27; mean = 0.23) were lower than those of the Japanese rice cultivars (0.23–0.26; mean = 0.25).

The starches in the rice cultivars grown under low temperatures have a significantly higher amylose contents than those of the rice cultivars grown under high temperature, whereas the SLC contents of the amylopectin were lower [23–27]. Thus, we consider that the starch properties of Japanese rice cultivars were influenced by the ambient temperatures during the development of the grain, which yielded high "new λ_{max}" values.

5.1.3.2. Physical properties of cooked rice grains

There were differences in the values of physical properties of the cooked rice grains. The balance of -H1/H1, -H2/H2, A3/A1, and A6/A4 are important indices when evaluating the palatability of rice [16].

The hardness of the surface layer (H1) was higher in the Chinese rice cultivars than the Japanese rice varieties, and the hardness of the overall layer (H2) of Chinese rice cultivars was higher than that of the Japanese rice varieties. The stickiness of the surface layer (-H1) was significantly lower in the Chinese rice cultivars than the Japanese rice cultivars, whereas the stickiness of overall layer (-H2) was higher in the Chinese rice varieties than the Japanese rice cultivars. The balance degree of the surface layer (-H1/H1) was significantly lower in the Chinese rice varieties than the Japanese rice varieties, and that of the surface layer (A3/A1) was significantly lower in the Chinese rice varieties than the Japanese rice varieties.

Low amylose rice cultivars are stale-resistant according to Takami et al., who previously reported the staling characteristics of cooked low amylose rice [28]. In the staling test, we stored the cooked rice at 10°C for 16 h (**Figure 6**). The staling rate for starch based on the surface layer hardness (H1) was significantly higher in the Chinese rice cultivars than the Japanese rice cultivars. In particularly, Jinongda 878 (2.3 times), Daohuaxiang (1.9 times), and Nanjing 9108 (1.9 times) had very high values. The staling rates for starch based on the hardness of the overall layer (H2) were almost the same in the Chinese and Japanese rice cultivars. The staling rate for starch based on the hardness of the surface layer (H1) in Chinese rice varieties had a positive correlation with the intensities of the 13 kDa prolamin spots and the protein content at $p < 0.01$.

5.1.3.3. SDS-PAGE

The intensities of the bands of 13 kDa prolamin were correlated positively with $\lambda_{max}/A\lambda_{max}$ (r = 0.66) and "new λ_{max}" (r = 0.65) in Chinese rice varieties, and with the glutamic acid (r = 0.85), whereas with λ_{max} (r = -0.65) had negative correlation.

http://dx.doi.org/10.5772/66398

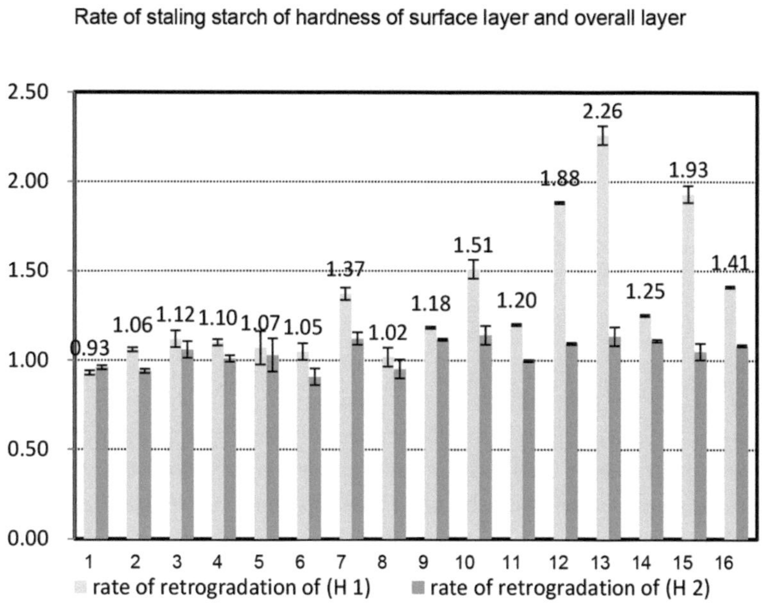

1, Koshihikari; 2,Tsuyahime; 3, Yumepirika; 4, Sagabiyori; 5, Kinumusume;

6, Kenjing 5; 7, Shendao 529; 8, Jinyuan 45; 9, Changyou 5; 10, Lianjing7; 11,Longjing 31;

12, Nanjing 9108; 13, Jinongda 878; 14, Shennong 265; 15, Daohuaxiang; 16, Jinchuan 1

Figure 6. Results of the staling starch test of the physical properties of cooked rice grains.

The rice seed storage proteins mainly comprise glutelins and prolamins. PB-I is highly enriched with prolamin and it constitutes approximately 20% of the milled rice protein contents. PB-II mainly contains glutelins and it constitutes 60–65% of the milled rice protein [29]. The protein content of rice grains is influenced by the weather conditions, as it is increased by high air temperature or high water temperature, but decreased by low water temperature or sun shading during the ripening stage [30]. Matsui et al. [31] showed that the Final vis. and Cons values of near-isogenic line pairs for the low glutelin gene (*Lgc1*) locus were significantly higher in low glutelin lines, and the surface stickiness was also significantly lower in the low glutelin lines. Protein production also tended to increase with higher levels of nitrogenous fertilizer at any planting density [32].

5.1.4. Conclusion

In this study, we evaluated 16 Japanese and Chinese rice cultivars in terms of their main chemical components, iodine absorption curve, apparent amylose content (AAC), pasting property, resistant starch content, physical properties, and SDS-PAGE.

Based on these quality evaluations, we can conclude that Chinese rice cultivars are characterized by their high protein content. The hardness of the surface layer (H1) and overall layer

(H2) were higher in the Chinese rice cultivars than the Japanese rice cultivars, whereas the stickiness of the surface layer, and the balance degree of the surface layer were lower in the Chinese rice cultivars than the Japanese rice cultivars, although the stickiness of the overall layer (-H2) was higher in the Chinese rice cultivars than the Japanese rice cultivars. In addition, the texture of cooked rice was strongly correlated with the iodine absorption factors.

In a previous study, we developed a novel formula for estimating AAC based on the iodine absorption curve. The validation test showed a determination coefficient of 0.996 for estimating AAC of Chinese rice cultivars as unknown samples.

5.2. Palatability evaluation of the Japanese premium rice "Koshihikari" in Sado Island

As described in Section 3, in the case of physicochemical evaluation, we can get the common results for the rice samples produced in the different regions if we use the same method. As an example, we report the results of physicochemical measurements of rice palatability using rice samples produced by the different cultivation conditions, and rice samples grown in the mountainous region and the field region in Sado Island, Japan.

5.2.1. Introduction

Rice is one of the most important staple foods in the world. "Koshihikari" is the most famous premium rice cultivar, and it shares about 37% of all rice production in Japan. "Koshihikari" is cultivated all over Japan, but that grown in Niigata Prefecture is considered the best "Koshihikari" in Japan with regards to its quality and quantity. Sado Island is part of the Niigata Prefecture and the "Koshihikari" cultivated in Sado, Iwafune, and Uonuma is evaluated as the highest quality rice.

Sado Island is located off the Niigata City seashore and it is the second largest island in Japan after Okinawa Island. Sado Island is famous for its Toki (crested ibis, *Nipponia nippon*), which is an endangered species in Japan. The inhabitants of Sado city and its farmers are trying to make an ideal habitat for Toki. The feed for Toki are small fish and insects, such as the loach, grasshoppers, and worms as well as small river crabs, and frogs. The farmers are making an effort to keep the environment agriculturally sound to protect the ibis from extinction.

Sado city awards the farmers a harmonizing environment certificate known as the "homeland to live with crested ibis" if the farmers adopt an ibis-friendly agricultural method. Farmers create a narrow pond, preserve bio-diversity with winter-flooding, and reduce the use of chemical fertilizers and agricultural chemicals by over 50% compared with conventional cultivation.

However, the effects of this ibis-friendly agricultural method, which also include abolishing drying paddy field after transplanting, does not equate to palatability of the rice grains. Therefore, the relationship between the palatability of rice and agronomical condition to preserve the bio-diversity for ibis was investigated. Furthermore, the quality of rice grown in Sado Island was compared with that of rice grown in mountainous areas and in the field.

5.2.2. Materials and methods

5.2.2.1. Comparison of Sado Koshihikari and the other premium rice in Japan

Ten premium rice cultivars, including "Koshihikari" from Sado Island, "Sagabiyori," "Yumepirika," "Akitakomachi," "Tsuyahime,""Hitomebore," and "Koshihikari from Sado" were subjected to physicochemical measurements and sensory test.

5.2.2.2. Comparison of mountainous and field regions

Rice samples grown in eight regions, including the mountainous and field regions, were subjected to the physicochemical measurements to evaluate rice quality.

5.2.2.3. Effects of the ibis-friendly cultivations on the palatability of the rice grains

Rice grains were harvested after an ordinary or ibis-friendly cultivation and were subjected to quality evaluations. Ibis-friendly cultivation meant that farmers adopted winter-flooding (keeping water even during the winter season to save the habitat for Toki in winter) or deletion of "Nakaboshi" (drainage of paddy field for approximately a week after transplanting to activate the roots of rice plant).

5.2.2.4. Preparation of polished rice samples

After de-husking, brown rice was polished to a milling yield of 90–91%, using an experimental TM05 rice milling machine (Satake, Co. Ltd. Higashihiroshima, Japan). Rice flours were prepared by a cyclone mill (SFC-S1, Udy, Fort Colins, CO, USA).

5.2.2.5. Amylose content

Amylose content was measured by the iodine colorimetric method. Calibration samples were prepared by mixing the standard potato amylose (Type III, Sigma Chemical Co., St. Louis, MO, USA) and standard amylopectin (waxy rice removed from fat and proteins).

5.2.2.6. Measurements of protein content, Mido, and quality score

Protein contents and quality scores were measured with a near-infrared spectrometer (AN820, Kett Electric Laboratory Co. Ltd., Tokyo, Japan) (Near Infrared Spectrometer, Foss Japan Co. Ltd, Tokyo, Japan). Mido was measured using a Midometer (Toyorice, Wakayama, Japan).

5.2.2.7. Measurement of the physical properties of the boiled rice grains

Polished rice grains (10 g) were added with 16 mL of distilled water in an aluminum cup. The rice was in the water for 1 h, cooked for 25 min in the automatic electric rice cooker (SR-SW182, Panasonic Co. Ltd., Kadoma, Japan) and kept warm for 10 min in it. After the rice was kept warm, the boiled rice grains were moved into a plastic bag and kept in the bag for 2 h at 25°C. Thereafter, the physical property (hardness and stickiness) of each boiled rice grain (20 samples) was measured by a Tensipresser (My Boy System, Taketomodenki Co. Ltd., Tokyo, Japan), as reported by Okadome et al. (single grain low-compression high-compression test).

5.3. Results and discussion

5.3.1. Comparison of palatability of premium rice cultivars in Japan

"Koshihikari" from the Sado Island received extremely high value for the quality value, Mido value, and low value for protein. "Koshihikari" from Sado Island had an extremely low hardness in surface layer of boiled grains, whereas extremely low amylose rice such as "Yumepirika" had the lowest hardness in surface layer of boiled rice gains. The ratio of stickiness to hardness of surface layer of boiled grains was the highest for "Yumepirika" followed by "Koshihikari" from Sado Island. Palatability of the "Koshihikari" from Sado Island seems to be one of the best premium rice cultivars in Japan due to its low protein content.

5.3.2. Comparison of mountainous and field regions

As a result of measurement of physical properties, five rice samples from mountainous regions showed a lower pasting temperature and a higher ratio of adhesiveness to hardness (A6/A4, overall). It was reported that the ratio of adhesiveness to hardness is correlated with palatability of boiled rice grains. Rice samples grown in the mountainous region seems to be more palatable than those in the field region because the former rice is more adhesive.

5.3.3. Effect of winter-flooding and Nakaboshi on palatability of rice

Rice samples grown using winter-flooding (without "Nakaboshi") showed higher amylose contents, than those grown with winter-flooding and "Nakaboshi." Rice grains grown using winter-flooding and Nakaboshi had lower amylose contents implying that the boiled rice grains of this group were softer and stickier than the rice grains subjected to winter-flooding alone. In case of cultivation using winter-flooding, the rice grains grown using "Nakaboshi" became stickier after boiling compared with those cultivated without "Nakaboshi." In case of cultivation without Nakaboshi, rice samples grown using winter-flooding became less adhesive after boiling compared with those grown without winter-flooding. These results are shown in **Figure 7**.

To summarize, farmers should adopt Nakaboshi in case of winter-flooding, and avoid winter-flooding in case of not performing Nakaboshi. Although adoption of winter-flooding and not performing Nakaboshi are recommended to maintain biodiversity, simultaneously performing of both of these measures negatively affects the palatability of the boiled rice grains.

5.4. Conclusions

Farmers in Sado city make an effort to harmonize the environment with sound agriculture to protect the ibis from extinction. The palatability of "Koshhikari" from Sado Island is among

http://dx.doi.org/10.5772/66398

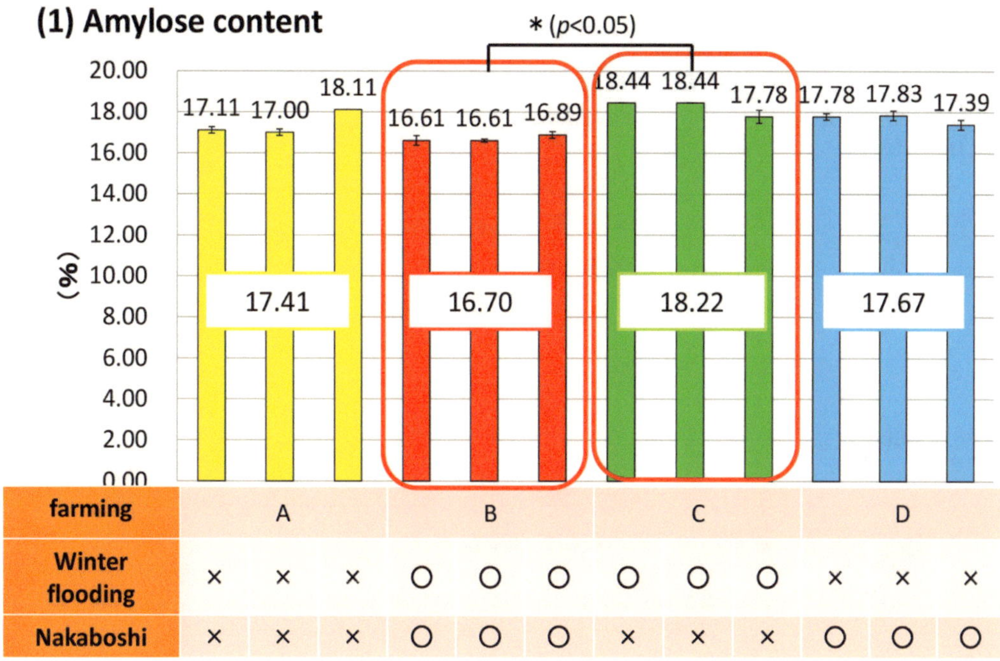

Figure 7. The effect of winter-flooding and Nakaboshi on amylose contents of rice grains.

the premium rice cultivars in Japan. The rice quality from Sado Island grown in mountainous areas and those in field regions were examined along with the palatability of rice and the agronomical conditions to preserve the bio-diversity for the ibis.

The results show that the palatability of "Koshihikari" from Sado Island seems to be one of the best among the premium rice cultivars in Japan because of its low protein contents. Five rice samples from mountainous regions showed better physical properties compared with three rice samples grown in the fields. Adoption of both winter-flooding and not performing Nakaboshi is recommended to maintain biodiversity; however, simultaneously performing both of these measures negatively affects the palatability of the boiled rice grains.

6. Summary

There are various kinds of rice cultivars of which qualities are diversified, such as hard Indica rice and soft Japonica rice in the world. Consumers in southern Asia prefer hard rice grains and people in North-eastern Asia like soft and sticky rice grains. Novel method to evaluate the quality of the cooked rice is necessary to breed high-quality rice cultivars and to select the suitable rice for each consumer and for each purpose. We try to develop the novel method to evaluate the rice quality using various kinds of apparatus, such as Tensipresser, RVA, NIR, and spectrophotometer. Simple, rapid, and accurate method to evaluate the quality of rice grains is very valuable.

Author details

Ken'ichi Ohtsubo* and Sumiko Nakamura

*Address all correspondence to: ohtsubok@nupals.ac.jp

Niigata University of Pharmacy and Applied Life Sciences, Niigata City, Japan

References

[1] Ohtsubo K, Toyoshima H, Okadome H: Quality assay of rice using traditional and novel tools. Cereal Foods World. 1998;43:203–206.

[2] Juliano BO: A simplified assay for milled rice amylose. Cereal Sci. Today. 1971;16:334–340, 360.

[3] Toyoshima H, Okadome H, Ohtsubo K, Suto M, Horisue N, Inatsu O, Narizuka A, Aizaki M, Inouchi N, Fuwa H: Cooperative test on the small-scale rapid method for the gelatinization properties test of rice flours with a rapid visco analyser (in Japanese). Nippon Shokuhin Kogakukaishi. 1997;44:579–584.

[4] Ohtsubo K, Ishima T, Yanase H: Colorimetric method for fat acidity measurement of rice grains (in Japanese). Rep. Natl. Food Res. Inst. 1987;50:59–65.

[5] Batcher OM, Deary PA, Dawson EH: Cooking quality of 26 variety of milled white rice. Cereal Chem. 1957; 34:277–285.

[6] Okadome H, Toyoshima H, Ohtsubo K: Multiple measurements of physical properties of individual cooked rice grains with a single apparatus. Cereal Chem. 1999;76:855–860.

[7] Chikubu S, Watanabe S, Sugimoto T, Manabe M, Sakai F, Taniguchi Y: Establishment of palatability estimation formula of rice by multiple regression analysis (in Japanese). Denpun Kagaku 1985;32:51–60.

[8] Iwamoto M, Suzuki T, Kongseree N, Uozumi J, Inatsu O: Analysis of protein and amino acid contents in rice flour by near-infrared spectroscopy (in Japanese). Nippon Shokuhin Kogyo Gakkaishi. 1986;12:848–853.

[9] Inatsu O: Studies on improving the eating quality of Hokkaido rice (in Japanese). Rep. Hokkaido Prefect. Agric. Exp. Station. 1988;66:3–7.

[10] Shimizu N, Katsura J,Yanagisawa T, Inoue S, Withey RP, Cowe IA, Eddison CG, Withey RP, Blakeney A, Kimura T, Yoshizaki S, Okadome H, Toyoshima H, Ohtsubo K: Determination of apparent amylose content in Japanese milled rice using near-infrared transmission spectroscopy. Food Sci. Technol. Res. 1999;5:337–342.

[11] Villareal CP, Dela C, Normita M, Juliano BO: Rice amylose analysis by near-infrared transmittance spectroscopy. Cereal Chem. 1994;71:292–296.

[12] Delwiche SR, McKenzie K, Webb BD: Quality characteristics in rice by near-infrared reflectance analysis of whole grain milled rice samples. Cereal Chem. 1996;73:257–263.

[13] Yamagata I, Andoh M, Yanase H: Characteristics of eating quality evaluation systems (in Japanese). Seimai Kogyo. 1990;No123:10–20.

[14] Nakamura S, Satoh H, Ohtsubo K: Development of formulae for estimating amylose content, amylopectin chain length distribution, and resistant starch content based on the iodine absorption curve of rice starch. Biosci. Biotechnol. Biochem. 2015;79:443–455.

[15] Juliano BO, Onate LU, Mundo AM: A simplified assay for milled rice amylose. Food Technol. 1965;19:1006–1011.

[16] Okadome H, Kurihara M, Kusuda O, Toyoshima H, Kim JI, Shimotsubo K, Matsuda T, Ohtsubo K: Multiple measurements of physical properties of cooked rice grains with different nitrogenous fertilizers (in Japanese). Jpn. J. Crop Sci. 1999;68:211–216.

[17] Zhang X, Cui Z, Cui J, Matsue Y, Ogata T, Kusutani A: Sensory test for the palatability of Japanese rice cultivars by Chinese and Japanese panels (in Japanese). Jpn. J. Crop Sci. 2015;84:176–181.

[18] Laemmi UK: Cleavage of structural proteins during the assembly of the head of bacterio phage T4. Nature. 1970;227:680–685.

[19] Gallant DJ, Bouchet B, Baldwin PM: Microscopy of starch evidence of a new level of granule organization. Carbohydrate Polym. 1997;32:177–191.

[20] Nakamura Y, Sakurai A, Inaba Y, Kimura K, Iwasawa N, Nagamine T: The fine structure of amylopectin in endosperm from Asian cultivated rice can be largely classified into two classes. Starch/Starke. 2002;54:117–131.

[21] Hizukuri S: Polymodal distribution of the chain lengths of amylopectins and its significance. Carbohydr. Res. 1986;147:342–347.

[22] Robin JP, Mercier C, Charbonniere R, Guilbot A: Gel filtration and enzymatic studies of insoluble residues from prolonged acid treatment of potato starch. Cereal Chem. 1974;51:389–406.

[23] Inouchi N, Hibiu H, Li T, Horibata T, Fuwa H, Itani T: Structure and properties of endosperm starches from cultivated rice of Asia and other countries. J. Appl. Glycosci. 2005;52:239–246.

[24] Asaoka M, Okuno K: Effect of environmental temperature at the milky stage on amylose content and fine structure of amylopectin of waxy and nonwaxy endosperm starches of rice. Agric. Biol. Chem. 1985;49:373–379.

[25] Inouchi N, Ando H, Asaoka M, Okuno K, Fuwa H: The effect of environmental temperature on distribution of unit chains of rice amylopectin. Starch/Strake. 2000;52:8–12.

[26] Adam A, Karen AKM, Anna MM, Donn HB, Brucc RH: Effect of growth location in the United States on amylose content, amylopectin fine structure, and thermal properties of starches of long grain rice cultivars. Cereal Chem. 2006;83:93–98.

[27] Umemoto T, Nakamura Y, Satoh H, Terashima K: Differences in amylopectin structure between two rice varieties in relation to the effects of temperature during grain-filling. Starch/Starke. 1999;51:58–62.

[28] Takami K, Koriyama T, Ohtsubo K: Staling characteristics of cooked low-amylose rice and a proposal of evaluation method. Nippon Shokuhin Kagaku Kogaku Kaishi. 1998;45:469–477.

[29] Mitsukawa N, Konishi R, Uchiki M, Masumura T, Tanaka K: Molecular cloning and characterization of a cystein-rich 16.6 kDa prolamin in rice seeds. Biosci. Biotechnol. Biochem. 1998;45:469–477.

[30] Honjyo K: Variation of protein content between rice varieties and the influences of environmental factors on the protein content. Jpn. J. Crop Sci. 1971;40:183–189.

[31] Matsui T, Ishizaki K, Nakamura S, Ohtsubo K: Differences in physical properties of boiled rice and gelatinization properties of rice flour between pairs of near-isogenic lines for low glutelin gene (Lgc1) locus. Nippon Shokuhin Kagaku Kogaku kaishi. 2013;60: 204–211.

[32] Honjyo K: Effect of the fertilization on protein content and protein production in paddy grain. Jpn. J. Crop Sci. 1971;40:190–196.

Chapter 7

Comparison on Grain Quality Between Super Hybrid and Popular Inbred Rice Cultivars Under Two Nitrogen Management Practices

Min Huang, Shuanglü Shan, Jiana Chen,
Fangbo Cao, Ligeng Jiang and Yingbin Zou

Additional information is available at the end of the chapter

http://dx.doi.org/10.5772/66386

Abstract

This study was conducted to determine the differences in grain quality traits between super hybrid and popular inbred rice cultivars grown under two nitrogen (N) management practices. Field experiments were done at the Experimental Farm of Guangxi University, Guangxi Province, China in early and late rice-growing seasons in 2014. Two representative super hybrid cultivars Liangyoupeijiu (LYPJ) and Y-liangyou 1 (YLY1) and a popular inbred rice cultivar Huanghuazhan (HHZ) were grown under fixed-time N management (FTNM) and site-specific N management (SSNM) practices in each season. Grain quality traits and N uptake were measured for each cultivar. LYPJ and YLY1 had higher milling efficiency, poorer appearance and palatability, and equal nutritional value than HHZ. The higher milling efficiency and poorer appearance in LYPJ and YLY1 were associated with their higher rice width compared with HHZ. Total N application rate was reduced by 15–20% under SSNM than under FTNM, whereas there was nearly no significant difference in grain quality between SSNM and FTNM. Our results suggest that (1) strategies for grain quality improvement in super hybrid rice should be focused on appearance and palatability, and (2) replacing FTNM with SSNM can reduce N input without sacrificing grain quality in rice production.

Keywords: grain quality, grain shape, nitrogen management, super hybrid rice, temperature

1. Introduction

Rice is the staple food for about 65% of the population of China and therefore rice productivity is critical to the national food security [1]. Although Chinese rice production has increased more than threefold in the past five decades [2], rapid population growth and economic development have been posing a growing pressure for increased food production [3]. It is projected that China will need to produce about 20% more rice by 2030 in order to fulfil its domestic needs [4]. To achieve this goal, great efforts should be made to breed new rice cultivars with higher yield potential [5]. In 1996, China established a nationwide mega-project on the development of super rice based on the ideotype concept [6]. In 1998, Prof. Longping Yuan proposed a strategy for developing super hybrid rice by combining an ideotype approach with the use of inter-subspecific heterosis [7]. The main target of super hybrid rice breeding is to develop high-yielding cultivars. However, the demand for high quality rice increases in China as living standards improve [8]. In 2000, 40% of the rice area was planted with rice cultivars with high grain quality and most of these cultivars were inbreds [9]. Hence, whether the super hybrid rice cultivars can be widely planted in China to some extent depends on their grain quality.

In rice, grain quality is generally classified into four components: milling efficiency, appearance, cooking and eating characteristics and nutritional value [10]. Milling efficiency is typically assessed as brown rice percentage (BRP), milled rice percentage (MRP) and head rice percentage (HRP) [11]. Appearance is often judged by the percentage of chalky rice grains (PCRG) and degree of chalkiness (DC) [12]. Cooking and eating characteristics are mostly determined by gelatinization temperature (GT), amylose content (AC) and gel consistency (GC) of the grain endosperm [13]. A nutritional value is commonly evaluated by protein content (PC) [14]. Among these, milling and appearance traits are highly correlated with grain shape. In general, rice length (RL), rice length-to-width ratio (RLWR), or length-to-thickness ratio are negatively associated with grain milling efficiency, while increased rice width (RW) and thickness tend to result in increased milling efficiency [15] but poorer appearance [16, 17]. In addition, there are strong relationships between some of the grain quality traits. Chalkiness reduces grain resistance to forces applied during the milling process, causing a decrease in HRP [18, 19]. Brown rice with high PC is more resistant to abrasive milling than that with low PC [20].

Rice grain quality depends not only on cultivars but also on crop management practices and environmental conditions [21, 22]. N application has been reported as a common management practice that affects rice grain quality. Leesawatwong et al. [20] observed that the N application increased the milling efficiency and nutritional value in four Thai extra-long grain rice cultivars. Ning et al. [23] found that contents of four proteins (albumin, globulin, prolamin and gutelin) were increased with the increased N level in 31 japonica cultivars. Similarly, Wang et al. [24] reported that PC increased with an increase in N rates in two hybrid rice cultivars and there was a significantly positive relationship between PC and the total N uptake (TNU) for each cultivar. The temperature, especially daily mean temperature during grain-filling period, is one of the dominant climatic factors affecting rice grain quality [25]. High temperature may significantly accelerate the grain-filling rate, but correspondingly shorten its duration, thus resulting in loosely packed starch granules, decreased grain weight and HRP and increased PCRG.

Recently, there has been a report describing the grain quality of a super hybrid rice cultivar Liangyoupeijiu (LYPJ) grown under different N rates [24]. Their results showed that LYPJ had higher HRP, AC and GC but lower PCRG and PC than an ordinary hybrid rice cultivar, Shanyou 63, across a wide range of N rates. However, this study was conducted using only one super hybrid cultivar in comparison with a check cultivar. Furthermore, although rice breeders in China often use Shanyou 63 as a check cultivar because of its superior yield stability, its grain quality is not good enough to suit the preference of consumers. Therefore, it is difficult to conclude whether the grain quality of super hybrid rice is good or not. In our current study, we compared two super hybrid rice cultivars with a popular inbred rice cultivar with good quality under two N management practices in two seasons. The objectives of this study were to (1) determine the differences in grain quality traits between super hybrid and popular inbred rice cultivars and (2) the effects of N management practices on rice grain quality.

2. Materials and methods

2.1. Site and soil

Field experiments were conducted at the Experimental Farm of Guangxi University (22°51′ N, 108°17′ E, 78 m a.s.l.), Guangxi Province, China in early and late rice-growing seasons in 2014. The soil of the experimental field was an Ultisol (USDA taxonomy) with the following properties: pH 6.75, 32.3 g kg^{-1} organic matter, 120 mg kg^{-1} alkali-hydrolysable N, 31.6 mg kg^{-1} available P and 126 mg kg^{-1} available K. The soil test was based on samples taken from the 0 to 20 cm soil layer.

Cultivar	Type	Year of release	Female parent	Male parent
Liangyoupeijiu	Indica-japonica hybrid	1999	Peiai64S	9311
Y-liangyou 1	Indica hybrid	2006	Y58S	9311
Huanghuazhan	Indica inbred	2005	Huangxinzhan	Fenghuazhan

Table 1. Information about rice cultivars used in the study.

2.2. Plant and treatments

Three rice cultivars, including Liangyoupeijiu (LYPJ), Y-liangyou 1 (YLY1) and Huanghuazhan (HHZ), were used in this study. LYPJ and YLY1 are two representatives of high-yielding cultivars developed from China's super hybrid rice breeding project [26]. HHZ has been widely grown by rice farmers in southern China because of its good grain quality. Detailed information about rice cultivars is given in **Table 1**. Two N management practices were imposed: fixed-time N management (FTNM) and site-specific N management (SSNM) (**Table 2**). A

N management practice	Season	N application timing and rate (kg ha^{-1})[a]				Total N application rate (kg ha^{-1})
		Basal	Early tillering	Panicle initiation	Spikelet differentiation	
FTNM	Early and late	112.5	45	45	22.5	225
SSNM[b]	Early	56	60	45	15	176
	Late	56	60	60	15	191

[a] Basal was defined as 1 day before transplanting, early tillering as 7 days after transplanting, panicle initiation as the first appearance of differentiated apex and spikelet differentiation as the appearance of glumous flower primordia at the tips of elongating primary rachis-branches [27].

[b] A chlorophyll meter (SPAD-502, Soil-Plant Analysis Development Section, Minolta Camera Co., Osaka, Japan) was used to determine top dressings. SPAD value was measured on 10 topmost fully expanded leaves per plot. At panicle initiation, if SPAD < 37, apply 60 kg ha^{-1}; if between 37 and 39, apply 45 kg ha^{-1}; if > 39, apply 30 kg ha^{-1}. At spikelet differentiation, if SPAD < 37, apply 45 kg ha^{-1}; if between 37 and 39, apply 30 kg ha^{-1}; if between 39 and 42, apply 15 kg ha^{-1}; if > 42, apply 0 kg ha^{-1} [28].

Table 2. N application timing and rate and the total N application rate for fixed-time N management (FTNM) and site-specific N management (SSNM) in early and late seasons.

split-plot design was used with N management practices as main plots and cultivars as subplots. The experiment was replicated three times and subplot size was 20 m^2.

Pre-germinated seeds were sown in a seedbed. Twenty-two-day-old seedlings were transplanted on 12 April and 31 July in early and late seasons, respectively. Transplanting was carried out at a hill spacing of 20 cm × 27 cm with two seedlings per hill. In addition to the N fertilizer, plants received 112.5 kg P_2O_5 ha^{-1} and 157.5 kg K_2O ha^{-1}. P fertilizer was applied at basal. K fertilizer was split equally at basal and panicle initiation. The regimen for water management was in the sequence of flooding, midseason drainage, re-flooding and moist intermittent irrigation. Pests and weeds were controlled using chemicals.

2.3. Sampling and measurements

Ten hills were sampled diagonally from a 5 m^2 harvest area (excluding plants in the borders) for each replication at maturity. These samples were separated into straw and panicles. Panicles were threshed by hand and the filled grains were separated from unfilled grains by submerging them in tap water. Dry weights of straw, rachis and filled and unfilled grains were determined after oven drying at 70°C to constant weight. The N content of straw, rachis and filled and unfilled grains were determined by an autoanalyzer (Integral Futura, Alliance Instruments, Frépillon, France) to calculate N uptake in filled grains (NUFG) and the total N uptake (TNU). Plants were harvested from the 5 m^2 area and grains were threshed and sun-dried. Filled grains were separated from unfilled grains and debris by winnowing. Around 500 g of the filled grains were taken from each sample and stored at room temperature for 4 months to ensure stable grain quality [29]. A weight of 130 g of the stored grains was dehulled with a roller sheller and polished in a polishing machine according to the National Standard NT 147-88 of China. Brown rice percentage (BRP), milled rice percentage (MRP) and head rice percentage (HRP) were calculated based on the rough rice weight. Percentages of chalky rice grains (PCRG), degree of chalkiness (DC), gelatinization temperature (GT), gel consistency (GC), amylose content (AC) and protein content (PC) were determined according to the descriptions of Huang et al. [30]. Rice length (RL), rice width (RW) and rice length-to-width ratio (RLWR) were measured on 10 head rice grains using a photoenlarger magnified at 10×. Temperature data were collected from the local weather station.

2.4. Statistical analysis

Data were analyzed following analysis of variance (Statistix 8, Analytical software, Tallahassee, FL, USA). The statistical model used included sources of variation due to replication, cultivar, N management practice, season and interactions of cultivar × N management practice, cultivar × season, N management practice × season and cultivar × N management practice × season. Means of cultivars were compared based on the least significant difference (LSD) test at the 0.05 probability level.

3. Results

Daily mean temperature trended to increase during the early season period, whereas a decrease trend was observed during the late season period (**Figure 1a**). Average daily mean temperatures during the grain-filling period were 28.6°C for HHZ and 29.2°C for LYPJ and YLY1 in early season (**Figure 1b**), as well as 24.0°C for HHZ and 23.3°C for LYPJ and YLY1 in the season (**Figure 1c**).

Significant cultivar effects were observed for all measured grain quality traits except for PC (**Tables 3** and **4**). LYPJ had slightly higher BRP and MRP than HHZ, while the differences were not significant between YLY1 and HHZ. MRP was higher in LYPJ and YLY1 than in HHZ by 13 and 30%, respectively. LYPJ and YLY1 showed 14- and 9-fold, respectively, higher PCRG than HHZ. DC was 20-fold higher in LYPJ and 11-fold higher in YLY1 compared with HHZ. LYPJ showed 12% lower GT than HHZ, while the difference between YLY1 and HHZ was relatively small. AC was 29% higher in LYPJ but 12% lower in YLY1 than in HHZ. GC was 8% higher in LYPJ, but 13% lower in YLY1 than in HHZ. N management practice had no significant effects on any of the grain quality traits except for GT. SSNM showed slightly higher GT than FTNM. Seasonal effects were significant on all the grain quality traits except for BRP. Compared to early season, late season had lower MRP, PCRG, DC and GC but higher HRP, GT, AC and PC. The interactive effects between cultivar and season were significant on all the grain quality traits except for PC, while the other interactive effects were generally insignificant.

There were significant cultivar effects on all three grain shape traits (**Tables 3** and **4**). RL was slightly lower in LYPJ than in YLY1 and HHZ. LYPJ and YLY1 had 12 and 11%, respectively, higher RW than HHZ. RLWR was lower in LYPJ and YLY1 than in HHZ by 13 and 10%, respectively. N management practice showed no significant effects on RL and RLWR, whereas SSNM had a slightly but significantly higher RW than FTNM. A significant seasonal effect was observed for RL but not for either RW or RLWR. RL was slightly lower in early season than in late season.

Cultivar and N management practice had no significant effects on TNU and NUFG (**Figure 2a** and **b**). Significant seasonal effects were observed for TNU and NUFG. Late season showed higher TNU and NUFG than early season by 5 and 22%, respectively.

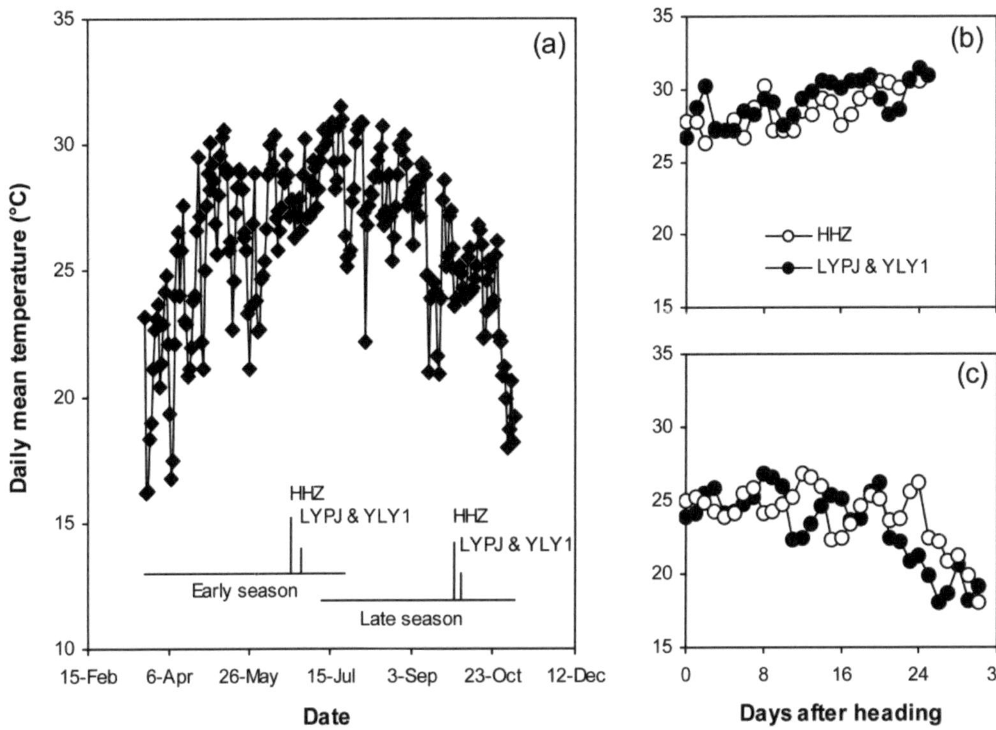

Figure 1. Daily mean temperature during early and late rice-growing seasons (a) and grain-filling period (from heading to maturity) in the early (b) and late (c) seasons. In (a), horizontal lines denote the duration (from sowing to harvesting) of early and late seasons and vertical lines represent the heading dates for three rice cultivars (HHZ, Huanghuazhan; LYPJ, Liangyoupeijiu; and YLY1, Y-liangyou 1).

4. Discussion

Our results indicate that the super hybrid rice cultivars LYPJ and YLY1 had higher milling efficiency, poorer appearance and equal nutritional value compared with the popular inbred rice cultivar HHZ. Wang et al. [24] also observed a higher milling efficiency in LYPJ than in an ordinary hybrid rice cultivar Shanyou 63 and they pointed out that the higher milling efficiency in LYPJ was associated its lower PCRG compared with Shanyou 63. In this regard, it is suggested that chalkiness reduces grain resistance to forces applied during the milling process, causing a decrease in milling efficiency [18, 19]. However, this is not the case in the present study, in which LYPJ and YLY1 had higher PCRG and DC than HHZ. In fact, in addition to the chalkiness, grain shape and PC are also correlated with grain milling efficiency in

http://dx.doi.org/10.5772/66386

Source	Grain quality trait[a]									Grain shape trait[b]		
	BRP	MRP	HRP	PCRG	DC	GT	AC	GC	PC	RL	RW	RLWR
Cultivar (C)	6.02**	7.69**	57.14**	102.48**	116.69**	132.72**	983.19**	85.11**	2.29ns	15.40**	307.62**	330.44**
N management practice (N)	0.14ns	0.75ns	0.19ns	2.45ns	2.97ns	7.98**	2.71ns	2.06ns	0.07ns	0.64ns	6.07*	2.35ns
Season (S)	4.04ns	11.20**	33.85**	12.92**	54.92**	56.15**	55.11**	18.54**	163.28**	5.75*	0.00ns	0.59ns
C × N	1.27ns	0.13ns	0.75ns	0.38ns	0.57ns	8.03**	0.05ns	0.31ns	0.07ns	0.64ns	2.66ns	1.91ns
C × S	8.32**	9.32**	19.10**	26.19**	44.71**	13.46**	23.67**	6.72**	1.09ns	15.54**	1.14ns	5.43*
N × S	2.69ns	0.94ns	0.74ns	1.44ns	4.46*	5.45*	0.36ns	0.07ns	7.60*	1.77ns	6.07*	2.35ns
C × N× S	0.67ns	0.14ns	0.81ns	0.22ns	0.69ns	5.49*	0.01ns	2.51ns	1.40ns	0.50ns	2.66ns	1.91ns

[a] BRP, brown rice percentage; MRP, milled rice percentage; HRP, head rice percentage; PCRG, percentage of chalky rice grains; DC, degree of chalkiness; GT, gelatinization temperature; AC, amylose content; GC, gel consistency; PC, protein content.

[b] RL, rice length; RW, rice width; RLWR, rice length-to-width ratio.

* Significance at the 0.05 probability level.

** Significance at the 0.01 probability level.

ns Non-significance at the 0.05 probability level.

Table 3. F values of analysis of variance for effects of cultivar, N management practice and season on grain quality and shape traits in rice.

Season	N management practice	Cultivar	Grain quality trait[a]									Grain shape trait[b]		
			BRP (%)	MRP (%)	HRP (%)	PCRG (%)	DC (%)	GT	AC (%)	GC (mm)	PC (%)	RL (mm)	RW (mm)	RLWR
Early	FTNM	LYPJ	82.1(0.4)[c]	73.1(0.4)	51.9(1.7)	50.0(3.2)	8.43(0.81)	5.43(0.07)	23.2(0.2)	85.3(1.3)	8.1(0.2)	6.57(0.05)	2.24(0.04)	2.93(0.19)
		YLY1	80.2(0.3)	71.3(0.6)	64.6(0.7)	21.0(2.1)	3.53(0.77)	6.60(0.05)	14.8(0.2)	71.0(1.0)	8.9(0.2)	6.67(0.03)	2.22(0.01)	3.00(0.14)
		HHZ	80.5(0.2)	71.8(0.3)	42.2(2.0)	3.3(0.7)	0.33(0.07)	6.93(0.04)	17.5(0.2)	83.7(2.0)	8.3(0.1)	6.80(0.04)	2.00(0.02)	3.40(0.34)
	SSNM	LYPJ	81.0(0.2)	72.8(0.2)	50.6(1.0)	42.7(2.2)	7.07(0.27)	6.10(0.17)	23.4(0.2)	86.0(2.0)	8.6(0.1)	6.60(0.03)	2.25(0.03)	2.93(0.17)
		YLY1	79.6(0.5)	70.8(0.6)	63.7(0.2)	14.7(4.1)	2.13(0.48)	6.73(0.07)	14.9(0.2)	76.0(0.2)	9.1(0.2)	6.60(0.02)	2.20(0.02)	3.00(0.10)
		HHZ	80.8(0.5)	71.5(0.4)	40.0(1.7)	2.0(0.4)	0.13(0.04)	6.83(0.05)	17.7(0.4)	81.7(1.5)	8.9(0.1)	6.80(0.04)	2.00(0.02)	3.40(0.65)
Late	FTNM	LYPJ	80.4(0.4)	71.1(0.4)	60.0(0.7)	22.7(5.9)	2.43(0.69)	6.37(0.09)	23.0(0.2)	84.7(1.8)	10.5(0.3)	6.63(0.04)	2.21(0.02)	3.00(0.48)
		YLY1	80.5(0.4)	71.5(0.6)	60.7(1.0)	23.3(4.4)	2.27(0.44)	7.00(0.05)	16.9(0.2)	66.0(2.7)	10.6(0.3)	6.80(0.04)	2.19(0.02)	3.11(0.43)
		HHZ	79.5(0.8)	70.9(0.6)	54.5(4.9)	2.3(0.7)	0.23(0.09)	6.93(0.07)	18.4(0.2)	75.0(3.6)	10.9(0.6)	6.67(0.07)	2.00(0.02)	3.34(0.37)
	SSNM	LYPJ	80.5(0.1)	71.4(0.2)	56.6(1.2)	20.7(5.0)	2.37(0.50)	6.47(0.09)	23.4(0.2)	88.3(0.3)	10.3(0.4)	6.67(0.03)	2.30(0.02)	2.90(0.14)
		YLY1	80.8(0.3)	71.6(0.3)	63.1(1.0)	23.7(0.9)	2.57(0.38)	6.90(0.10)	17.2(0.3)	65.0(0.6)	10.4(0.1)	6.83(0.07)	2.25(0.03)	3.04(0.11)
		HHZ	79.9(0.6)	70.6(0.4)	57.0(1.7)	2.0(0.6)	0.30(0.06)	7.00(0.05)	18.8(0.2)	77.7(1.9)	10.2(0.3)	6.73(0.07)	2.00(0.02)	3.37(0.10)
Mean of cultivar														
LYPJ	81.0a	72.1a	54.8b	34.0a	5.08a	6.09c	23.3a	86.1a	9.4b	6.62b	2.25a	2.94c		
YLY1	80.3b	71.3b	63.0a	20.7b	2.63b	6.81b	16.0c	69.5c	9.8a	6.73a	2.21b	3.03b		
HHZ	80.2b	71.2b	48.4c	2.4c	0.25c	6.93a	18.1b	79.5b	9.6ab	6.75a	2.00c	3.38a		

Note: Means of cultivars with the same letters for each parameter are not significantly different according to LSD at the 0.05 probability level.

[a] BRP, brown rice percentage; MRP, milled rice percentage; HRP, head rice percentage; PCRG, percentage of chalky rice grains; DC, degree of chalkiness; GT, gelatinization temperature; AC, amylose content; GC, gel consistency; PC, protein content.

[b] RL, rice length; RW, rice width; RLWR, rice length-to-width ratio.

[c] Values in parenthesis are SE ($n = 3$).

Table 4. Grain quality and shape traits in three rice cultivars, including two super hybrid cultivars Liangyoupeijiu (LYPJ) and Y-liangyou 1 (YLY1) and a popular inbred cultivar Huanghuazhan (HHZ), grown under fixed-time N management (FTNM) and site-specific N management (SSNM) in early and late seasons.

http://dx.doi.org/10.5772/66386

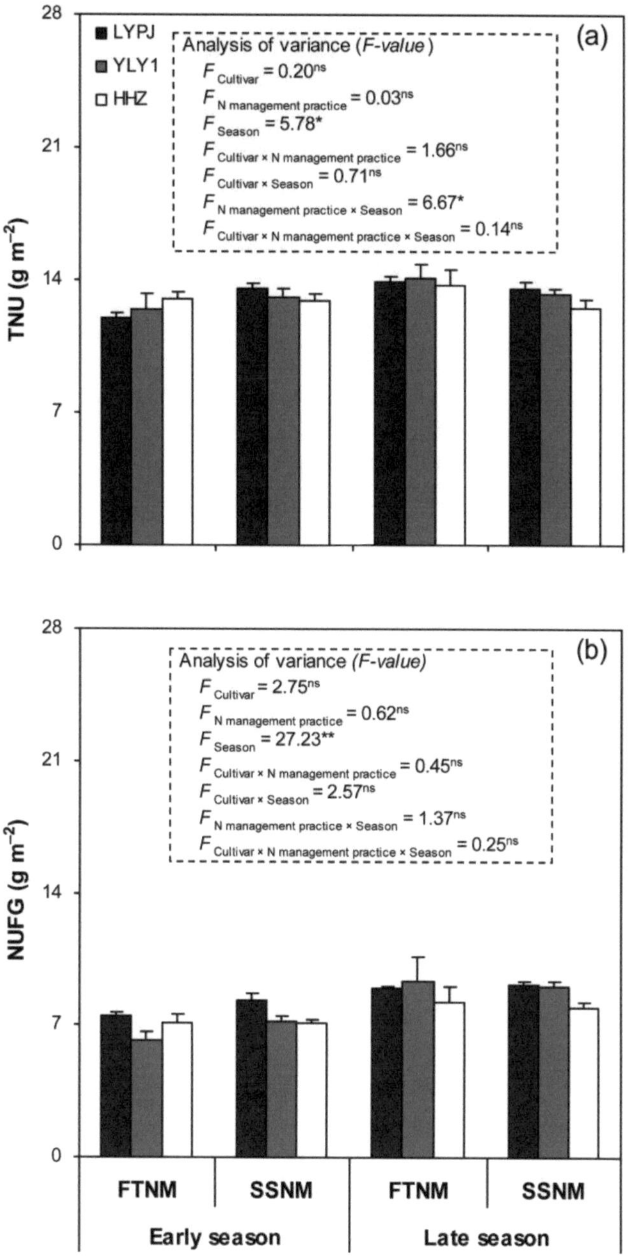

Figure 2. Total N uptake (TNU) and N uptake in filled grains (NUFG) in three rice cultivars, including two super hybrid cultivars Liangyoupeijiu (LYPJ) and Y-liangyou 1 (YLY1) and a popular inbred cultivar Huanghuazhan (HHZ), grown under fixed-time N management (FTNM) and site-specific N management (SSNM). Vertical bars represent SE ($n = 3$). *Significance at the 0.05 probability level. **Significance at the 0.01 probability level. nsNon-significance at the 0.05 probability level.

rice [15, 20]. Generally, increased RW or PC tends to result in increased milling efficiency. In this study, the former seems to be responsible for the higher milling efficiency in LYPJ and

YLY1 than in HHZ, because LYPJ and YLY1 had equal PC compared with HHZ. Moreover, it has been reported that there are positive relationships between RW with PCRG and DC [16, 17]. Therefore, the poorer appearance (higher PCRG and DC) in LYPJ and YLY1 might also be associated with their higher RW. These results also reveal that it may be difficult to achieve a synchronous improvement in milling efficiency and appearance in super hybrid rice. Among the cooking and eating quality traits, GT is often considered as an indicator affecting the cooking time of rice, for rice with higher GT requires a longer time to cook [31]. This point, however, is uncertain. Bhattacharya and Sowbhagya [32] observed that water uptake and hence the cooking time was strongly influenced by the surface area per unit weight. Other workers have also observed that water uptake of rice at boiling temperature is not related to GT but it is related to grain size and shape [33, 34]. In the present study, significant differences were observed both in GT and in grain shape between LYPJ and YLY1 with HHZ, suggesting that further studies are required to understand the difference in cooking characteristics between them. AC and GC have been considered the key indicators relating to palatability [31, 35]. Generally, consumers, especially in China, favor the rice with moderate AC and GC [25]. HHZ is such a cultivar. In the present study, LYPJ and YLY1 had largely higher or lower AC and GC than HHZ, implying that the palatability of LYPJ and YLY1 was not good enough to suit the preference of consumers. Our study suggests that strategies for grain quality improvement in super hybrid rice should be focused on appearance and palatability.

Previous studies showed that the N application could increase milling efficiency and nutritional value (PC) in rice grains and there was a positive relationship between the PC and N application rate [20, 23, 24]. Furthermore, Wang et al. [24] stated that the increased PC from the higher N application rate depended on increased TNU. However, in the present study, although the total N application rate was lower under SSNM than under FTNM by 20 and 15% in early and late season, respectively (**Table 2**), no significant differences were observed in N uptake (TNU and NUFG) in rice crops as well as milling efficiency and nutritional value in rice grains between SSNM and FTNM. Our results imply that replacing FTNM with SSNM can reduce N input without sacrificing grain quality in rice production.

It is well known that early season rice generally has poorer grain quality than late season rice and the poorer grain quality of early season rice is to a great extent attributed to the higher daily mean temperature during grain-filling period [25]. Consistently, in the present study, average daily mean temperature during grain-filling period across three rice cultivars was about 5°C higher in early season than in late season and grain quality was generally poorer in early season than in late season. More interestingly, we found that TNU and NUFG were significantly lower in early season than in late season. As mentioned above, higher N uptake in rice crops can lead to a higher grain PC, which may consequently result in a higher grain milling efficiency [20, 24]. Therefore, in this study, variation in N uptake was partly responsible for the seasonal differences in milling efficiency and nutritional value. Daily temperature is one of the factors influencing rice N uptake under favorable growth conditions and adequate N supply [36]. In general, higher daily temperature leads to a higher rice N uptake rate. Our previous study showed that daily temperature during the early growth stage was lower in early season than in late season, which resulted in that N uptake was lower in early season than that in late season [37]. This might be one reason why early season had lower TNU and NUFG than late season in the present study. Another reason for the seasonal variation in N uptake might be the seasonal changes in the N application rate under SSNM, which was 8% lower in early season than in late season (**Table 2**). These results also indicate that SSNM can achieve a good match of N supply with crop demand.

In addition, we observed that the interactive effects between cultivar and season on grain quality traits were generally significant. It is not surprise because (1) the seasonal variation in average daily mean temperature during grain-filling period was different between the super hybrid rice cultivars (5.9°C) and the popular inbred rice cultivar (4.6°C) (**Figure 1b** and **c**) and (2) the temperature effects on grain quality traits are cultivar-dependent [25, 38]. For example, Zhong et al. [25] found that under high temperature, GC decreased or remained little changed for cultivars with higher amylase content and increased for cultivars with lower amylase content. A similar result was also observed in the present study (**Table 4**). Namely, high temperature in early season decreased GC for LYPJ, which had higher amylase content and increased GC for YLY1, which had lower amylase content.

5. Conclusions

It is concluded that (1) strategies for grain quality improvement in super hybrid rice should be focused on appearance and palatability and (2) replacing FTNM with SSNM can reduce N input without sacrificing grain quality in rice production.

Acknowledgements

This study was supported by the Earmarked Fund for China Agriculture Research System (Grant No. CARS-01).

Author details

Min Huang[1]*, Shuanglü Shan[1], Jiana Chen[1], Fangbo Cao[1], Ligeng Jiang[1, 2] and Yingbin Zou[1]

*Address all correspondence to: jxhuangmin@163.com

1 Southern Regional Collaborative Innovation Center for Grain and Oil Crops (CICGO), Hunan Agricultural University, Changsha, China

2 Key Laboratory of Crop Cultivation and Farming System, Guangxi University, Nanning, China

References

[1] Fan M, Lu S, Jiang R, Liu X, Zhang F. Triangular transplanting pattern and split nitrogen fertilizer application increase rice yield and nitrogen fertilizer recovery. Agronomy Journal. 2009;**101**:1421–1425. DOI: 10.2134/agronj2009.0009

[2] Peng S, Tang Q, Zou Y. Current status and challenges of rice production in China. Plant Production Science. 2009;**12**:3–8. DOI: 10.1626/PPS.12.3

[3] Zhang Q. Strategies for developing green super rice. Proceedings of the National Academy of Sciences of the United States of America. 2007;**104**:16402–16409. DOI: 10.1073/pnas.0708013104

[4] Cai H, Chen Q. Rice research in China in the early 21st century. Chinese Rice Research Newsletter. 2000;**8**:14–16.

[5] Peng S, Khush GS, Virk P, Tang Q, Zou Y. Progress in ideotype breeding to increase rice yield potential. Field Crops Research. 2008;**108**:32–38. DOI: 10.1016/j.fcr.2008.04.001

[6] Cheng S, Liao X, Min S. China's super rice research: background, goals and issues. China Rice. 1998;**1**:3–5.

[7] Yuan L. Breeding of super hybrid rice. In: Peng S, Hardy B, editors. Rice Research for Food Security and Poverty Alleviation. Los Baños: International Rice Research Institute; 2001. pp. 143–149.

[8] Zhang X, Wang D, Fang F, Zhen Y, Liao X. Food safety and rice production in China. Research of Agricultural Modernization. 2005;**26**:85–88.

[9] Liao X, Chen Q, Pan G. Status and strategies for improving high-quality rice production in China. Journal of Agrotechnical Economy. 2002;**5**:32–34.

[10] Li J, Xiao J, Grandillo S, Jiang L, Wan Y, Deng Q, Yuan L, McCouch SR. QTL detection for rice grain quality traits using an interspecific backcross population derived from cultivated Asian (*O. sativa* L.) and African (*O. glaberrima* S.) rice. Genome. 2004;**47**:697–704. DOI: 10.1139/g04-029

[11] Ordonez Jr SA, Silva J, Oard JH. Association mapping of grain quality and flowering time in elite japonica rice germplasm. Journal of Cereal Science. 2010;**51**:337–343. DOI: 10.1016/j.jcs.2010.02.001

[12] Wan XY, Wan JM, Weng JF, Jiang L, Bi JC, Wang CM, et al. Stability of QTLs for rice grain dimension and endosperm chalkiness characteristics across eight environments. Theoretical and Applied Genetics. 2005;**110**:1334–1346. DOI: 10.1007/s00122-005-1976-x

[13] Rabiei B, Valizadeh M, Ghareyazie B, Moghaddam M, Ali AJ. Identification of QTLs for rice grain size and shape of Iranian cultivars using SSR markers. Euphytica. 2004;**137**:325–333. DOI: 10.1023/B:EUPH.0000040452.76276.76

[14] Shewry PR. Improving the protein content and composition of cereal grain. Journal of Cereal Science. 2007;**46**:239–250. DOI: 10.1016/j.jcs.2007.06.006

[15] Zheng TQ, Xu JL, Li ZK, Zhai HQ, Wan JM. Genomic regions associated with milling quality and grain shape indentified in a set of random introgression lines of rice (*Oryza sativa* L.). Plant Breeding. 2007;**126**:158–163. DOI: 10.1111/j.1439-0523.2007.01357.x

[16] Tan YF, Xing YZ, Li JX, Yu SB, Xu CG, Zhang Q. Genetic bases of appearance quality of rice grains in Shanyou 63, an elite rice hybrid. Theoretical and Applied Genetics. 2000;**101**:823–829. DOI: 10.1007/s001220051549

[17] Jongkaewwattana S, Geng S. Inter-relationships amongst grain characteristics, grain-filling parameters and rice (*Oryza sativa* L.) milling quality. Journal of Agronomy and Crop Science. 2001;**187**:223–229. DOI: 10.1046/j.1439-037X.2001.00521.x

[18] Lisle AJ, Martin M, Fitzgerald MA. Chalky and translucent rice grains differ in starch composition and structure and cooking properties. Cereal Chemistry. 2000;**77**:627–632. DOI: 10.1094/CCHEM.2000.77.5.627

[19] Counce PA, Bryant RJ, Bergman CJ, Bautista RC, Wang YJ, Siebenmorgen TJ, Moldenhauer KAK, Meullenet JFC. Rice milling quality, grain dimensions and starch branching as affected by high night temperatures. Cereal Chemistry. 2005;**82**:645–648. DOI: 10.1094/CC-82-0645

[20] Leesawatwong M, Jamjod S, Kuo J, Dell B, Rerkasem B. Nitrogen fertilizer increases seed protein and milling quality of rice. Cereal Chemistry. 2005;**82**:588–593. DOI: 10.1094/CC-82-0588

[21] Tan YF, Sun M, Xing YZ, Hua JP, Sun XL, Zhang QF, Corke H. Mapping quantitative trait loci for milling quality, protein content and color characteristics of rice using a recombinant inbred line population derived from an elite rice hybrid. Theoretical and Applied Genetics. 2001;**103**:1037–1045. DOI: 10.1007/s001220100665

[22] Fitzgerald MA, McCouch SR, Hall RD. Not just a grain of rice: the quest for quality. Trends in Plant Science. 2008;**14**:133–139. DOI: 10.1016/j.tplants.2008.12.004

[23] Ning H, Liu Z, Wang Q, Lin Z, Chen S, Li G, Wang S, Ding Y. Effect of nitrogen fertilizer application on grain phytic acid and protein concentrations in japonica rice and its variations with genotypes. Journal of Cereal Science. 2009;**50**:49–55. DOI: 10.1016/j.jcs.2009.02.005

[24] Wang Q, Huang J, He F, Cui K, Zeng J, Nie L, Peng S. Head rice yield of "super" hybrid rice Liangyoupeijiu grown under different nitrogen rates. Field Crops Research. 2012;**134**:71–79. DOI: 10.1016/j.fcr.2012.05.001

[25] Zhong LJ, Cheng FM, Wen X, Sun ZX, Zhang GP. The deterioration of eating and cooking quality caused by high temperature during grain filling in early-season indica rice cultivars. Journal of Agronomy and Crop Science. 2005;**191**:218–225. DOI: 10.1111/j.1439-037X.2005.00131.x

[26] Zhang Y, Tang Q, Zou Y, Li D, Qin J, Yang S, Chen L, Xia B, Peng S. Yield potential and radiation use efficiency of 'super' hybrid rice grown under subtropical conditions. Field Crops Research. 2009;**114**:91–98. DOI: 10.1016/j.fcr.2009.07.008

[27] Jiang P, Xie X, Huang M, Zhou X, Zhang R, Chen J, Wu D, Xia B, Xu F, Xiong H, Zou Y. Comparison of yield performance and nitrogen response between hybrid and inbred

rice under different ecological conditions in southern China. Journal of Integrative Agriculture. 2015;**14**:1283–1294. DOI: 10.1016/S2095-3119(14)60929-1

[28] Jiang P, Xie X, Huang M, Zhou X, Zhang R, Chen J, Wu D, Xia B, Xiong H, Xu F, Zou Y. Potential yield increase of hybrid rice at five locations in southern China. Rice. 2016;**9**:11. DOI: 10.1186/S12284-016-0085-6

[29] Perez CM, Juliano BO, Liboon SP, Alcantara JM, Cassman KG. Effects of late nitrogen fertilizer application on head rice yield, protein content and grain quality of rice. Cereal Chemistry. 1996;**73**:556–560.

[30] Juliano BO, Onate LU, Del Mundo AM. Relation of starch composition, protein content and gelatinization temperature to cooking and eating qualities of milled rice. Food Technology. 1965;**19**:1006–1011.

[31] Huang M, Jiang L, Zou Y, Zhang W. On-farm assessment of effect of low temperature at seedling stage on early-season rice quality. Field Crops Research. 2013;**141**:63–68. DOI: 10.1016/j.fcr.2012.10.019

[32] Bhattacharya KR, Sowbhagya CM. Water uptake by rice during cooking. Cereal Science Today. 1971;**16**:420–424.

[33] Batcher OM, Deary PA, Dawson EH. Cooking quality of 26 varieties of milled white rice. Cereal Chemistry. 1957;**34**:277–285.

[34] Halick JV, Kelly VJ. Gelatinization and pasting characteristics of rice varieties as related to cooking behavior. Cereal Chemistry. 1959;**36**:91–96.

[35] Cagampang GO, Perez CM, Juliano BO. A gel consistency test for eating quality of rice. Journal of the Science of Food and Agriculture. 1973;**24**:1589–1594. DOI: 10.1002/jsfa.2740241214

[36] Peng S, Cassman KG. Upper thresholds of nitrogen uptake rates and associated nitrogen fertilizer efficiencies in irrigated rice. Agronomy Journal. 1998;**90**:178–185. DOI: 10.2134/agronj1998.00021962009000020010x

[37] Huang M, Yang C, Ji Q, Jiang L, Tan J, Li Y. Tillering responses of rice to plant density and nitrogen rate in a subtropical environment of southern China. Field Crops Research. 2013;**149**:187–192. DOI: 10.1016/j.fcr.2013.04.029

[38] Resurreccion AP, Hara T, Juliano BO, Yoshida S. Effect of temperature during ripening on grain quality of rice. Soil Science and Plant Nutrition. 1977;**23**:109–112. DOI: 10.1080/00380768.1977.10433027

Chapter 8

The Deep Purple Color and the Scent are Two Great Qualities of the Black Scented Rice (*Chakhao*) of Manipur

Ibemhal Devi Asem, Rajkumar Imotomba and
Pranab B. Mazumder

Additional information is available at the end of the chapter

http://dx.doi.org/10.5772/67193

Abstract

Specialty rice varieties with unique properties such as unique color, flavor, aroma and nutraceutical qualities are increasingly in demand other than the traditional white rice varieties. Black rice has various nutraceutical properties with high anthocyanin content and the anthocyanin antioxidants are very important in preventing various diseases. Black scented rice of Manipur, which are commonly known as *Chakhao*, are deep purple colored and scented, and are very glutinous, as well. Two *Chakhao* cultivars, *Chakhao Poireiton* and *Chakhao Amubi*, were shown to have high anthocyanin and phenolics content and strong antioxidant activity. The main anthocyanins of *Chakhao Poireiton* are delphinidin 3-galactoside, delphinidin 3-arabinoside, cyanidin 3-galactoside and cyanidin 3-glucoside and that of *Chakhao Amubi* are delphinidin 3-galactoside, delphinidin 3-arabinoside and cyanidin 3-galactoside. By GC-MS analysis, a cross mixture of 26 volatile compounds from *Chakhao Poireiton* and 11 volatile compounds from *Chakhao Amubi* were found to be responsible for emanating the aroma. Among the complex mixture of volatile oil components, n-hexadecanoic acid and octadec-9-enoic acid were the most abundant in *Chakhao Poireiton* and 17-pentatriacontene, 13-octadecenal (Z) and hexadecenoic acid eicosyl ester in *Chakhao Amubi*.

Keywords: black scented rice, *Chakhao* rice, Manipur, anthocyanin, nutraceutical properties, flavor (Scent) active compounds

1. Introduction

Rice is the staple food for over half of the world's population. Rice cultivation provides employment for over one billion people directly or indirectly. Besides traditional white or common rice varieties, specialty rice varieties with unique properties such as unique flavor, aroma (unique aromas), color (red, purple), nutrition (glossiness, stickiness and smooth texture), chemical composition, esthetic, waxy (very low amylose content) and superior processing qualities are increasingly in demand. Rice according to the pericarp color can be broadly classified into black (purple), red and white rice. Black rice has high anthocyanin content located in the pericarp layers, which gives it a dark purple color [1, 2]. The demand for various types of specialty rice is increasing in recent years, which are sold for as much as 50% more than traditional rice cultivars [3]. Black rice has been used in various traditional medicines; recently, many researchers have reported that they have several health benefits in various studies, and thus, black rice is being considered as the new superfood by US scientists. In Asian countries, black rice is often consumed after mixing with white rice to enhance flavor, color and nutritional value, which includes high protein, total essential amino acids, vitamin B1 and minerals Fe, Zn, Mn and P, and it is intensely colored because of anthocyanin [4]. One serving of black rice even though contains some calories, but offers a high amount of flavanoid phytonutrients, important fiber, mineral content such as iron and copper, and it is a good source of plant-based protein which is hard to get to plant-based eaters who rely on grains and legumes for protein [5]. Black rice is rich in antioxidant anthocyanin [6, 7]. A spoonful of black rice bran provides the same amount or more anthocyanin than a spoonful of blueberries [8]. Anthocyanin antioxidants are very important in the prevention of cardiovascular disease, protection against cancer, improving brain function, reducing inflammation, etc [9, 10].

Flavor is the primary importance of specialty rice, and superior flavor increases consumer satisfaction and repeatedly purchase [11, 12]. Flavor is composed of taste and aroma, while aroma is conferred by volatile compounds emanating from cooked rice [13]. From cooked rice, a number of compounds have been identified, but only a few make up the characteristic aroma [9, 14, 15]. Based on the aroma, rice cultivars can be separated generally into aromatic and non-aromatic types: Aromatic rice has a relatively diverse range of unique aromas [13] such as jasmine rice, which is characterized as having buttery, corn, dairy, starchy, cooked grain and nutty attributes, and basmati rice, which is characterized as having hay like and earthy attributes [16]. A complex mixture of odor-active compounds comprises the aroma of both aromatic and nonaromatic rice; approximately 300 volatile compounds have been identified from various cultivars of aromatic and nonaromatic rice [17]; and several odor-active compounds in cooked aromatic rice have been determined using odor units [18, 19]. In aromatic rice, 2-acetyl 1-pyrroline (2-AP) is described as having a "popcorn-like" odor by American and "pandan-like" odor by Asian consumers which is synthesized in aerial parts of aromatic rice during growth, and in some nonaromatic types, it is present at a very low negligible concentration [20, 21]. 2-AP is not only the compound responsible for the unique aromas, but their aromas are due to qualitative and quantitative variations in a diverse cross section of odor-active compounds [13]. In cooked rice, lipid-derived odor-active compounds were formed during the degradation of oleic (octanal, heptanal, nonanal, (*E*)-2-nonenal, decanal and 2-heptanone are formed from oleic acid), linoleic (hexanal, pentanol, pentanal, (*E*)-2-octenal, (*E*,*E*)-2,4-decadienal and 2-pentylfuran are formed from linoleic acid) and linolenic acid [22, 23]. The oxidation of lipid yields rancid odors and also induces various deteriorative reactions with proteins, amino acids and other components [24]. Examples of thermally derived flavor compounds formed in rice during cooking, which have seasoning-like and meaty-like aromas, are 3-hydroxy-4,5-dimethyl-2(5*H*)-furanone and bis (2-methyl-3-furyl)disulfide, respectively [19]. Thus, the volatile chemistry of rice grain could be due to a single predominant compound or a complex mixture of several compounds associated with a unique flavor and strength of aroma in the diverse set of fragrant rice.

Now, let us have a glance of Manipur, Manipur is a place of exquisite natural beauty and splendor. Manipur is a state of India laying on the northeastern corner of India bordering Myanmar (Burma). The state is rectangular in shape with a precious little valley in the center encircled by mountain ranges on all sides with salubrious climate [25]. The soil cover of Manipur can be divided into two broad types, viz. red ferruginous soil in hill area and alluvium in valley. Rice is the staple food for Manipuris. Agriculture mainly on rice and allied activities are the largest source of livelihood of majority of rural masses and backbone of the state's economy. Rice-based agriculture and allied services provided direct employment to about 70% of the total working population of the state. Rice crop is commonly grown during *kharif* season in Manipur. Rice is largely cultivated by small and marginal farmers. Manipur is endowed with several rice germplasms having special cultural values and unique characters, including colored, aromatic and quality rice landraces.

In this chapter, we reported the volatile oil components responsible for scent of two *Chakhao* cultivars (*Oryza sativa* cv. *Chakhao Poireiton* and *Oryza sativa* cv. *Chakhao Amubi*) of Manipur and reviewed our previous study on anthocyanin content and nutraceutical properties present in these two cultivars.

2. *Chakhao* rice of Manipur

Manipur has a large variety of indigenous rice germplasm which range their adaptation from low-lying lake areas to rainfed uplands of Manipur hills. There is a diverse set of locally adapted aromatic rice, ranging in color from white to red and purple, all of which are very glutinous in nature, also. They are commonly known as *Chakhao*. The literal meaning of *Chakhao* in Manipuri language is delicious rice (*Chak* means rice and *ahoaba* means delicious). *Chakhao* rice of Manipur are poor yielders which are found only in this state of India. Usually, they are grown during *Kharif* season, and they are low yielding, highly lodging, late maturing and also poorly studied as well. Some of the purple color *Chakhao* (black scented rice) cultivars are *Chakhao Poireiton, Chakhao Amubi, Wairi Chakhao, Khurkhul Chakhao, Pong Chakhao* and *Chakhao Sempak*. Red color *Chakhao* cultivars are *Chakhao Anganba* and *Langphou Chakhao,* and white color cultivars are *Chakhao Manam Nungshibi* and *Chakhao Angouba*.

The two great qualities of black scented rice of Manipur are their color and scent. These two great qualities make the *Chakhao* a unique specialty rice. They are deep purple in color and fragranced. *Chakhao* are used in community feast and ceremonial purposes as a delicacy, and they are also important in religious rituals. These are one of the high rated dishes served as desserts, flakes, bread, cakes, beverages, chapati and a special snack "Utong Chak" prepared within bamboo sticks [26]. They are sold in the local markets at a high rate. Manipuris *Chakhao* also have potential nutraceutical properties which is used by the traditional medical practitioners.

They have high potential for export to the international markets. However, as they are low yielding, the farmers of Manipur grow *Chakhao* very limitedly.

3. The two great qualities of black scented rice (*Chakhao* rice) of Manipur

3.1. Anthocyanin content and Nutraceutical properties of *Chakhao*

In our previous study [26], we had reported the anthocyanin and phenolics content and antioxidant activity of two black scented rice cultivars of Manipur, *Chakhao Poireiton* and *Chakhao Amubi*. Both cultivars have high anthocyanin, phenolics content and strong antioxidant activity. Using pH differential method, the total monomeric anthocyanin present in the two black scented rice cultivars (*Chakhao Poireiton* and *Chakhao Amubi*) and one non-scented white rice variety (CAUR1) was measured. Acidified methanol extract was used for the measurement, and total monomeric anthocyanin found in *Chakhao Poireiton* and *Chakhao Amubi* was 740 and 692 mg/kg, respectively, which was comparatively higher than non-scented white rice, CAUR1 (134 mg/kg). In our study, we have found out that anthocyanin content in both the *Chakhao* cultivars was higher than white rice variety, and our result is also consistent with the study on other colored rice. Colored rice varieties from Minahasa Regency, North Sulawesi, Indonesia, have higher anthocyanin content from the noncolored rice varieties [27]. Similarly, two colored rice of Thailand were reported to have a comparatively higher anthocyanin content than that of the Thailand non-pigmented rice variety [28]. Eight different pigmented varieties in Thailand were also studied and reported that rice varieties with dark purple color contained a higher amount of anthocyanin [29].

Total phenolics content was estimated for *Chakhao Poireiton* and *Chakhao Amubi* using Folin-Ciocalteu method from acidified methanol extracts, and the results were expressed as gallic acid equivalent. The total phenolics content in *Chakhao Poireiton* was 577 mg/100 g and in *Chakhao Amubi* was 500 mg/100 g of the powdered sample as gallic acid equivalent. The maximum DPPH-free radical-scavenging activity of *Chakhao Poireiton* and *Chakhao Amubi* extracts were 70.28 and 60.84%, respectively, and that of the standard ascorbic acid was 93.73% given by DPPH assay (**Figure 1**) [26]. Though scavenging activity is a little lower than the standard ascorbic acid, the antioxidant activity of *Chakhao Poireiton* and *Chakhao Amubi* extracts was strong enough, thus showing that higher phenolics content and antioxidant capacity are correlated with *Chakhao Poireiton* and *Chakhao Amubi*. Many researchers have reported about the phenolics content and antioxidant activity of other pigmented rice which are consistent with our study on *Chakhao Poireiton* and *Chakhao Amubi* of Manipur. One such study showed that anthocyanin extract of Korean black rice (*Heugjinjubyeo*) exhibited good free radical-scavenging activity [10]. Similarly, total phenol content using Folin-Ciocalteu method and antioxidant activities using thiocyanate method and DPPH-free radical-scavenging assay of different Thai rice white color, red color and black rice cultivars were conducted and reported that colored rice showed higher antioxidant activity from the white rice [30].

In our previous study, we had also reported the different types of anthocyanin present in Manipuris *Chakhao Poireiton* and *Chakhao Amubi*. Further, the acidified methanol extracts of

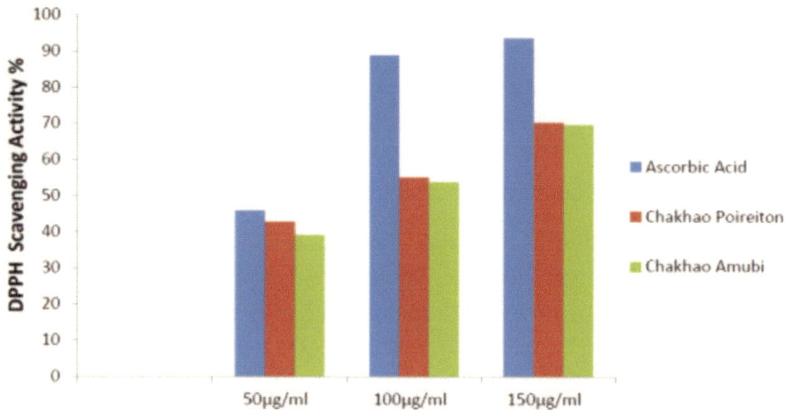

Figure 1. DPPH scavenging activity% of the methanol extracts of *Chakhao Poireiton* and *Chakhao Amubi.*

Chakhao Poireiton and *Chakhao amubi* have been used for HPLC analysis for the identification of anthocyanin components. HPLC analysis identified four main anthocyanins from *Chakhao Poireiton* and three main anthocyanins from *Chakhao Amubi*. The four main anthocyanins of *Chakhao Poireiton* were delphinidin 3-galactoside, delphinidin 3-arabinoside, cyanidin 3-galactoside and cyanidin 3-glucoside and that of *Chakhao Amubi* were delphinidin 3-galactoside, delphinidin 3-arabinoside and cyanidin 3-galactoside [26]. Among the identified anthocyanin, delphinidin 3-galactoside was found to be the most predominant in both the cultivars. Similarly, several researchers had also reported about the different types of anthocyanin present in other black rice rather than the Manipuris black scented rice. For example, cyanidin 3-O glucoside and peonidin 3-O glucoside were identified from Korean black rice (*Heugjinjubyeo*) by HPLC analysis [10]. Other researchers also performed and reported the HPLC and LC-MS analyses on black rice extract and identified two major anthocyanins: cyanidin-3-glucoside and peonidin-3-glucoside [31]. From the black rice of Osaka, Japan, by using HPLC method, two anthocyanins have also been identified [32].

In the present scenario, there is an increased interest in the alternative sources of anthocyanin due to a rising demand for economical sources of natural and stable pigments [33]. Anthocyanins have shown to be potent antioxidants which are superior to well-known antioxidants such as butylated hydroxyanisole (BHA), alpha-tocopherol, 6-hydroxy-2,5,7,8-tetramethychromane-2-carboxylic acid (Trolox), catechin and quercetin [34]. Antioxidant capacity is becoming a parameter to characterize food or medicinal plants and their bioactive components. In the human body, dietary antioxidants protect against reactive oxygen species [35]. Anthocyanins may reduce the risk of cardiovascular diseases and cancer with anti-inflammatory, antioxidant and chemoprotective properties [10], and also, if the antioxidants intake is increased, there may have a number of health effects, such as reducing the incidence of cancer and cardiovascular diseases [36]. Anthocyanins have been incorporated into the human diet, and due to their diverse physiological abilities to treat conditions such as hypertension, pyrexia, liver disorders, dysentery and diarrhea, urinary problems and the common cold, anthocyanins have been used as traditional herbal medicines [37]. Not only this, the dietary antioxidants can help to fight reactive oxygen species and free radicals and help to decrease the risk of chronic diseases such as coronary heart disease and certain cancers [38]. Our results showed that the black scented rice cultivars (*Chakhao Poireiton* and *Chakhao Amubi*) have better antioxidant properties than noncolored rice varieties; thus, it can be suggested that colored rice varieties could be used as a natural antioxidant source.

3.2. Volatile oil components for scent in *Chakhao*

A complex mixture of volatile compounds comprised the aroma of *Chakhao* rice. GC-MS analysis of the *Chakhao Poireiton* identified 26 volatile compounds (**Table 1**; **Figure 2**). They are in the decreasing order of n-hexadecanoic acid (22.92%), octadec-9-enoic acid (11.66%), 4-beta-H-Pregna (8.78%), 9, 12-octadecadienoic acid (Z.Z) (8.11%), benzene methyl (5.61%), 2-furancarboxaldehyde 5-methyl (3.92%), g-hexadecenoic acid octadecyl ester, (Z) (3.88%), 9-hexadecenoic acid, eicosyl ester, (Z) (3.75%), 2-furancarboxaldehyde (3.41%), pentadecane (2.90%), I 7-pentatriacontene z (2.59%), stigmast 5-EN-3-OL. (3.beta.24S)-(2.00%), tetradecanoic acid (1.91%), stigmast-5-EN.3-OL, oleat (1.68%), 9-octadecenoic acid (Z), 9-octadecenyl ester, (Z) (1.67%), dodecane (1.59%), furan, 2,2'-methylenebis[5-methyl (1.59%), 9-hexadecenoic acid, 9-octadecenyl ester, (Z.Z) (1.58%), furan 2-(2-furanylmethyl)-5-methyl (1.51%), benzene butyl (1.44%), benzofuran 4, 7-dimethyl (1.20%), benzofuran, 2-methyl-3 (0.78%), undecane (0.71%), butanenitrile, 3-methyl (0.25%).

Peak	R. Time	Name	Area	Area%
1.	3.720	Butanenitrile 3-methyl	104,009	0.25
2.	4.540	Benzene methyl	2,295,838	5.61
3.	6.873	2-Furancarboxaldehyde	1,393,622	3.41
4.	11.768	2-Furancarboxaldehyde 5-methyl	1,603,643	3.92
5.	15.012	Benzene butyl	588,613	1.44
6.	16.549	Undecane	289,797	0.71
7.	16.765	Benzofuran 2-methyl-3	18,639	0.78
8.	18.310	Benzene, pentyl	31,571	1.30
9.	18.943	Furan 2-(2-furanylmethyl)-5-methyl	619,648	1.51
10.	19.660	Dodecane	649,180	1.59
11.	20.133	Benzofuran 4, 7-dimethyl	491,294	1.20
12.	21.742	Furan, 2,2'-methylenebis[5-methyl]	649,017	1.59
13.	27.782	Pentadecane	1,188,621	2.90
14.	33.804	Tetradecanoic acid	781,382	1.91
15.	37.940	n-Hexadecanoic acid	9,379,913	22.92
16.	41.112	9, 12-Octadecadienoic acid (Z.Z)	3,317,895	8.11
17.	41.236	Octadec-9-enoic acid	4,771,521	11.66
18.	41.317	Oleic acid	1,336,655	3.27

Peak	R. Time	Name	Area	Area%
19.	48.343	g-Hexadecenoic acid octadecyl ester. (Z)	1,588,616	3.88
20.	48.818	9-Hexadecenoic acid, eicosyl ester, (Z)	1,535,517	3.75
21.	49.042	I 7-Pentatriacontene z	1,059,294	2.59
22.	51.890	9-Octadecenoic acid (Z)-, 9-octadecenyl ester, (Z)	684,629	1.67
23.	51.966	9-Hexadecenoic acid. 9-octadecenyl ester. (Z.Z)	647,022	1.58
24.	53.000	4-Beta-H-Pregna	3,591,642	8.78
25.	53.419	Stigmast-5-En.3-oL. Oleat	687,915	1.68
26.	56.337	Stigmast 5-En-3-oL. (3.Beta.24S)	818,755	2.00
			40,924,254	100.00

Table 1. Volatile oil profiling using GC-MS of *Chakhao Poireiton*.

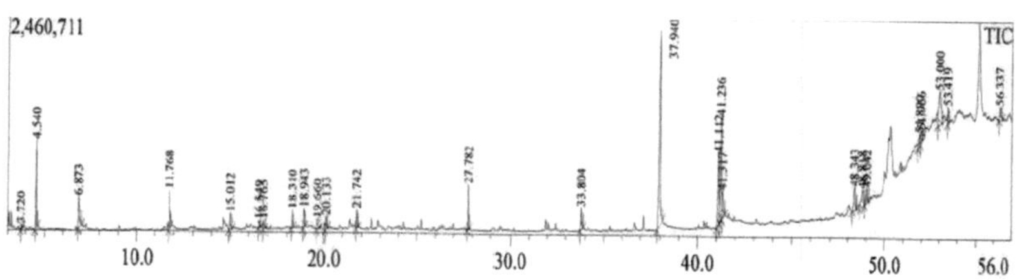

Figure 2. GC-MS chromatogram of *Chakhao Poireiton*.

GC-MS analysis of the *Chakhao Amubi* identified 11 volatile compounds (**Table 2**; **Figure 3**). The decreasing order of volatile compounds comprising the aroma of *Chakhao Amubi* analyzed by GC-MS is as follows: 17-pentatriacontene (40.60%), 13-octadecenal (Z) (12.03%), 9-hexadecenoic acid eicosyl ester-(Z) (11.98%), tetracosamethyl-cyclododecasiloxane (10.27%), Z-9-pentadecenol (9.79%), 9-octadecenoic acid (Z)-tetradecyl ester (0. 4.29%), toluene (3.38%), l-(+)-ascorbic acid 2,6-dihexadecanoate (2.21%), Z.Z-3-13-Octadecedien-I-o I (1.96%), 9-octadecenoic acid, 1,2,3-propanetriyl ester, (E,E,E) (1.93%), cyclononasiloxane octadecamethyl (1.56%). All the compounds together contributed to the unique aroma of their respective cultivars and their aroma increases during cooking, which may be due to the interaction of the compounds with the protein and lipids.

Among the complex mixture of volatile oil components, n-hexadecanoic acid and octadec-9-enoic acid were the most abundant compounds emanating the scent in *Chakhao Poireiton,* and in *Chakhao Amubi,* 17-pentatriacontene, 13-octadecenal (Z) and hexadecenoic acid eicosyl ester were the most abundant compounds. Each variety has a unique fragrance resulting from a number of volatile compounds which may vary from well-characterized popcorn-like aroma/2-AP-associated aroma although little is known about their relationships with aroma/flavor [39]. 2-AP has been found as one of the odor-active compounds which give aroma to the black rice [4]; however, in the present study of *Chakhao Poireiton* and *Chakhao Amubi,* no 2-AP has been detected. Similarly, 2-AP was not detected in black rice, instead reported 94 volatile compounds of which nonanal, butylated hydroxytoluene, 1-hexanol, naphthalene and 1-octen-3-ol were the main volatile compounds [28]. Earlier, concentration of 2-AP has been used as an indicator of aroma in the selection of aromatic lines due to its significant importance in aromatic rice; however, it does not hold good for all the aromatic rice. And also, the overall aroma during cooking may also be associated with dissociation of the starch-lipid complex and enhancement of the formation of a cross section of lipid-derived volatiles compounds.

Peak	R. Time	Name	Area	Area%
1.	4.550	Toluene	748,852	3.38
2.	37.880	l-(+)-ascorbic acid 2,6-dihexadecanoate	490,011	2.21
3.	48.618	Cyclononasiloxane octadecamethyl	346,471	1.56
4.	49.428	9-octadecenoic acid, 1,2,3-propanetriyl ester, (E,E,E)	427,185	1.93
5.	49.635	Z.Z-3-13-octadecedien-I-o I	434,892	1.96
6.	50.685	9-octadecenoic acid (Z)-tetradecyl ester	950,550	4.29
7.	50.866	13-octadecenal, (Z)	2,667,664	12.03
8.	51.096	9-hexadecenoic acid. eicosyl ester (Z)	2,655,953	11.98
9.	51.241	Tetracosamethyl-cyclododecasiloxane	2,277,003	10.27
10.	51.682	Z-9-pentadecenol	2,169,969	9.79
11.	55.156	17-pentatriacontene	9,001,555	40.60
			22,170,105	100.00

Table 2. Volatile oil profiling of *Chakhao Amubi*.

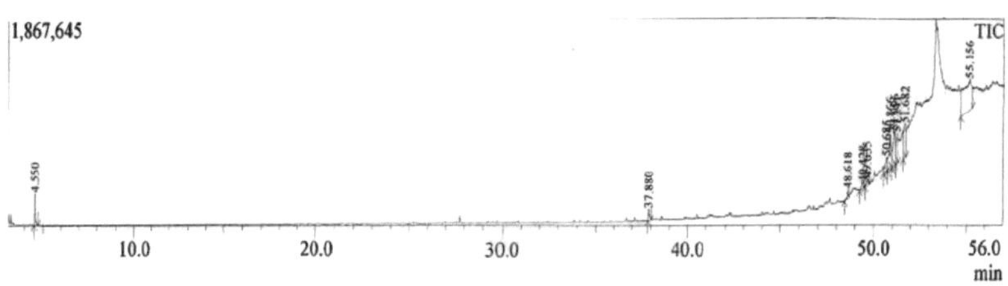

Figure 3. GC-MS chromatogram of *Chakhao Amubi*.

During the rice-breeding program, evaluation of rice is done in many ways according to consumer desires and references. Some of the traits are easy to assess, while some are very difficult. Flavor assessment is one among the difficult traits. Assessment of flavor requires time, cost and traditional sensory panels which are not accurate and also limit the number of progeny. Thus, the study of volatile oil components of different *Chakhao* rice would replace the sensory analysis and reduce time and cost as well. The potential of using the scent chemistry

http://dx.doi.org/10.5772/67193

would make a rapid progeny selection at the very initial and accurate screening of a large number of progeny for premium new cultivars with the desired traits increasing consumer satisfactions.

4. Conclusion

Nowadays, people are seeing forward more to the supplementation of natural antioxidant in the diet; thus, the consumption of pigmented rice will be a great thinking for the improvement in human health. Rice varieties with higher anthocyanin pigment have stronger scavenging activity than white rice varieties, and thus, these pigmented rice varieties are reducing agents and possess strong radical-scavenging activity. *Chakhao* rice extracts could be a potential source of antioxidative phytochemicals and useful ingredient for nutraceuticals or functional food products rather than the toxic synthetic colorants. The supplementation of *Chakhao* in the diet would have a great impact on human health. More publicity on the relationship between antioxidants and disease risk mechanisms would increase consumption of the anthocyanin-containing rice. Establishment of antioxidant capacity in rice crop will result in the betterment of whole as rice crop constitutes the main food for populations in different countries.

The approach of identification of individual rice aroma would make possible the selection of multiple flavor types and the development of superior new cultivars for a wide cross section of flavors without using sensory tests. The separation and characterization of these compounds may be of potential use in rice-breeding programs focusing on flavor. A better understanding of the flavor (taste and aroma) of *Chakhao* will increase consumer preference and will help selection of better *Chakhao* for breeding with great grain quality. The anthocyanin and volatile oil extract from Manipuri *Chakhao* rice have great future potential which can be used in the industries as a natural food coloring and odor-active substances in beverages and also in pharmaceutical industries.

Acknowledgements

The authors acknowledge Women Scientist Scheme A, Department of Science and Technology, Govt. of India (Reference No: SR/WOS-A/LS-632/2012(G)) for financial support for conducting the study.

Author details

Ibemhal Devi Asem[1]*, Rajkumar Imotomba[1] and Pranab B. Mazumder[2]

*Address all correspondence to: ibemasem@gmail.com

1 Krishi Vigyan Kendra, Utlou, Bishnupur, India

2 Department of Biotechnology, Assam University, Silchar, India

References

[1] Ryu SN, Park SZ, Ho CT. 1998. High performance liquid chromatographic determination of anthocyanin pigments in some varieties of black rice. J Food Drug Anal. 6: 729–36.

[2] Takashi I, Bing X, Yoichi Y, Masaharu N, Tetsuya K. 2001. Antioxidant activity of anthocyanin extract from purple black rice. J Med Food. 4:211–18.

[3] Chaudhary RC. 2003. Speciality rices of the world: effect of WTO and IPR units production trend and marketing. J Food Agric Environ. 1:34–41.

[4] Yang DS, Lee K, Jeong O, Kim K, Kays SJ. 2008. Characterization of volatile aroma compounds in cooked black rice. J Agric Food Chem. 56: 235–40.

[5] Suzuki M, Kimura T, Yamagishi K, Shinmoto H, Yamaki K. 2004. Comparison of mineral contents in 8 cultivars of pigmented brown rice. Nippon Shokuhin Kagaku Kogaku Kaishi. 51: 424–27.

[6] Dr. Axe. 2015. The Forbidden Rice: Black Rice Nutrition & Benefits. http://draxe.com/forbiddenrice/7/24/2015.

[7] Wolf M. 2015. Health benefits of black Rice. Demand Media, Jillian Michaels 7/25/2015.

[8] Xu Z. 2010. Whole Grain Council, Black Rice Rivals Blueberries as antioxidant source, Louisiana State University Agricultural Center Study. Presentation at the National Meeting of the American Chemical Society, Boston MA.

[9] Manach C, Scalbert A, Morand C, Remesy C, Jimenez L. 2004. Polyphenols: food sources and bioavailability. Am J Clin Nutr. 79(5): 727–4.

[10] Park YS, Kim SJ, Chang HI. 2008. Isolation of anthocyanin from black rice (*Heugjinjubyeo*) and screening of its antioxidant activities. Kor J Microbiol Biotechnol. 36(1): 55–60.

[11] Bhattacharjee P, Singhal R S, Kulkarni PR. 2002. Basmati rice: a review. Int J Food Sci Technol. 37: 1–12.

[12] Suwansri S, Meullenet JF, Hankins JA, Griffin K. 2002. Preference mapping of domestic/imported Jasmine rice for U.S.-Asian consumers. J Food Sci. 67: 2420–31.

[13] Yang DS, Lee K, Kim K, Kays SJ. 2008. Site of origin of volatile compounds in cooked rice. Cereal Chem. 85(5): 591–8.

[14] Grosch W, Schieberle P. 1997. Flavor of cereal products—a review. Cereal Chem. 74: 91–7.

[15] Maga JA. 1984. Rice product volatiles: a review. J Agric Food Chem. 32: 964–70.

[16] Limpawattana M. 2007. An integrated approach to sensory analysis of rice flavor. Ph.D. Dissertation, The University of Georgia, 2007.

[17] Widjaja R, Craske JD, Wootton M. 1996. Comparative studies on volatile components of non-fragrant and fragrant rices. J Sci Food Agric. 70: 151–61.

[18] Buttery RG, Ling LC, Juliano BO. 1982. 2-Acetyl-1-pyrroline: an important aroma component of cooked rice. Chem Ind (Lond). 12: 958–9.

[19] Jezussek M, Juliano BO, Schieberle P. 2002. Comparison of key aroma compounds in cooked brown rice varieties based on aroma extract dilution analysis. J Agric Food Chem. 50: 1101–5.

[20] Paule CM, Powers JJ. 1989. Sensory and chemical examination of aromatic and nonaromatic rices. J Food Sci. 54: 343–6.

[21] Yoshihashi T. 2002. Quantitative analysis of 2-acetyl-1-pyrroline of an aromatic rice by stable isotope dilution method and model studies on its formation during cooking. J Food Sci. 67(2): 619–22.

[22] Monsoor MA, Proctor A. 2004. Volatile component analysis of commercially milled head and broken rice. J Food Sci. 69: 632–36.

[23] Zhou Z, Robards K, Helliwell S, Blanchard C. 2002. Composition and functional properties of rice. Int J Food Sci Technol. 37: 849–68.

[24] Nawar WW. 1996. Lipids. In: Fennema OR, Editor. Food Chemistry. Dekker: New York, pp. 225–319.

[25] Singh RK, Baghel SS. 2003. Aromatic rices of manipur, a treatise on the scented rices of India (1st. ed.). New Delhi: Kalyani Publishers, p. 347.

[26] Asem ID, RK Imotomba, PB Mazumder, JM Laishram. 2015. Anthocyanin content in the black scented rice (*Chakhao*): its impact on human health and plant defense. Symbiosis. 66: 47–54, doi: 10.1007/s13199-015-0329-z.

[27] Moko EM, Purnomo H, Kusnadi J, Ijong FG. 2014. Phytochemical content and antioxidant properties of colored and non colored varieties of rice bran from Minahasa, North Sulawesi, Indonesia. Int Food Res J. 21(3): 1053–59.

[28] Wu L, Zhai M, Yao Y, Dong C, Shuang S, Ren G. 2013. Changes in nutritional constituents, anthocyanins, and volatile compounds during the processing of black rice tea. Food Sci Biotechnol. 22(4): 917–23.

[29] Yodmanee S, Karrila TT, Pakdeechanuan P. 2011. Physical, chemical and antioxidant properties of pigmented rice grown in Southern Thailand. International Food Research Journal. 18(3): 901–6.

[30] Muntana M and Prasong S. 2010. Study on total phenolic contents and their antioxidant activities of Thai white, red and black rice bran extracts. Pak J Biol Sci. 13(4): 170–74.

[31] Xia X, Ling W, Ma J, Xia M, Hou M, Wang Q, Zhu H and Tang Z. 2006. An anthocyanin-rich extract from black rice enhances atherosclerotic plaque stabilization in apolipoprotein E–deficient mice. J Nutr. 136: 2220–5.

[32] Yawadio R, Tanimori S, Morita N. 2007. Identification of phenolic compounds isolated from pigmented rices and their aldose reductase inhibitory activities. Food Chem. 101: 1616–25.

[33] Hu C, Zawistowski J, Ling WH, Kitts DD. 2003. Black rice (*Oryza sativa* L. *indica*) pigmented fraction suppresses both reactive oxygen species and nitric oxide in chemical and biological model systems. J Agric Food Chem. 51:5271–7.

[34] Kahkonen MP, Heinonen M. 2003. Antioxidant activity of anthocyanins and their aglycones. J Agric Food Chem 51: 628–33.

[35] Saenkod C, Liu Z, Huang J and Gong Y. 2013. Anti-oxidative biochemical properties of extracts from some Chinese and Thai rice varieties. Afr J Food Sci. 7(9): 300–5.

[36] Diplock AT, Charleux JL, Crozier-Willi G, Kok FJ, Rice-Evans C, Roberfroid M, Stahl W, Vina Ribes J. 1998. Functional food science and defence against reactive oxidative species. Brit J Nutr. 80:77–112.

[37] Konczak I, Zhang W. 2004. Anthocyanins—more than nature's colours. J Biomed Biotechnol. 5: 239–40.

[38] Abdel-Aal E-S, Abou-Arab AA, Gamel TH, Hucl P, Young JC, Rabalski I. 2008. Fractionation of blue wheat anthocyanin compounds and their contribution to antioxidant properties. J Agric Food Chem. 56: 11171–77.

[39] Champagne ET. 2008. Rice aroma and flavor: a literature review. Cereal Chem. 85: 445–54.

Section 3

Breeding for Biotic and Abiotic Stress

Chapter 9

Salt Stress Tolerance in Rice: Emerging Role of Exogenous Phytoprotectants

Anisur Rahman, Kamrun Nahar,
Jubayer Al Mahmud, Mirza Hasanuzzaman,
Md. Shahadat Hossain and Masayuki Fujita

Additional information is available at the end of the chapter

http://dx.doi.org/10.5772/67098

Abstract

Excess salinity in soil is one of the major environmental factors that limit plant growth and yield of a wide variety of crops including rice. On the basis of tolerance ability toward salinity, rice is considered as salt-sensitive crop, and growth and yield of rice are greatly affected by salinity. In general, rice can tolerate a small amount of saltwater without compromising the growth and yield. However, it greatly depends on the types and species of rice and their growth stage. Salinity-induced ionic and osmotic stresses reduce rate of photosynthesis and consequently cause oxidative stress, which is also responsible for growth reduction. The negative effects of salt stress that mentioned ultimately reduced yield of most crops including rice, except some halophytes. In recent decades, researchers have developed various approaches toward making salt-tolerant rice varieties. Using phytoprotectants is found to be effective in conferring salt tolerance to rice plants. In this chapter, we reviewed the recent reports on different aspects on salt stress tolerance strategies in light of using phytoprotectants.

Keywords: *Oryza sativa*, abiotic stress, oxidative stress, ROS, phytohormones, calcium

1. Introduction

In a changing world, plants experience various kinds of environmental stresses (salinity, drought, heat, cold, flooding, heavy metals, ozone, UV radiation, etc.), which affect plant growth, yield and productivity that challenges the food security of ever-growing population all over the world [1, 2]. Of the environmental factors, salinity is one of the most brutal abiotic stresses, because most crop plants are sensitive to salt stress [1, 3, 4]. The condition of soil characterized by high concentrations of soluble salt is called salinity. About 6% of world's total land area and one-third of world's irrigated land area are affected by salinity directly or by secondary salinity. Among the cultivable area, 20% of irrigated land and 2% of the dry land area are affected by salinity directly or by secondary salinity [5]. The problem of salinization is increasing day by day, often due to bad agricultural practices and climate changes. It is assumed that 50% of the cultivable land will be salt affected by the middle of the twenty-first century [6].

Excess salinity in soil is one of the major environmental factors that limit growth and yield of a wide variety of crops including rice [3, 4, 7]. Many studies revealed the negative effect of salinity on plant growth, development and yield [8–13]. The damages by higher salinity in plant start from germination and exist till death of plant [13]. It is evident that salt stress has negative correlation with seed germination and vigor of wide variety of crops [3]. In the seedling stage, salt stress affects plant growth by osmotic and ionic stress. Salinity-induced ionic and osmotic stresses reduce rate of photosynthesis and consequently cause oxidative stress, which is also responsible for growth reduction [3, 4, 14]. The negative effects of salt stress that mentioned ultimately reduced yield of most crops including rice, except some halophytes.

On the basis of tolerance ability toward salinity, rice is considered as salt-sensitive crop, and growth and yield of rice are greatly affected by salinity [12]. Salinity-induced yield reduction of rice is alarming for the food security of ever-growing population of the world, especially in Asia, because 90% of the world's rice is produced and consumed in Asia and more than 3 billions of Asian intake their 50–80% daily calorie from rice [15]. So, it is an urgent task of plant biologists to develop salt-tolerant rice cultivar to ensure food security of rising population, since expansion of rice-growing areas is limited because of industrialization and various stresses including salinity [16].

Along with developing tolerant cultivar by breeding and genetic engineering, detoxification of reactive oxygen species (ROS) and methylglyoxal (MG), maintenance of nutrient homeostasis and reduction of salt uptake comprise some stress tolerance mechanisms within the plant under salt stress condition [11, 14, 17, 19]. Considering the factors that discussed, use of exogenous phytoprotectants become one of the important approaches for improving salt stress tolerance by osmoregulation, ROS and MG detoxification and ion homeostasis in rice [11, 14, 17, 18]. Considering the above-mentioned factors, in this chapter, we discuss salinity-induced damages and alleviation of salt stress by using exogenous phytoprotectants in rice plant.

2. Effect of salinity on rice plants

Soil salinity is not a new problem for rice production. In the coastal area, rice has been growing in saline soils since long time. The adjacent rivers, canals, streams and other water bodies are always contaminated with salt to a certain extent. With the initiation of green revolution, most of the rice production is being depending on irrigation water and salt intrusion in coastal areas becomes a common phenomenon recently. In general, rice can tolerate a small amount of saltwater without compromising the growth and yield. However, it greatly depends on the types and species of rice and their growth stage [3]. Lee et al. [19] reported that the tolerance level of indica is higher than that of japonica at seedlings stage. At early stage of growth, rice is grouped as salinity susceptible cereal and confines its efficiency of production at mature stage [20, 21]. According to IRRI [22], soil salinity beyond EC ~4 dS m^{-1} is considered as moderate salinity for rice, while more than 8 dS m^{-1} is high. However, it is not absolute measure, and it depends on other soil factors because they are interacted with each other. Excess salt caused both ionic toxicity and osmotic stress in rice plants. Under high salinity, rice plants show various morphological, physiological or biochemical alterations and symptoms and even may die when the salt stress becomes very high (**Figure 1**). Sodium ion itself causes direct cellular injury to plants, and additionally, higher amount of Na^+ in root zone inhibits K^+ uptake because of their antagonistic effect [23]. This shortage of K^+ inside the cell unavoidably leads to decrease in plant growth because K^+ has vital role in preserving membrane potential, enzyme activities and cell turgor [21, 23]. Apart from Na^+ or other anions, some of the cations like Cl^- also show toxicity in rice. Ionic stress causes chlorosis and necrosis which either accelerate senescence or impair growth and development. Due to the excess accumulation of Na^+ in the cytoplasm during NaCl salinity, cellular metabolisms such as protein synthesis and enzyme activities are hampered, and therefore, source-sink relationship and photosynthesis are disrupted [24]. Many reports showed that Na^+ accumulation in shoots is relatively well correlated with the survival of rice plants under salinity stress [24], and hence, keeping a lower cytosolic Na^+ is considered as one of the vital strategies for salt tolerance in glycophytes [24]. As a result of salinity-induced osmotic stress, water uptake by plant is hampered and plant suffers from physiological drought. This also led to the interruption of nutrient uptake. Under osmotic stress, regulation of water transport becomes a vital adaptive strategy of rice plants because a sufficient amount of water is indispensable for the cells to maintain their growth and vital cellular functions such as photosynthesis and metabolisms. This situation also induces stomatal closure, which results in the reduction of evaporation and water transport. A set of hormonal regulations also associated with these processes is upregulated many-fold under salt stress [24]. In the next subsections, we provided a description of the specific responses of plants under salinity.

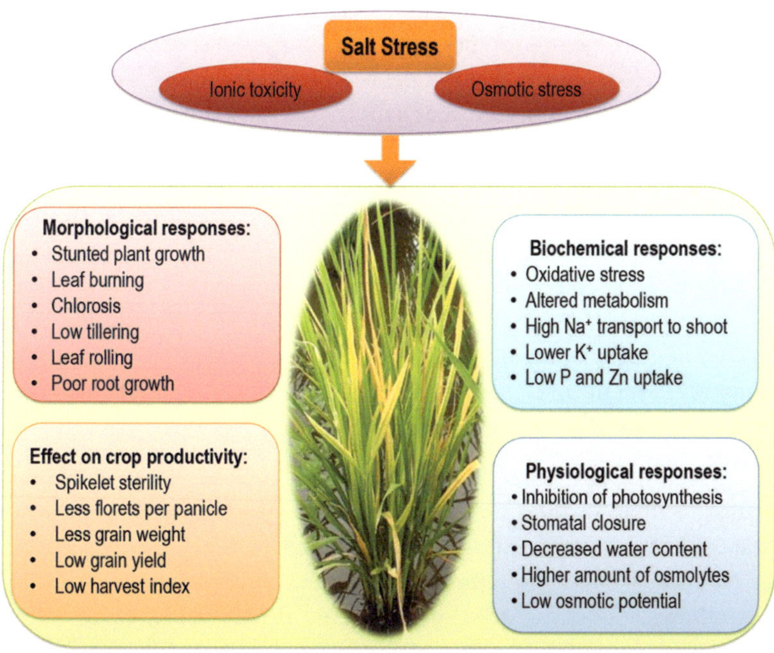

Figure 1. Salinity-induced major responses in rice plants.

2.1. Seed germination

Germination of seed is the starting and one of the most important phases of plant life cycle that determines the better establishment of seedlings as well as growth, development and yield of plant. It is well documented that the presence of salt in germination media hampers germination and seedling establishment by osmotic and ionic stress [3]. Salinity hampers seed germination by affecting major events of germination such as imbibition, metabolism activation, emergence of embryonic tissues and seedling establishment [4, 25]. Salinity-induced osmotic and ionic stresses inhibit and delay seed germination by limiting hydrolysis of seed through water imbalance, cell membrane destruction and enzyme activity reduction. Poor hydrolysis also limits translocation of food reserve from storage tissue to develop embryo that also negatively affects germination and seedling establishment. Salinity inhibits amylase activity, which is the major reason of poor hydrolysis or imbibitions of stored substances [26]. However, the negative effect of salinity on seed germination may vary due to various factors, such as varietal variation, level of salinity and other external factors [4]. Like other crop plants, rate of germination, speed of germination and seedling establishment of rice are greatly affected by salinity [26, 28]. Many studies revealed that salinity reduces germination percentage, germination index, germination speed, germination energy percentage (GE%) and mean germination time along with seedlings establishment in rice. Rajakumar [29] reported that salinity decreased germination percentage gradually with the increase of salt stress (0–300 mM NaCl) by higher accumulation of proline (Pro) and lower starch and protein content. Higher salinity (100 and 200 mM) delayed germination (3–6 days) and decreased germination percentages (upto 61%) by increasing solute leakage and decreasing α-amylase activity [30]. Islam and Karim [31] reported that germination percentage and germination index decreased in 17 rice genotypes with increasing the level of salinity in germination media. Germination percentage and mean germination time decreased in six rice genotypes with increasing the concentration of salinity [32]. Kazemi and Eskandari [33] observed the effect of salinity (0–8 dS m^{-1}) on three rice cultivars (Anbar, LD and Hamar) and reported that salinity decreased germination percentage (up to 96%), germination speed, and plumule and radical growth with increasing the level of salinity. Effect of six levels of salinity (0–20 dS m^{-1}) was studied by Hakim et al. [28] on 12 rice cultivars and noticed that the level of damage on seed germination and early seedling growth increased with increasing the level of salinity. They found that salinity decreased germination percentage, speed of germination and GE% up to 88, 99 and 100%, respectively, with highest level of salinity (20 dS m^{-1}). Ologundudu et al. [34] also found that germination percentage, speed of germination and GE% decreased in eight rice cultivars under different levels of salinity (0–15 dS m^{-1}).

2.2. Plant growth

It is well documented that seedling and early vegetative growth stages are most susceptible to salinity during the entire life cycle of plant [3, 4]. On the basis of tolerance ability, rice is salt-sensitive crop and the sensitivity to salinity varies with the growth stages. Among the growth stages, early seedling stages of rice are also considered as most sensitive to salinity compared with other stages [35]. Aref [36] subjected rice plant to salinity at tillering, panicle initiation, panicle emergence and ripening stage where salt-induced damages were higher at tillering and panicle initiation stages compared with other two stages.

The immediate response of plant seedlings to high concentration of salt around the root is osmotic effect that reduces water uptake capacity. Osmotic stress disturbs the water balance and causes water loss from the cell that gradually reduces cell elongation, cell division, stomatal closure and leaf area as well as photosynthesis and growth [3, 5]. The presence of high concentration of salt (NaCl) around the root zone also reduces plant growth by ionic toxicity through over accumulation of Na^+ and Cl^-. In addition, Na^+ influx causes chlorosis, necrosis and premature senescence of adult leaves and thus limits the photosynthetic area available to support continued growth of salt-affected plants [5, 16, 37]. In later stage, salt-induced ionic and osmotic stresses and lower rate of photosynthesis cumulatively cause oxidative stress by overproduction of ROS, which is also responsible for growth reduction under salt stress condition [3, 12].

However, many studies revealed the salt-induced growth reduction in rice. Exposure of higher salinity (150 mM NaCl) in rice seedlings reduces plant height and biomass by salt-induced osmotic, ionic and oxidative stresses [17]. Rahman et al. [18] also reported that salt stress (200 mM NaCl) reduced plant growth by creating ionic and water imbalance, and oxidative stress. Salt-induced growth reduction by oxidative stress in rice seedlings was also reported by Özdemir et al. [27]. Under salt stress condition, growth reduction and water loss were higher in susceptible cultivar compared with tolerant cultivar [3, 31]. Kumar and Khare [38] found that salt stress (100 mM NaCl $\approx$10 dS m^{-1}) reduced root length, root dry weight, shoot length and shoot dry weight both in sensitive and in tolerant cultivar where growth reduction is higher in sensitive cultivar compared with tolerant. On the other hand, growth of rice seedlings decreased with increasing the level of salinity [31, 33]. Ologundudu et al. [34] conducted experiment with eight rice cultivars under different level of salinity (0–15 dS m^{-1}) and reported that root and shoot length, root and shoot dry weight, and total dry matter production decreased with increasing the level of salinity.

2.3. Physiological attributes

Among different physiological processes, photosynthesis is a vital physiological attribute related to plant growth and development that is affected by salinity. Photosynthesis is a complex process depending on gas-exchange characteristics, photosynthetic pigments, photosystems, components of electron transport system and activities of different enzymes involved in carbon metabolism. Therefore, damage to any of those components affects photosynthesis negatively [39, 40]. Rate of photosynthesis declines under saline condition primarily due to osmotic stress that results in stomatal closure and secondarily by higher accumulation of Na^+ and Cl^- that can damage thylakoid membrane in the choloroplast [3]. Salinity-induced limitation of CO_2 diffusion causes inactivation of RuBisCo. Moradi et al. [41] demonstrated a remarkable decrease in photosynthetic CO_2 fixation, transpiration and stomatal conductance (gs) in three rice cultivars due to salt stress. Salt-sensitive genotype IR29 showed greater reduction of these attributes than tolerant genotypes IR652 and IR632. In another study, Cha-um et al. [42] reported that in dark reaction, net photosynthetic rate (NPR), gs and transpiration rate (E) decreased in both varieties, Homjan (HJ) and Pathumthani 1 (PT1) varieties, under salt-stressed condition. Photosynthetic pigments, chl *a* and chl *b*, are greatly affected by different abiotic stresses including salinity. Accumulation of toxic Na^+ reduces the content of precursor of chl biosynthesis (such as glutamate and 5-aminolevullinic acid) and thus interrupts chl biosynthesis under saline condition [40]. Salinity-induced chlorophyll (chl) reduction is observed in rice like other crops. Amirjani [43] found a dose-dependent reduction of chl content against salt stress (0, 25, 50, 100 and 200 mM NaCl). At 200 mM NaCl, chl *a* and chl *b* reduced by 44 and 27%, respectively, compared to control. Rahman et al. [17] reported that 12-day-old rice seedlings exposed to 150 mM NaCl for 3 days caused 23 and 19% reduction in chl *a* and chl *b*, respectively, compared to control. As the duration of salt stress extended for further 3 days, chl *a* and chl *b* reduced by 46 and 48%, compared to control.

Reduction of photosynthesis due to stomatal closure and subsequent CO_2 shortage affects source to sink translocation of photosynthates and carbohydrate metabolism in leaves. Photosynthates may contribute to salt stress tolerance by serving as osmolytes. Pattanagul and Thitisaksakul [44] observed carbohydrate metabolism in salt-tolerant and salt-susceptible rice genotypes. They mentioned increased total soluble sugar and sucrose content in susceptible genotype under salt stress, while starch accumulation increased in salt-tolerant genotype. To explain this result, they assumed that salt-susceptible rice genotype could not use carbohydrate for plant growth, thus growth reduced and total soluble sugar and sucrose content increased.

2.4. Osmotic stress

Salt stress imposes hyperosmotic stress in plants. Osmotic stress is physiological dysfunction resulted from alteration of solute concentration around a cell. Salt imposition in the growing media increases solute concentration around the root zone and lowering of the soil water potential for which plants cannot uptake water as a result plants suffer from osmotic stress. Salt-induced osmotic stress is often termed as physiological drought which is accountable for physiological disorders and injuries within the plants [45, 46]. Several research findings demonstrated osmotic stress and its damage effects induced by salinity. Osmotic potentials in roots and in the oldest and youngest leaves were measured. Salt-resistant cultivars (Nona Bokra and IR 4630) were characterized by lower osmotic potential, in contrast to the salt-sensitive cultivars (I Kong Pao and IR 3 1785) [47]. Redillas et al. [48] reported that trehalose (Tre) producing transgenic *Oryza sativa* expressing *TPSP (Ubi1:TPSP)* accumulated more Tre under salt stress and showed less osmotic stress, in contrast to nontransgenic cultivar. Two rice cultivars [IR651 (salt-tolerant) and IR29 (salt-sensitive)] were grown under salt stress (100 mM NaCl), where total soluble sugars accumulation was higher in shoot of tolerant cultivar, compared to sensitive cultivar. Increased total soluble sugar was suggested to regulate osmotic potential and water uptake capacity under salt stress [49]. Starch degradation and sugar accumulation were also reported in salt-affected rice plant and described as a strategy to improve osmotic status and plant survival. Salt-sensitive rice cultivar (Pathumthani 1) showed lower starch degradation and reduced sugar accumulation, compared to Homjan (salt-tolerant cultivar expressing starch metabolism related genes, *AGPL1, AGPS2b* and *SBEIIb*) under salt stress (150 mM NaCl, 7 days). Water use efficiency of Pathumthani 1 significantly reduced by 35%, whereas water use efficiency in salt-tolerant cultivar did not change. Starch metabolism capacity of sensitive cultivar was lower in salt-sensitive cultivar that decreased the water use efficiency [50]. Salt stress due to lowering the soil water potential causes reduction of water uptake and as a secondary effect stomata closes. This stomatal closure reduces CO_2 availability and fixation which reduce photosynthesis [51]. Reduction of tissue water content, modulation of osmotic potential and accumulation of compatible solutes are the most common indications of salt-induced osmotic stress as studied in several research findings. Two different levels of salt stresses (150 and 300 mM NaCl, 48 h) were imposed to two different cultivars of rice, namely BRRI dhan49 and BRRI dhan54. Salt stress imposition decreased leaf RWC and increased Pro content in both cultivars. Being a salt-sensitive cultivar, BRRI dhan49 showed decrease in leaf RWC by 19 and 29%, under 150 and 300 mM salt stresses, whereas in salt-tolerant cultivar BRRI dhan54, RWC decreased by 12 and 28% under 150 and 300 mM NaCl stress, respectively, compared to control [14]. Cha-Um and Kirdmanee [42] observed decreased water use efficiency in salt-sensitive rice cultivar exposed to salt (150 mM NaCl) stress. The KDML105 rice callus was grown in salt (250 mM NaCl) containing media for 2, 4, 6, 8 and 10 days. With the increase of salt stress duration, the water content of rice cells progressively decreased. Salt stress increased accumulation of Pro. The Pro content increased with the increase of duration of stress. Proline level reached to the highest level at the 10th day of salt stress [52]. Rahman et al. [18] reported reduction of leaf RWC and increase of Pro as stress marker due to exposure of rice (*Oryza sativa* L. cv. BRRI dhan47) seedlings to 200 mM NaCl stress. Rice cultivars differing in salt tolerance capacity were subjected to different levels of salt stresses (20, 30, 40 or 50 mM NaCl) [47].

2.5. Ionic toxicity/imbalance

Salt stress generally involves with ionic toxicity/injury, and a high salt content in the growing media is able to break the ion homeostasis, destroy the ionic balance and induce nutritional disorder in plant cells [53]. The nutritional disorders may be associated with the effect of salinity on nutrient availability, competitive uptake, distribution or transport within the plant [2]. Generally, Na^+, K^+, Ca^{2+}, Mg^{2+}, Cl^- and sometimes SO_4^{2-} and CO_3^{2-} ions are responsible for salinization. As Na^+ predominates in the soils, the common problem for crop production is soil sodicity. In addition, sodic soils are important because they create very poor soil structure due to disaggregation of soil properties which impair water movement [3]. So, most of the plant researches focuses on the transport mechanism of Na^+ ion and its action to the plant cell. The ionic imbalances, which are developed by the accumulation of toxic ions, such as Na^+ and Cl^-, and the depletion of ions, such as K^+ and Ca^{2+}, are directly affected by salinity [54]. The maintenance of Na^+ and K^+ homeostasis becomes more vital under salt stress. Plant exposed to salt stress is affected by Na^+ influx, which causes K^+ efflux and triggers K^+ and Ca^{2+} leakage from plant cells because plants uptake higher amount of Na^+ and show increasing Na^+/K^+ ratio [17, 18, 55]. Entrance and accumulation of Na^+ to a high level become toxic to many enzymes because it competes with K^+ for binding sites of important enzymes [56]. Moreover, Na^+ influx causes chlorosis and necrosis in adult leaves and premature senescence of mature leaves by upsetting protein synthesis and interfering with enzyme activity [5, 57]. Micronutrient deficiencies are very common under salt stress because the availability of micronutrients in saline soils is dependent on its solubility, pH and redox potential of the soil solution and the nature of binding sites on the inorganic and organic particle surfaces [58]. Rahman et al. [18] carried out an experiment with rice plant and found NaCl-induced stress disturbed ion homeostasis through increase of Na^+ content and decreased of K^+ content in the shoots and roots of rice seedlings, which might be due to entrance of higher amount of Na^+ into plant by nonselective cation channel (NSCC) that caused K^+ efflux or leakage through NSCC and guard cell outward rectifying potassium channels (GORK). Importantly, they reported that higher Na^+ accumulation also resulted in a higher Na^+/K^+ ratio which disrupted homeostasis by decreasing Mg, Mn and Zn contents. In rice, substantial variation in uptake and accumulation of Na^+ between genotypes was repeatedly observed, and tolerant genotypes tended to accumulate less Na^+ and maintain higher ratios of K^+/Na^+ and Ca^{2+}/Na^+ in plant tissues under salt stress [5, 59]. Azarin et al. [60] demonstrated that salt treatment strongly stimulated accumulations of Na^+, Na^+/K^+ ratio, Cl^- in root and shoot and reduced K^+, NO_3^- contents in both organs under 1.2% NaCl stress condition. They also stated that Cl^- ions negatively affected the biomass growth and survival of rice plants. So, it can be concluded that salinity initially causes disruption of ion homeostasis and increases ionic toxicity and nutritional disorder in plant by increasing Na^+ and Cl^- uptake which finally hampers the growth and development of rice plant.

2.6. Oxidative stress

One of the common consequences of salt stresses in plant is the accelerated production of ROS [51]. Exposure of plant to high salinity induces closure of stomata. As a result, CO_2 availability as well as fixation in the leaf tissues becomes reduced. At the same time, a decreased reduction of CO_2 by Calvin cycle and a state of excessive excitation energy occur, of which chloroplasts become exposed, and sequentially, photosynthetic electron transport system becomes impaired. Thus, salinity causes the excessive synthesis of ROS including superoxide ($O_2^{\bullet -}$), singlet oxygen ($^1O_2$), hydroxyl radicals ($OH\bullet$) and hydrogen peroxide (H_2O_2) and results in oxidative stress [3, 4, 61]. A water deficit condition also occurs under salinity which further contributes to the generation of ROS [62]. In plant cells, ROSs are continuously produced as a consequence of aerobic metabolism in all the intracellular organelles, particularly in chloroplast, mitochondria and peroxisomes [63] (**Figure 2**). The reduction of the CO_2/O_2 ratio of chloroplast increases the synthesis of H_2O_2 by accelerating the photorespiration process [64]. These ROSs are very reactive in nature and interfere with plants normal metabolism by causing the peroxidation of lipid, oxidation of protein, nucleic acid and DNA [65]. However, plants try to detoxify ROS by their well-established antioxidant defense system, which includes both the enzymatic antioxidants, that is, catalase (CAT), superoxide dismutase (SOD), ascorbate peroxidase (APX), monodehydroascorbate reductase (MDHAR), dehydroascorbate reductase (DHAR), glutathione *S*-transferase (GST) and glutathione reductase (GR), and nonenzymatic antioxidants, that is, glutathione (GSH), ascorbate (AsA), tocopherols and carotenoids [2]. These act coordinately in scavenging ROS and protecting cells from oxidative stress [2, 9].

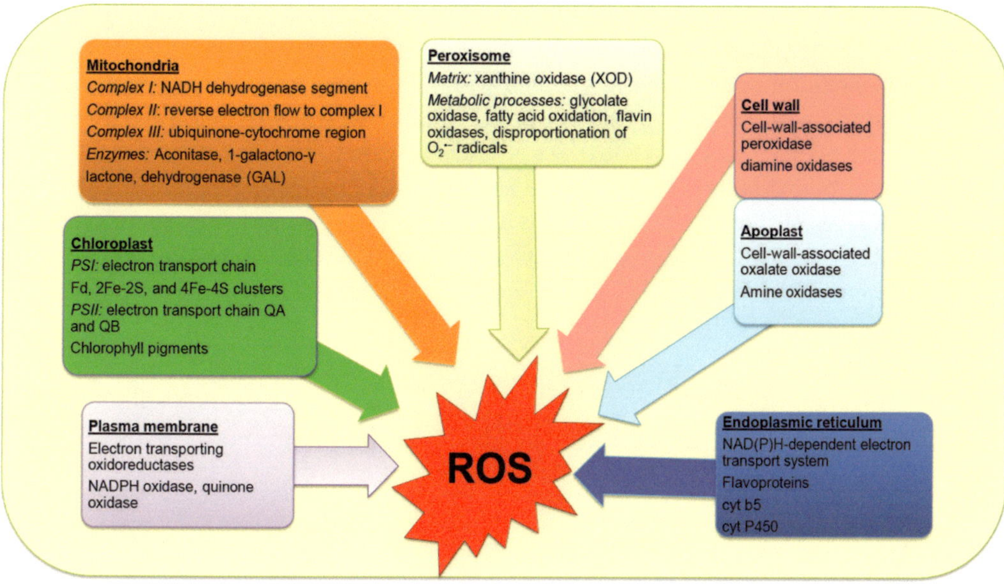

Figure 2. Sites of production of reactive oxygen species (ROS) in plants [63].

Vaidyanathan et al. [66] studied with two rice cultivars, namely salt-sensitive Pusa Basmati 1 (PB) and salt-tolerant Pokkali (PK), and observed that the lipid peroxidation and H_2O_2 production were lower in PK than those of PB with concomitant enhancement of the activities of ROS-detoxifying enzymes as well as elevated levels of AsA and GSH. According to Rahman et al. [17], salt stress (150 mM NaCl) increased oxidative damage in time-dependent manner. Lipid peroxidation increased by 80 and 203%, H_2O_2 content by 74 and 92%, and LOX activity increased by 69 and 95% after 3 and 6 days of treatment, respectively. Accumulation of H_2O_2 and $O_2^{\bullet-}$ in leaves was much higher than control, especially after 6 days of stress (**Figure 3**). Results confirmed that the enhanced activities of MDHAR, DHAR, SOD and CAT in salt-stressed rice plant reinforced the salt stress tolerance. Salinity-induced oxidative stress in two rice varieties viz. Swarna (susceptible) and Nonabokra (tolerant) was investigated by Ghosh et al. [67]. The two rice varieties were subjected to 200 mM NaCl, and both varieties showed increased accumulation of ROS and higher lipid peroxidation. Tolerant variety Nonabokra showed reduced damage of biomolecules including lipid, protein, enzymes and membranes with significantly higher accumulation of protective phenolic compounds compared to sensitive variety Swarna. Under high salinity, salt-sensitive varieties, that is, Hitomebore and IR28 showed decreased SOD activity and increased POX activity. The research result of Abdallah et al. [68] suggested that salt stress increased the level of SOD, CAT and Peroxidase (POX) in the shoot of two studied rice varieties, that is, Giza 178 and Giza 177. They found the increased SOD and POX activities in the NaCl-stressed treatment. Hong et al. [69] demonstrated that upon NaCl exposure, in rice roots, the OsGR2 and OsGR3 isoforms of GR were expressed and triggered by the increased H_2O_2 level under salt stress. The two important enzymes those scavenge ROS under salt stress in rice are APX and GR

[70]. Although the excess generation of ROS under salt stress is very common, it depends on the types of the plant genotype, dose and duration of stress [2]. Tolerant plants show lower ROS generation rate which is correlated with their higher antioxidant defense capacities. In our study with two rice genotypes grown under salt stress (150 and 300 mM NaCl, 48 h), we observed that compared with the salt-tolerant variety, the salt-sensitive one showed increased lipid peroxidation (76 and 159%) and H_2O_2 (35 and 69%) generation at 150 and 300 mM NaCl, respectively [14]. These were due to the increased activities of APX, DHAR, MDHAR, GPX, GR, CAT and glyoxalase I (Gly I) in tolerant (BRRI dhan54) variety than the sensitive variety (BRRI dhan49) [14].

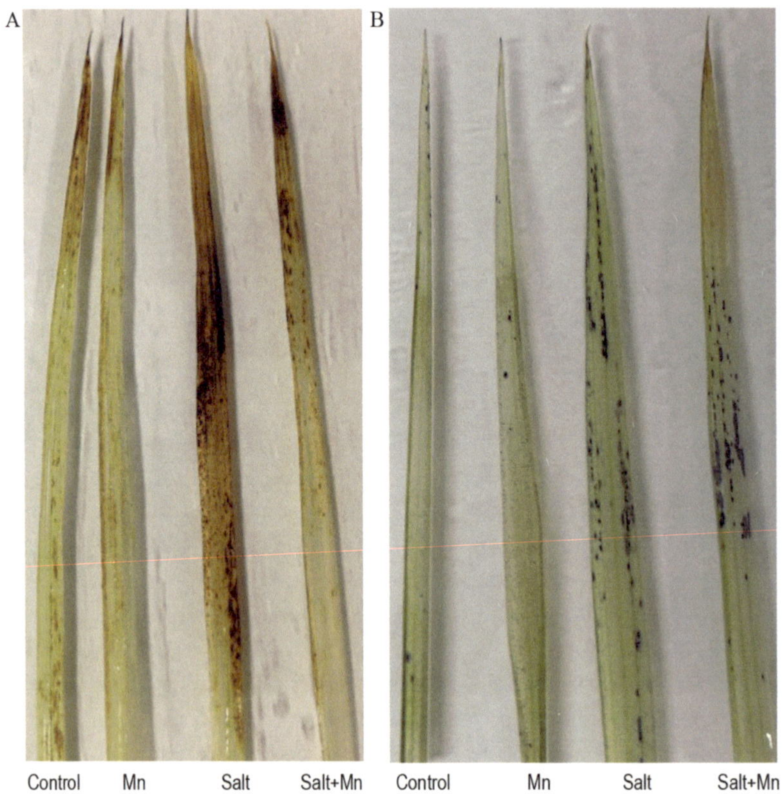

Figure 3. Histochemical localization of H_2O_2 (A) and $O_2^{\bullet -}$ (B) in leaves of rice seedlings after 6 days of salt stress (150 mM NaCl) with or without exogenous Mn (0.5 mM $MnSO_4$). Hydroponically grown 12-day-old seedlings were subjected to both salt and Mn treatments both individually and combinedly (adapted from Rahman et al. [17], with permission from Springer).

2.7. Yield and grain quality

Salt-induced damages not only affect the vegetative growth but also affect reproductive development and yield of plant. Yield and grain quality of rice greatly influenced by salinity as salinity hampers growth, photosynthesis and net assimilation rate. Along with the vegetative stages, salt stress affects reproductive stage of rice that reduces yield, yield contributing parameters and grain quality. Zheng and Shanon [71] reported that salt stress affected reproductive stage of rice due to lower growth and lower survival percentage of seedlings under different levels of salinity (19–11.7 dS m^{-1}). They noticed that salinity decreased grain yield by decreasing tiller number, pollen viability, fertility percentage and 1000-grain weight where the level of yield reduction increased with increasing the level of salinity. Kumar and Khare [38] subjected tolerant and sensitive cultivar to 100 mM NaCl (≈10 dS m^{-1}) and noted that salinity reduced number of grain per panicle, filled grain percentage, 1000-grain weight and grain yield both in sensitive and in tolerant cultivar where yield reduction was higher in sensitive cultivar. They also noted that grain quality of rice deteriorated by salinity through reduction of protein and starch content of grain. Saleethong et al. [72] reported that salt stress reduced grain yield both in tolerant (32%) and in sensitive cultivars (56%) by affecting yield contributing parameters where yield reduction is higher in sensitive cultivar. They also reported that salt stress decreased grain quality by reducing N, P, K and Mg content in rice grain. Ali et al. [73] noticed that salinity (8.5 dS m^{-1}) reduced grain yield of rice by reducing photosynthesis, leaf area index (LAI) and productive tiller. Chunthaburee et al. [74] conducted experiment with four different cultivars and found that salt stress (25 mM NaCl) reduced yield and harvest index by decreasing 1000-grain weight, filled grain percentage and panicle fertility. Arsa et al. [75] reported that salt stress (2.5% NaCl) along with drought stress reduced grain yield and aroma of rice grain. The level of yield reduction increased with increasing the level of salinity. Yield reduction also depends on stages of plant growth when it is affected by salinity. A green house experiment was conducted by Aref [36] with different levels of salinity (2–8 dS m^{-1}) where salt stress exposed at different stages (tillering, panicle initiation, panicle emergence and ripening) of plant. He reported that salinity decreased grain yield, biomass yield and harvest index when salt stress was exposed at tillering and panicle initiation stage and yield reduction increased with increasing the level of salinity.

3. Use of phytoprotectants in conferring salt stress tolerance in rice

3.1. Osmoprotectants

Plants to cope with salt-induced osmotic stress synthesize and accumulate compatible solutes or osmoprotectants. These are electrically neutral nontoxic and highly soluble molecules. Osmotic balance, stabilization of proteins and membranes are vital functions of osmoprotectants. Plants overexpressing genes for biosynthesis or metabolic genes for osmoprotectants or their precursors showed enhanced salt stress tolerance. Induction of salt stress enhances biosynthesis of osmoprotectants such as Pro, glycine betaine (GB), trehalose (Tre), sorbitol and ectoine, which contribute hyperosmotic stress tolerance generated from salt stress (**Table 1**) [51, 76]. Several research findings evidenced the role of osmoprotectants to confer salt stress tolerance in rice seedlings (**Table 1**). Salt stress (150, 200, 250 and 300 mM) was imposed to in vitro rice shoot apices cultures of two Malaysian rice cultivars MR 220 and MR 253. Salt stress decreased plant height, root length, biomass and chl content, and these stress effects were alleviated by exogenous Pro application. Supplementation of 5 mM Pro increased endogenous Pro level and plant height of both cultivars. Application of different levels of

Pro (5, 10, 15 and 20 mM) increased fresh weight under 150 mM NaCl stress in both cultivars [77]. Exogenous Pro reduced the Na^+/K^+ ratio and contributed an improved ionic homeostasis and reduced ion toxicity in salt-affected rice plant [78]. Effect of exogenous Pro (25 and 50 mM) was studied in salt-sensitive (BRRI dhan29) and moderately salt-tolerant (BRRI dhan47) cultivars of rice grown under salt stress (50 and 100 mM NaCl) condition. Exogenous Pro (either doses) reversed salt-induced growth reduction of both cultivars under 25 mM NaCl stress. Proline also increased yield of salt-sensitive rice at same salt stress. Exogenous Pro increased chl and endogenous Pro levels and AsA contents. Increasing K^+/Na^+ ratio exogenous Pro contributes to reduce ionic toxicity. Increasing AsA content and activity of antioxidant enzyme GPX, and Pro imparted oxidative stress tolerance [79]. Improved germination, growth and photosynthetic pigment levels were featured due to Pro (1, 5 and 10 mM) supplementation under salt stress (NaCl 100, 200, 300 and 400 mM) [80]. Abdallah et al. [68] demonstrated that application of Tre (25 mM) alleviated salt damage effects resulted from 30 and 60 mM NaCl. They observed the beneficial effects on different physiological parameters. Trehalose supplementation with salt stress increased the content of photosynthetic pigments including chl *a*, chl *b* and carotenoid. Decreasing total carbohydrates exogenous Tre increased total soluble sugar and cellular Tre and decreased Pro level. Trehalose application also modulated antioxidant enzymes such as CAT, SOD and POX activities. Increased fresh weight, dry weight and RWC were also attributed by Tre application under salt stress. Exogenous GB sprays improved water use efficiency, structure of chl, CO_2 assimilation and rate of photosynthesis and growth under salt stress condition [42]. Increasing activities of GST and SOD and upregulating the levels of AsA and GSH exogenous Pro (5 mM) and GB (5 mM) conferred oxidative stress tolerance in both salt-sensitive (BRRI dhan49) and salt-tolerant (BRRI dhan54) rice cultivars where the performance of BRRI dhan54 was better [14]. Nounjan et al. [78] also reported exogenous Tre application-modulated antioxidant enzymes activities including SOD, POX and APX. The increase of Gly II activity by exogenous Pro and GB application is an indication of improving MG toxicity tolerance in salt-affected rice seedlings [14]. Sorbitol (5 and 10 mM) and Tre (5 and 10 mM) supplementation improved oxidative stress tolerance of salt (170 mM) stress-affected rice (*Oryza sativa* L. cv. KDML105) plant by preventing H_2O_2 generation, lipid peroxidation and membrane electrolyte leakage [81].

3.2. Plant hormones

Plant hormones or phytohormones are chemicals produced within the plants at low concentration and function as signaling molecules those regulate growth of plant. Plant hormones are essential endogenous molecules involved in regulation of plant development and tolerance toward various stresses including salinity [82]. Recently, various kinds of plant hormones such as abscisic acid (ABA), auxin, cytokinins (CK), brassinosteroids, jasmonates and gibberellins (GA_3) are exogenously applied for alleviating various kind of abiotic stresses including salinity (**Table 2**). Li et al. [83] reported that ABA pretreatment conferred salt stress tolerance in salt-treated rice seedlings by abundant energy supply and active anabolism of nitrogen, nucleotide acid and carbohydrate through upregulating protein and energy metabolism (**Table 1**). Seed treatment with ABA played potential role in alleviating salt stress damages by reducing Na^+ and Cl^- concentrations, Na^+/K^+ ratio and increasing K^+ and Ca^{2+} contents [84].

Species and cultivars	Salinity dose and duration	Protectants	Protective effects	References
O. sativa ssp. *indica*	150 mM NaCl	50 mM GB	Increased chl and carotenoid contents Increased water use efficiency (WUE) Increased seed weight and yield	Cha-Um and Kirdmanee [42]
O. sativa cv. Nipponbare	25 mM NaCl, 12 h	1 and 5 mM Pro and GB	Reduced Na^+ uptake Increased K^+/Na^+ ratio	Sobahan et al. [127]
O. sativa cv. KDML 105	100 mM NaCl, 6 days	10 mM Pro	Increased FW and DW Reduced Na^+/K^+ ratio Increased endogenous Pro and transcript levels of P5CS and P5CR Upregulated transcription of genes encoding several antioxidant enzymes	Nounjan et al. [78]
O. sativa cv. Nipponbare	150 mM NaCl, 5 days	5 mM GB	Prevented salt-induced swelling of thylakoids, disintegration of grana stacking and intergranal lamellae, and disruption of mitochondria	Rahman et al. [128]
O. sativa cv. KDML 105	170 mM NaCl, 24 h	5–10 mM Sorbitol and Tre	Enhanced growth Reduced H_2O_2 and MDA contents and electrolyte leakage	Theerakulpisut and Gunnula [81]
O. sativa L. cv. BRRI dhan29 and BRRI dhan47	50 and 100 mM NaCl	25 and 50 mM Pro	Increased K^+/Na^+ ratio Increased chl, intracellular Pro Increased AsA content and activity of GPX Increased growth and yield	Bhusan et al. [79]
O. sativa cv. Giza 177 and Giza 178	30 and 60 mM NaCl	25 mM Tre	Increased photosynthetic pigments and total carbohydrate Increased FW, DW and RWC Decreased total carbohydrates but increased total soluble sugar and Tre, decreased Pro level Increased CAT activity and decreased SOD and POX activities	Abdallah et al. [68]

Table 1. Beneficial effects of exogenous osmoprotectants in mitigating salt stress-induced damages in rice.

The plant growth regulator auxin, namely indole-3-acetic acid (IAA), plays potential role in improving stress tolerance and grain quality of rice by regulating physiological and biochemical attributes under salt stress condition. Javid et al. [85] reported that exogenous application of IAA improved grain yield of rice by improving 1000-grain weight and filled grain percentage through increasing sucrose and glucose content under salt stress condition (**Table 1**). Seed priming with IAA improves germination percentage by improving α-amylase activity (**Table 1**) [86]. The phytohormone cytokinin, namely kinetin, improves grain yield and grain quality of rice under salt stress condition by improving sucrose and glucose content in rice grain [85].

It is well documented that exogenous brassinosteroids play pivotal role in enhancing salt stress tolerance. Seed priming with 24-epibrassinolide (EBL) improved chl content and stress tolerance by activating antioxidant enzymes SOD, CAT, GPX and DHAR and reducing lipid peroxidation in salt-treated rice seedlings [87]. Exogenous application of EBL and 28-homobrassinolide on salt contaminated growing media (150 mM NaCl) reverse inhibitory effect of salt stress and improve germination percentage, seedling growth by enhancing levels of nucleic acid and soluble proteins [88]. Jasmonate is involved in plant developmental processes and the defense response by acting as signaling molecule. Kang et al. [89] reported that the exogenous application of jasmonic acid (JA) after 24 and 48 h of salt stress (20 and 40 mM NaCl) exposure recovered salt-induced damages in terms of reduction in root and shoot dry weight, especially in salt-sensitive cultivar compared with tolerant. Exogenously applied JA inhibited salt-induced damage by decreasing Na content and increasing K, Ca and Mg levels through upregulating plant hormone such as ABA and JA under salt stress condition. Seed priming with GA_3 played positive role in germinating seed and on yield of rice under salt stress condition. Priming with 10 μM GA_3 improves growth of seedling by increasing starch and total soluble and reducing sugar contents and α-amylase activity in germinating seed under salt stress condition [86]. Misratia et al. [90] reported that supplementation with GA_3 improved grain yield by increasing panicle plant^{-1}, filled grains plant^{-1} and weight of 1000 grains under salt stress condition.

3.3. Signaling molecule

Similar to the nervous system of human and other animals, plants also sense different situation through the complex interplay of signal transduction networks and machinery. Accordingly, plants develop behavioral changes or develop cognition and storage of processed information to adapt in rapidly changing or variable environment. In recent years, considerable attempts have been taken to elucidate the signaling network in plants through physiological and molecular approaches. Moreover, exogenous application of some signaling molecules has also established their role in abiotic stress tolerance, and eventfully, these molecules became suitable candidate to be used as exogenous protectants (**Table 3**). Since last few years, NO has showed multifarious roles in regulating redox-related gene expression, scavenging $O_2^{\bullet-}$ and terminating chain reaction of lipid oxidation. Therefore, role of NO in salt stress tolerance in different crop plants is well documented [91, 92]. Habib et al. [93] reported that among different doses of NO (0, 0.05, 0.1, 0.2, 0.3, 0.4 and 0.5 mM SNP as NO donor), 0.1 and 0.2 mM SNP improved germination under 80 mM salt stress in four rice genotypes. Uchida et al. [94] found that pretreatment with NO (1 mM SNP, 48 h) upregulated the activity of antioxidant enzymes such as SOD, CAT and APX, and thus, NO alleviated injuries caused by salt stress (100 mM NaCl, 8 days) in rice seedlings.

Species and cultivars	Salinity dose and duration	Protectants	Protective effects	References
O. sativa ssp. *Indica*	150 mM NaCl, 48 h	5 μM ABA pretreatment, 48 h	Upregulated energy metabolism, primary metabolism, secondary metabolism, protein and defense Increased FW and decreased Pro content	Li et al. [83]
O. sativa cv. IR-26	0, 50 and 75 mM NaCl Transplanting to harvest	10 μM ABA priming, 24 h	Reduced Na^+ and Cl^- concentrations, Na^+/K^+ ratio Increased K^+ and Ca^{2+} contents Increased grain yield	Gurmani et al. [84]
O. sativa cv. IR 651, IR29	6 dS m^{-1} Panicle initiation to harvest	50 μM IAA, foliar spray	Increased sucrose and glucose contents Increased filled grain percentage, 1000-grain weight and grain yield	Javid et al. [85]
O. sativa cv. Hwangyoungbyeo	20 mM NaCl	20 μM IAA priming, 12 h	Increased α-amylase activity Increased germination percentage	Kim et al. [86]
O. sativa cv. IR 651, IR29	6 dS m^{-1} Panicle initiation to harvest	50 μM kinetin, foliar spray	Increased sucrose and glucose content Increased filled grain percentage, 1000-grain weight and grain yield	Javid et al. [85]
O. sativa cv. Pokkali	50 and 100 mM NaCl	150 ppm GA_3 foliar spray	Increased panicle $plant^{-1}$, filled grains $plant^{-1}$ and weight of 1000 grains Increased grain yield	Misratia et al. [90]
O. sativa cv. Pusa Basmati-1	NaCl 75–125 mM, 12 days	10^{-11}, 10^{-9}, 10^{-7} M EBL, priming 8 h	Increased chl a, chl b and total chl content Decreased MDA content Upregulated SOD, CAT, GPX and DHAR activities	Sharma et al. [87]
O. sativa cv. IR-64	150 mM NaCl, 7 days	0.5, 1 and 3 μM EBL and 28-homobrassinolide	Increased germination percentage Increased seedlings length, fresh weight and dry weight Enhanced nucleic acid and soluble protein level	Anuradha and Rao [88]
O. sativa cv. Dongjinchalbyeo and Dongjinbyeo	20, 40 and 80 mM NaCl, 8 days	30 μM JA	Decreased Na content Increased K, Ca and Mg contents Increased JA and ABA contents Increased quantum yield of PS II	Kang et al. [89]

Table 2. Beneficial effects of exogenous photohormones in mitigating salt stress-induced damages in rice.

Species and cultivars	Salinity dose and duration	Protectants	Protective effects	References
O. sativa cv. IRRI-6	80 mM NaCl	0.1 mM SNP, spray	• Improved growth and biomass production • Increased Pro • Enhanced the activities of antioxidant enzymes • Increased yield	Habib and Ashraf [98]
O. sativa cv. Shaheen, PB-95, IRRI-6 and KS-282	80 mM NaCl	Seed soaking with 0.05–0.5 mM SNP	• Improved seed germination • Enhanced seedling growth and vigority	Habib et al. [93]
O. sativa cv. Nipponbare	100 mM NaCl, 8 days	H_2O_2 (<10 μM), pretreatment, 48 h	• Enhanced the activities of ROS scavenging enzymes	Uchida et al. [93]
O. sativa cv. Pokkali and KDML 105	25 mM NaCl	1 mM Spd	• Increased fresh weight, dry weight and grain yield	Paweena et al. [110]
O. sativa cv. Pokkali and KDML 105	150 mM NaCl, 7 days	1 mM Spd	• Increased Plant height, fresh and dry weight • Scavenging free radicals Spd stabilized membrane, and maintained K^+/Na^+ status	Saleethong et al. [109]
O. sativa cv. Nonabokra and Swarna	200 mM NaCl, 7 days	1 and 2 mM Put	• Decreased $O_2^{\bullet-}$ and H_2O_2 contents • Increased anthocyanin and flavonoid contents, CAT, GPX and GR activities	Ghosh et al. [67]
O. sativa cv. IKP and Pokkali	50 and 100 mM NaCl, 5 and 12 days	1 mM Put	• Decreased root Na^+ content • Modulated ornithine decarboxylase and arginine decarboxylase involved in Put synthesis • Increased endogenous Put biosynthesis	Quinet et al. [108]
O. sativa M-1-48 and Pokkali	150 mM NaCl, 6 h	1 mM Spd or Spm	• Prevented chl loss • Improved photosystem activity • Decreased Na^+ content, and increased K^+, Ca^{2+} and Mg^{2+} contents	Chattopadhayay et al. [107]
O. sativa cv. I Kong Pao	150 and 300 mM	1 and 10 mM Put, Spd and Spm	• Decreased Na^+ and Cl^- accumulation • Improved cell viability • Inhibition of the alternative respiratory pathway	Ndayiragije and Lutts [106]
O. sativa cv. I Kong Pao	50 or 100 mM NaCl, 5, 12 and 19 days	1 mM Put, Spd and Spm	• Increased K^+/Na^+ ratio in the shoots	Ndayiragije and Lutts [106]

http://dx.doi.org/10.5772/67098

Species and cultivars	Salinity dose and duration	Protectants	Protective effects	References
O. sativa cv. M-1-48, Nonabokra and Gobindobhog	200 mM NaCl	1 mM Spd or 1 mM Spm	• Alleviation of growth inhibition, cellular damages • Decrease of H_2O_2, MDA, LOX activity, protein oxidation and protease activity • Maintenance of K^+/Na^+ balance, modulation osmolytes and improvement of activity of antioxidant enzymes	Roychoudhury et al. [112]
O. sativa cv. Niewdam Gs. no. 00621 and KKU-LLR-039	150 mM NaCl, 10 days	1 mM Spd	• Improved growth, anthocyanin and chl contents • Decreased Na+/K^+ ratio, Pro and H_2O_2 contents, particularly in the sensitive cultivar	Chunthaburee et al. [111]
Oryza sativa cv. IKP and Pokkali	50 and 100 mM NaCl, 5 days	1 mM Put, 5 days	• Reduced Na^+ accumulation • Increased PA content	Quinet et al. [108]
O. sativa cv. Pokkali and KDML105	150 mM NaCl, 7 days	1 mM Spd, 24 h	• Improved growth • Increased membrane stabilization • Efficient scavenging of free radicals and decreased MDA • Maintained K^+/Na^+ status	Saleethong et al. [109]
O. sativa cv. M-1-48, Nonabokra and Gobindobhog	200 mM NaCl, 15 days	1 mM Spd or 1 mM Spm, 15 days	• Increased Chl content • Decreased Na^+ content • Decreased MDA and H_2O_2 levels • Increased antioxidant metabolism	Roychoudhury et al. [112]

Table 3. Beneficial effects of exogenous signaling molecules and polyamines in mitigating salt stress-induced damages in rice.

Unlike NO, H_2O_2 is a long-lasting ROS, is able to cross organelle membrane and thus contributes to regulate redox-related gene expression. Depending on concentration, H_2O_2 itself causes damages to cell at higher concentration like other ROS or alleviates damaging effect of different abiotic stresses including salt stress at low concentration in various crop plants [2, 95]. However, triggering programmed cell death (PCD) during biotic stress is a well-known fact regulated by H_2O_2 signaling [96]. Use of H_2O_2 (at low concentration) as protectant has been well studied under salt stress. Uchida et al. [94] found that exogenous application of H_2O_2 upregulated activities of antioxidant defense enzymes (CAT, SOD, POX, APX and GR), and thus, H_2O_2 mitigated salt-induced damages in rice. In another study, Wang et al. [97] investigated potential role of H_2O_2 in salt stress tolerance in two rice varieties (Zhenghan-2 and Yujing-6). Exogenous H_2O_2 enhanced APX, CAT, peroxidases (POD), SOD and G6PDH activities in roots of both rice cultivars in a dose-dependent manner. Using a plasma membrane (PM), NADPH oxidase inhibitor, diphenyleneiodonium (DPI) and NaCl-induced H_2O_2 accumulation were reduced, and therefore, the activity of above enzymes was downregulated. They concluded that PM NADPH oxidase-mediated reduction of H_2O_2 is responsible for salt stress tolerance in Zhenghan-2 variety of rice by modulating the cellular antioxidant level. In a recent study, Habib and Ashraf [98] found that exogenous nitric oxide donor (0.1 and 0.2 mM SNP) could enhance salt (80 mM NaCl) stress tolerance in rice by enhancing the activities of antioxidant enzymes (SOD, POD and CAT). However, the effect was dose dependent where 0.1 mM SNP was found to be more effective. They studied four different cultivars and found variable amount of decrease in growth and biomass production which was associated with increase in Pro, AsA, H_2O_2 and MDA contents. However, spraying with NO donor effectively upregulated the antioxidant defense and improved growth and grain yield [98].

3.4. Polyamines

Polyamines (PAs) are organic cations distributing in broad range of organism perform a range of physiological functions [99]. Modulation of PAs interacting with other biomolecules (hormones/signaling molecule/amino acids) plays pivotal roles during plant developmental and stress adaptation process. PAs being cations bind to negatively charged surfaces of membrane and biomolecules and protect those from damage [100]. PAs stimulate antioxidant system which reduces ROS production and oxidative damages to plants [51, 101]. PAs have some direct action under salt stress which alleviates salt-induced damage in plants. As polycation, PAs can block a variety of cation through different pathways like K^+-selective channels, vacuolar-type channels and ammonium channels [102]. Salt-induced Na^+ influx, K^+ efflux, shoot-to-root K^+ recirculation and disruption of K^+/Na^+ homeostasis are directly modulated by PAs due to cationic nature which have been reported as obvious positive effects of PAs under salt stress [51, 103, 104]. Some previously reported roles of PAs in rice plant under salt stress condition have been discussed here (**Table 3**).

Ndayiragije and Lutts [105] reported that 1 mM Put, Spd or Spm application increased K^+/Na^+ ratio in the shoots of salt-sensitive *Oryza sativa* L. cv. I Kong Pao. In their further study, they reported that exogenous Put, Spd or Spm (1 and 10 mM) application decreased Na^+ and Cl^- accumulation, improved cell viability and inhibited the alternative respiratory pathway in salt-sensitive *Oryza sativa* L. cv. I Kong Pao [106]. Chattopadhayay et al. [107] described the roles of exogenous PAs to improve cation anion balance and photosynthesis. Exogenous Spd and Spm supplementation prevented salt-induced chl loss, inhibition of photochemical reactions of photosynthesis, improved PS I and PS II activities. Spd and Spm supplementation also decreased leakage of electrolytes and amino acids from roots and shoots caused by salinity. Exogenous Spd and Spm on endogenous decreased Na^+ content and increased K^+, Ca^{2+} and Mg^{2+} contents in salt-affected rice seedlings. Exogenous PAs were more effective to improve the performance of salt-sensitive M-1–48 plants, compared to tolerant Pokkali [107]. Exogenous Put (1 mM) addition increased endogenous Put level in both cultivars under salt stress. Exogenous Put-induced endogenous PAs metabolism was correlated with improved performance of rice cultivars under salt stress. Salt stress modulated the activities of PAs biosynthesis enzymes. Activities of ornithine decarboxylase and arginine decarboxylase (involved in Put synthesis) were higher in salt-treated plants and after exogenous Put addition in salt-resistant Pokkali than in salt-sensitive IKP cultivars of rice. Exogenous Put reduced Na^+ content in root of IKP cultivar [108]. Spermidine pretreatment (1 mM Spd, 24 h) improved K^+/Na^+ homeostasis in salt-tolerant (Pokkali) and salt-sensitive (KDML 105) rice cultivars in exposure with salt stress (150 mM NaCl, 7 days). Increased chl content and improved plant height, fresh and dry weight of both cultivars were confirmed by Spd pretreatment under salt stress [109]. Ghosh et al. [67] demonstrated that exogenous Put (1 and 2 mM) improved antioxidant defense system that reduced oxidative stress in salt-affected rice seedlings. Salt stress (200 mM NaCl) increased generation of $O_2^{\bullet-}$, H_2O_2 in tissues and caused membrane lipid peroxidation and protein carbonylation. Exogenous Put application increased the contents of nonenzymatic antioxidants such as anthocyanin and flavonoid. Activities of antioxidant enzymes GPX, CAT and GR were enhanced by exogenous Put application. Thus, enhanced antioxidant system conferred oxidative stress tolerance and overall salt stress tolerance in rice [67]. Two cultivars of rice differing in their salt tolerance capacity [Pokkali (salt-tolerant) and KDML 105 (salt-sensitive)] were examined to evaluate their performance after Spd pretreatment. Spermidine pretreatment decreased H_2O_2 accumulation. Exogenous Spd pretreatment also prevented lipid peroxidation and decreased electrolyte leakage by stabilizing membrane. Prevention of chl breakdown was also demonstrated in Spd pretreated salt-affected rice seedlings. These results indicated the roles of exogenous Spd to improve oxidative stress tolerance under salt stress [109]. Paweena et al. [110] noted the roles of Spd to improve yield attributes which ultimately increased the yield of salt-affected rice cultivars. Pretreatment with Spd increased grain yield per plant of KDML 105 by 62% and that of Pokkali by 16% under salt stress (25 mM NaCl) condition. Seed priming with Spd improved growth, anthocyanin and chl contents and decreased Na^+/K^+ ratio, Pro and H_2O_2 contents, particularly in the sensitive cultivar [111]. Alleviation of growth inhibition, cellular damages, decrease of H_2O_2, MDA, LOX activity, protein oxidation and protease activity, maintenance of K^+/Na^+ balance, modulation of osmolytes and improvement of activity of antioxidant enzymes (GPX, APX and CAT) were attributed by exogenous Spd and Spm supplementation those conferred salt stress tolerance in different rice cultivars. However, among different rice cultivars [(M-1-48 (salt-sensitive), Nonabokra (salt-tolerant) and Gobindobhog (highly sensitive)], salt-sensitive rice cultivars responded better compared to salt-tolerant cultivars in response to exogenous PAs application under salt stress [112].

3.5. Trace elements

The effect of trace elements in mitigating the effect of abiotic stress in plants including salinity has become a matter of interest in recent decades. However, the mechanism is still unclear, and their responses are mostly dose dependent [3, 113]. In recent decades, plenty of researches have been carried on the positive effect of trace element on salt stress tolerance in rice (**Table 4**), while many of them are focused on silicon (Si) as rice is a Si accumulator and its essentiality is well documented in rice.

Species and cultivars	Salinity dose and duration	Protectants	Protective effects	References
O. sativa cv. BRRI dhan47	150 mM NaCl, 3 and 6 days	0.5 mM $MnSO_4$, 3 and 6 days	• Decreased Na^+ uptake and the Na^+/K^+ ratio, and increased K^+ uptake • Increased the content of phenolic compounds, flavonoids and AsA • Increased RWC and decreased osmotic potential and Pro content • Decreased ROS production, lipid peroxidation and LOX activity • Increased the activities of MDHAR, DHAR, SOD, CAT, Gly I and Gly II	Rahman et al. [17]
O. sativa cv. Khazar and Zayandehrood	100 mM NaCl	3 mM Si as Na_2SiO_3	• Decreased Na^+ and K^+ concentrations and increased Si concentration • Increased Fv/Fm ratio	Mahdieh et al. [119]
O. sativa cv. IR29	10 dS m^{-1}	150 mg kg^{-1} Na_2SiO_3	• Increased FW and DW of shoot • Increased photosynthesis rate • Increased activities of APX and GPX and decreased CAT activity • Decreased Na^+ content	Farooq et al. [117]
O. sativa cv. KS-282 and IRRI-6	10 dS m^{-1}	150 mg kg^{-1} Na_2SiO_3	• Increased APX and GPX activities and decreased CAT activity • Reduced Na^+ uptake and transpiration rate • Increased RWC and photosynthetic efficiency	Farooq et al. [117]
O. sativa cv. Dongjin	100 mM NaCl; 6, 12 and 24 h	0.5, 1 and 2 mM $Na_2SiO_3$5 H_2O; 6, 12 and 24 h	• Increased shoot length, Shoot fresh weight and chl content • Decreased electrolytic leakage and Na^+ uptake	Kim et al. [129]
O. sativa cv. GR4, CSR10 and IR36	50 mM NaCl	3 mM Na_2SiO_3	• Increased root and shoot dry weight • Decreased concentrations of Cl^-, Na^+ and K^+ and increased K^+/Cl^- and K^+/Na^+ ratio • Increased net assimilation rate, stomatal conductance, intercellular CO_2 and decreased transpiration rate	Shi et al. [130]

Species and cultivars	Salinity dose and duration	Protectants	Protective effects	References
***O. sativa* cv. Pokkali and Peta**	200 mM NaCl, 6–8 days	0.065 mM GSH, 6–8 days	• Decreased H_2O_2 and MDA contents • Increased the activities of SOD, APX and GR • Increased levels of endogenous AsA and GSH	Wang et al. [122]
O. sativa cv. MRQ74	200 mM NaCl, 3 months	0.5 mM AsA, 3 months	• Decreased Na^+ uptake and the Na^+/K^+ ratio; increased K^+ and Ca^{2+} uptake • Increased RGR, fresh mass and dry mass • Increased Pro content • Increased POD, CAT and SOD	Alhasnawi et al. [123]
***O.sativa* cv. MRQ74 and MR269**	200 mM NaCl, 14 days	1.0 mM AsA, 14 days	• Decreased Na^+ uptake and the Na^+/K^+ ratio; increased N and K^+ uptake • Increased length, FW, DW of shoot and root • Increased chl content	Alhasnawi et al. [124]
***O. sativa* cv. Ciherang, IR 64, Lambur, Banyuasin, IR 42, Inpara 10, Margasari**	5.99 dS m^{-1} (EC), whole life cycle	1500 ppm AsA, applied in 4 times for whole life cycle	• Increased plant height and leaf area • Increased number of productive tiller • Increased length of panicle	Barus et al. [125]
***O. sativa* cv. Pokkali and Peta**	200 mM NaCl, 6–8 days	5 mM AsA, 6–8 days	• Decreased H_2O_2 and MDA contents • Increased the activities of SOD, APX and GR • Increased levels of endogenous AsA and GSH	Wang et al. [122]
O. sativa cv. Pokkali	120 mM NaCl, 72 h	0.75 and 1.5 mM gallic acid (phenolic compounds), 72 h	• Significantly decreased H_2O_2 and TBARS contents • Enhanced the activities of SOD, CAT, POX and APX • Increased of RGR, osmotic potential (ΨΠ), Fv/Fm and Pro	Ozfidan-Konakci et al. [126]

Table 4. Beneficial effects of exogenous trace elements and antioxidants in mitigating salt stress-induced damages in rice.

Zinc (Zn)-supplemented rice plants showed enhanced salt tolerance as reported by Iqbal and Aslam [114]. In their study, both salt-sensitive and salt-tolerant genotypes showed less plant height tiller, dry weight and fresh weight under salinity (70 mM NaCl), while these effects were reversed by the application of Zn (1–1000 nM). However, the effect of Zn was dose dependent. Mehmood et al. [115] observed that boron (B) could ameliorate the negative effect of salt stress in rice grown in the saline and saline sodic soils (9 dS m^{-1}). However, the effect was mostly dose dependent. Boron fertilized at 1.5 kg B ha^{-1} showed higher yield of rice which was due to reduced shoot Na^+ and Cl^- concentration and better ratio of K^+ and Na^+ in shoot. However, B application at 6 kg B ha^{-1} showed adverse affect on grain and straw production in saline sodic soils. Boron supplementation also improved seed setting of rice [115]. Zayed et al. [116] studied the effect of manganese (Mn), Zn and iron (Fe) on the performance of rice in saline soil. They observed that the application of the all micronutrients either alone or combined improved rice growth, dry matter production, leaf area index and chl content under saline condition. Moreover, yield attributes like panicle numbers, panicle weight, filled grains $panicle^{-1}$ and 1000-grain weight were markedly increased when plants received the micronutrient which resulted in higher yield under saline soil [116]. Farooq et al. [117] investigated the beneficial effects of exogenous Si in rice grown under salinity. Both salt-sensitive (IRRI-6) and salt-tolerant (KS-282) cultivars were used in this study. Their study showed that salt stress (10 dS m^{-1}) reduced plant biomass by damaging the membrane and reducing special products analysis division values and photosynthetic efficiency which was accompanied by higher accumulation of Na^+ and lower K^+. However, the damages were higher in IRRI-6 than KS-282. Importantly, Si (150 mg kg^{-1}) application counteracted the adverse effects of salt stress by reducing the uptake Na^+, lowering transpiration rate and improving water content and photosynthetic efficiency [117]. They also found the Si-induced upregulation of the activities of antioxidant enzymes viz. APX, GPX and CAT, and thus, the oxidative stress was mitigated [118]. Mahdieh et al. [119] investigated the relative performance of Si-supplemented salt-sensitive and salt-tolerant rice cultivars (25-day-old) under saline condition (100 mM NaCl, 96 h). Investigation revealed that both rice cultivars showed stunted growth, reduced photosynthetic efficiency (Fv/Fm) and increased chlorosis which were associated with increased Na^+ accumulation. On the other hand, Si supplementation decreased shoot Na^+ concentration, but increased Si uptake which in turn increased the growth, chl content and Fv/Fm ratio in salt-treated plants. In our laboratory, we studied the protective effect of exogenous Mn (0.5 mM $MnSO_4$) in mitigating short-term salt stress (150 mM NaCl, 3 and 6 days) in rice seedlings cv. BRRI dhan47. Any level of salinity not only diminished the ion homeostasis due to Na^+ influx and K^+ efflux but also caused osmotic stress due to limited amount of water and increased compatible solute content. As a result, salt-treated seedlings showed inhibited growth, chlorosis and excess generation of ROS (H_2O_2 and $O_2^{\bullet-}$) and higher amount of lipid peroxidation which was due to the disruption of antioxidant defense system [17]. However, the seedlings supplemented with exogenous Mn showed significant recovery of growth and chlorosis (**Figure 4**) by improving ionic and osmotic homeostasis through decreasing Na^+ influx and increasing water status, respectively. Exogenous Mn also enhanced the activities of antioxidant enzymes (MDHAR, DHAR, SOD and CAT) and maintained higher level of nonenzymatic antioxidants (AsA and GSH) which were able to detoxify ROS. Exogenous Mn also upregulated the glyoxalase system and detoxified the level of toxic MG. Moreover, exogenous Mn effectively maintained the ion homeostasis and improved the growth of seedlings under salinity. The beneficial role of Se in mitigating salt stress has been studied in many plants but hardly found in rice. Recently, we completed a pot experiment with three rice varieties viz. BRRI dhan45, BRRI dhan47 and Nipponbare grown under different concentration of saltwater (50–150 mM). Salt stresses reduced the plant height and tillers hill^{-1}, leaf relative water content and chl content in dose-dependent manner. Salt stress also reduced the effective tillers hill^{-1}, number of filled grains panicle^{-1}, 1000-grain weight, grain yield and straw yield. However, when the plants were supplemented with 0.5 mM Se (sodium selenite, Na_2SeO_3), these observed parameters were significantly increased compared to salt treatment alone. Importantly, the beneficial effect of Se on salt stress tolerance was prominent up to 100 mM NaCl, while it could not be more beneficial above this level of salt concentration [120]. The response of rice plant to Se and salt stress greatly varied in different cultivars (**Figure 5**).

Figure 4. Phenotypic appearance of rice seedlings under salt stress with and without exogenous Mn. Here, Mn and salt indicate 0.5 mM $MnSO_4$ and 150 mM NaCl, respectively (adapted from Rahman et al. [17], with permission from Springer).

3.6. Antioxidants

Mitigation of abiotic stresses including salinity by using antioxidants as exogenous protectants is becoming familiar for plant biologists in recent time. There are many antioxidants including AsA, GSH, tocopherol, nonprotein amino acid, phenolic compounds and alkaloids that play a significant role in mitigation of excessive cellular reactive oxygen species activities caused by salt stresses [2, 121]. Moreover, these protectants have diverse roles in plant growth and development. Several research results have indicated that exogenous application of antioxidants provided significant protection against salt-induced damages in plants [3]. However, very few researches have been carried out on the positive effect of antioxidant under salt stress in rice plant (**Table 4**).

Figure 5. Effect of Se supplementation in mitigating adverse effects of salt stress in different rice cultivars. Plants were subjected to 50 mM NaCl throughout its life cycles with or without 0.5 mM Na_2SeO_3.

Wang et al. [122] carried out an experiment to identify the effects of exogenous AsA and GSH on two rice cultivars, the salt-tolerant cultivar Pokkali and the salt-sensitive cultivar Peta, under salt stress (200 mM NaCl). They found that salt stress hampered the antioxidant enzyme activities and increased MDA and H_2O_2 contents of the chloroplasts. But, addition of exogenous AsA and GSH reduced H_2O_2 and MDA contents and increased SOD, APX, GR activities and endogenous levels of AsA and GSH in the chloroplasts of both cultivars under salt stress, but the effects were significantly more prominent in cv. Pokkali. On the other hand, GSH played more significant role than the AsA on the plastidial reactive oxygen scavenging systems. These results indicated that exogenous AsA and GSH differentially increased salinity tolerance and mitigate the damages of salt-induced stress. Alhasnawi et al. [123] investigated the effect of exogenous AsA application on embryogenic callus of indica rice cv. MRQ74 cultured under saline conditions (200 mM NaCl). They reported that salt stress reduced relative growth rate (RGR), callus fresh and dry masses, K^+ and Ca^{+2} content, and increased Na^+ content and Na^+/K^+ ratio. Application of AsA (0.5 mM) alleviated these effects of salinity. The activities of SOD, CAT and POX, as well as content of Pro, were increased due to NaCl treatment, and these parameters were mostly further increased by AsA application. Accordingly, AsA was able to enhance callus tolerance to NaCl stress. Similarly, AsA lessened the damaging effect of salt stress in two rice cultivars (MRQ74 and MR269) under NaCl stress [124]. They demonstrated that AsA (1.0 mM) considerably influenced a number of metabolic processes, leading to enhanced ability for seedling survival and growth through balancing or changing in the Na^+ and K^+ levels as well as the intracellular level of AsA. Exogenous AsA had effects on Na^+ and K^+ uptake and improving chl which have vital roles in number of metabolic processes. Ascorbic acid decreased Na^+ uptake and the Na^+/K^+ ratio and increased the uptake of Na^+ and K^+ in contrast to salt stress alone. It also increased the length, FW and DW of shoot and root and chl content. Therefore, it may be concluded that exogenous application of AsA has low external osmotic potential and ion toxicity and might be effective in mitigating the effects of salt stress. Barus et al. [125] studied the effect of salt stress and its interaction with AsA on some morphological traits of eight varieties of rice (Ciherang, IR 64, Lambur, Batanghari, Banyuasin, IR 42, Inpara 10 and Margasari). AsA (0, 500, 1000 and 1500 ppm) was applied in four times at the age of 15, 35, 55 and 75 days after sowing. All concentrations of ascorbic acid generally had a positive effect on morphological characters. The leaf area, productive tiller and panicle length increased in all cultivars after application of AsA. However, the best response was found at 1500 ppm concentration of AsA on Banyuasin variety. Besides the AsA and GSH, gallic acid (GA) had a remarkable biological effect on plants growth, including an improvement in plants' tolerance under salinity stress conditions. Ozfidan-Konakci et al. [126] conducted a study to examine the effect of exogenously applied GA on the tolerance of two rice cultivars (tolerant cultivar Pokkali and sensitive cultivar IR-28) to salt (NaCl) stress. Salt stress reduced the maximum photochemical efficiency (Fv/Fm) ratio, photochemical quenching coefficient (qP) and actual quantum yield (ΦPSII) of two rice cultivars, but the effect was more distinct in IR-28. Also, a significant increase in H_2O_2 content and lipid peroxidation was observed in NaCl treatment. The salt stress sensitivity of plants was higher in IR-28 than in Pokkali. However, addition of GA in NaCl-stressed Pokkali rice markedly decreased H_2O_2 and thiobarbituric acid reactive substances (TBARS) contents, and enhanced the activities of SOD, CAT, APX and POX as well as increase of relative growth rate (RGR), osmotic potential (ΨΠ), Fv/Fm ratio and Pro level in comparison with salt stress alone. So, it can be said that GA is able to mitigate NaCl toxicity of rice plant by increasing the level of antioxidants activities and photosynthetic efficiency.

4. Conclusion and perspectives

Salinity is one of the major abiotic stresses and constraints for agriculture worldwide because most of the crop plants are sensitive to salt stress [131]. Salt stress adversely affects physiology and biochemistry of plants primarily by creating water stress, ionic imbalance and toxicity, nutritional disorders, alteration of metabolic processes, oxidative stress, membrane disorganization and reduction of cell division and expansion. However, plants have well-organized antioxidative defense mechanisms at the cellular level to minimize the toxicity. But in severe stress condition plant cannot cope with stress by its own mechanism. So, use of exogenous protectants like osmoprotectants, plant hormones, signaling molecules, polyamines, trace elements, antioxidants, etc., are now popular in research aimed at enhancing abiotic stress tolerance. On the other hand, rice is one of the world's most important cereal crops with exceptional agricultural and economic importance as being a staple food for more than 50% population worldwide. This crop is highly susceptible to salt stress, and the productivity of rice decreased to a great extent under salinity. In recent time, the biochemical responses of rice plants to salt stress have been studied intensively. The use of exogenous protectants in rice plant under salt stress condition has been found to be very much effective to mitigate salt-induced damages. The mechanisms by which protectants regulate activities of enzymes are enhanced and are undoubtedly interesting and demand insightful studies in rice plant.

The clear mechanism of defense and signal transduction pathways is still in dark and should be discovered. Furthermore, the exact dose, duration and proper methods of application of exogenous protectants in rice plant under salt condition should be studied more precisely for complete elucidation of mechanism of protection.

Author details

Anisur Rahman, Kamrun Nahar[1,3], Jubayer Al Mahmud[1,4], Mirza Hasanuzzaman[2], Md. Shahadat Hossain, and Masayuki Fujita[1*]

*Address all correspondence to: fujita@ag.kagawa-u.ac.jp

1 Laboratory of Plant Stress Responses, Department of Applied Biological Science, Faculty of Agriculture, Kagawa University, Miki-cho, Japan

2 Department of Agronomy, Faculty of Agriculture, Sher-e-Bangla Agricultural University, Dhaka, Bangladesh

3 Department of Agricultural Botany, Faculty of Agriculture, Sher-e-Bangla Agricultural University, Dhaka, Bangladesh

4 Department of Agroforestry and Environmental Science, Faculty of Agriculture, Sher-e-Bangla Agricultural University, Dhaka, Bangladesh

References

[1] Cramer GR, Urano K, Delort S, Pezzotti M, Shinozaki K. Effects of abiotic stress on plants: a systems biology perspective. BMC Plant Biology 2011; 11:163. doi:10.1186/1471-2229-11-163

[2] Hasanuzzaman M, Hossain MA, da Silva JAT, Fujita M. Plant responses and tolerance to abiotic oxidative stress: antioxidant defense is a key factor. In: Bandi V, Shanker AK, Shanker C, Mandapaka M (eds) Crop stress and its management: perspectives and strategies. Berlin: Springer; 2012. pp. 261–316.

[3] Hasanuzzaman M, Nahar K, Fujita M. Plant response to salt stress and role of exogenous protectants to mitigate salt-induced damages. In: Ahmad P, Azooz MM, Prasad MNV (eds) Ecophysiology and responses of plants under salt stress. New York: Springer; 2013. pp. 25–87.

[4] Hasanuzzaman M, Nahar K, Fujita M, Ahmad P, Chandna R, Prasad MNV, Ozturk M. Enhancing plant productivity under salt stress: relevance of poly-omics. In: Ahmad P, Azooz MM, Prasad MNV (eds) Salt stress in plants: signaling, omics and adaptations. New York: Springer; 2013. pp. 113–156.

[5] Munns R, Tester M. Mechanism of salinity tolerance. Annual Review of Plant Biology 2008; 59:651–681.

[6] Mahajan S, Tuteja N. Cold, salinity and drought stresses: an overview. Archives of Biochemistry and Biophysics 2005; 444:139–158.

[7] Tester M, Davenport R. Na^+ tolerance in higher plants. Annals of Botany 2003; 91: 503–507.

[8] Hasanuzzaman M, Fujita M, Islam MN, Ahamed KU, Nahar K. Performance of four irrigated rice varieties under different levels of salinity stress. International Journal of Integrative Biology 2009; 6:85–90.

[9] Hasanuzzaman M, Hossain MA, Fujita M. Nitric oxide modulates antioxidant defense and the methylglyoxal detoxification system and reduces salinity-induced damage of wheat seedlings. Plant Biotechnology Reports 2011; 5:353–365.

[10] Hasanuzzaman M, Hossain MA, Fujita M. Selenium-induced up-regulation of the antioxidant defense and methylglyoxal detoxification system reduces salinity-induced damage in rapeseed seedlings. Biological Trace Element Research 2011; 143:1704–1721.

[11] Wu GQ, Wang SM. Calcium regulates K^+/Na^+ homeostasis in rice (*Oryza sativa* L.) under saline conditions. Plant, Soil and Environment 2012; 58:121–127.

[12] Mishra P, Bhoomika K, Dubey RS. Differential responses of antioxidative defense system to prolonged salinity stress in salt-tolerant and salt-sensitive Indica rice (*Oryza sativa* L.) seedlings. Protoplasma 2013; 250:3–19.

[13] Peng K, Chunling L, Wuxin Y, Chunlan L, Xiangdong L, Shen Z. Manganese uptake and interactions with cadmium in the hyperaccumulator-*Phytolacca americana* L. Journal of Hazardous Materials 2008; 154:674–681.

[14] Hasanuzzaman M, Alam MM, Rahman A, Hasanuzzaman M, Nahar K, Fujita M. Exogenous proline and glycine betaine mediated upregulation of antioxidant defense and glyoxalase systems provides better protection against salt-induced oxidative stress in two rice (*Oryza sativa* L.) varieties. BioMed Research International 2014. doi:10.1155/2014/757219

[15] Khush, G. What it will take to feed 5.0 billion rice consumers in 2030. Plant Molecular Biology 2005; 59:1–6. doi:10.1007/s11103-005-2159–2155

[16] Munns R. Comparative physiology of salt and water stress. Plant Cell and Environment 2002; 25:239–250.

[17] Rahman A, Hossain MS, Mahmud J, Nahar K, Hasanuzzaman M, Fujita M. Manganese-induced salt stress tolerance in rice seedlings: regulation of ion homeostasis, antioxidant defense and glyoxalase systems. Physiology and Molecular Biology of Plants 2016. doi:10.1007/s12298-016-0371-1

[18] Rahman A, Nahar K, Hasanuzzaman M, Fujita M. Calcium supplementation improves Na^+/K^+ ratio, antioxidant defense and glyoxalase systems in salt-stressed rice seedlings. Frontiers in Plant Science 2016. doi:10.3389/fpls.2016.00609

[19] Lee KS, Choi WY, Ko JC, Kim TS, Gregoria GB. Salinity tolerance of japonica and indica rice (*Oryza sativa* L.) at the seedling stage. Planta 2003; 216:1043–1046.

[20] Todaka D, Nakashima K, Shinozaki K, Yamaguchi-Shinozaki K. Towards understanding transcriptional regulatory networks in abiotic stress responses and tolerance in rice. Rice 2012; 5:1–9.

[21] Das P, Nutan KK, Singla-Pareek SN, Pareek A. Understanding salinity responses and adopting 'omics-based' approaches to generate salinity tolerant cultivars of rice. Frontiers in Plant Scince 2015. doi:10.3389/fpls.2015.007

[22] IRRI. Stress and disease tolerance. In: Rice Knowledge bank. International Rice Research Institute, Manila, Philippines (2006). Available from: http://www.knowledgebank.irri.org/ricebreedingcourse/Breeding_for_salt_tolerance.htm

[23] Xiong L, Zhu JK. Molecular and genetic aspects of plant responses to osmotic stress. Plant, Cell and Environment 2002; 25: 131–139.

[24] Horie T, Karahara I, Katsuhara M. Salinity tolerance mechanisms in glycophytes: an overview with the central focus on rice plants. Rice 2012; 5:11. doi:10.1186/1939-8433-5-11

[25] Wahid A, Farooq M, Basra SMA, Rasul E, Siddique KHM. Germination of seeds and propagules under salt stress. In: Pessarakli M (ed) Handbook of plant and crop stress. Boca Raton: CRC Press; 2011, pp. 321–337.

[26] Hua-long L, Han-jing S, Jing-guo W, Yang L, De-tang Z, Hong-wei Z. Effect of seed soaking with exogenous proline on seed germination of rice under salt stress. Journal of Northeast Agricultural University 2014; 21(3):1–6.

[27] Özdemir F, Bor M, Demiral T, Türkan I. Effects of 24-epibrassinolide on seed germination, seedling growth, lipid peroxidation, proline content and antioxidative system of rice (*Oryza sativa* L.) under salinity. Plant Growth Regulation 2014; 42: 203–211.

[28] Hakim MA, Juraimi AS, Begum M, Hanafi MM, Ismail MR, Selamat A. Effect of salt stress on germination and early seedling growth of rice (*Oryza sativa* L.). African Journal of Biotechnology 2010; 9:1911–1918.

[29] Rajakumar R. A study on effect of salt stress in the seed germination and biochemical parameters of rice (*Oryza sativa* l.) under *in vitro* condition. Asian Journal of Plant Science and Research 2013; 3:20–25.

[30] Shereen A, Ansari R, Raza S, Mumtaz S, Khan MA, Ali Khan M. Salinity induced metabolic changes in rice (*oryza sativa* l.) seeds during germination. Pakistan Journal of Botany 2011; 43:1659–1661.

[31] Islam MM, Karim MA. Evaluation of rice (*Oryza sativa* L.) genotypes at germination and early seedling stage for their tolerance to salinity. The Agriculturists 2010; 8(2): 57–65.

[32] Abbas MK, Ali AS, Hasan HH, Ghal RH. Salt tolerance study of six cultivars of rice (*Oryza sativa* L.) during germination and early seedling growth. Journal of Agricultural Scince 2013. doi:10.5539/jas.v5n1p250

[33] Kazemi K, Eskandari H. Effects of salt stress on germination and early seedling growth of rice (*Oryza sativa*) cultivars in Iran. African Journal of Biotechnology 2011; 10(77): 17789–17792.

[34] Ologundudu AF, Adelusi AA, Akinwale RO. Effect of salt stress on germination and growth parameters of rice (*Oryza sativa* L.). Notulae Scientia Biologicae 2014; 6(2): 237–243.

[35] Makihara D, Tsuda M, Morita M, Hirai Y, Kurod T. Effect of salinity on the growth and development of rice (*Oryza sativa* L.) varieties. Japanese Journal of Tropical Agriculture 1999; 43(4):285–294.

[36] Aref F. Effect of saline irrigation water on yield and yield components of rice (*Oryza sativa* L.). African Journal of Biotechnology 2013; 12:3503–3513.

[37] Hasegawa P, Bressan RA, Zhu JK, Bohnert HJ. Plant cellular and molecular responses to high salinity. Annual Review of Plant Physiology and Plant Molecular Biology 2000; 51:463–499.

[38] Kumar V, Khare T. Differential growth and yield responses of salt-tolerant and susceptible rice cultivars to individual (Na^+ and Cl^-) and additive stress effects of NaCl. Acta Physiologiae Plantarum 2016; 38:170. doi:10.1007/s11738-016-2191-x

[39] Chaves MM, Flexas J, Pinheiro C. Photosynthesis underdrought and salt stress: regulation mechanisms from wholeplant to cell. Annals of Botany 2009; 103:551–560.

[40] Ashraf M, Harris PJC. Photosynthesis under stressful environments: an overview. Photosynthetica 2013; 5:163–190.

[41] Moradi F, Abdelbagi M, Ismail. Responses of photosynthesis, chlorophyll fluorescence and ROS-scavenging systems to salt stress during seedling and reproductive stages in rice. Annals of Botany 2007; 99: 1161–1173.

[42] Cha-Um S, Kirdmanee C. Effect of glycinebetaine on proline, water use, and photosynthetic efficiencies, and growth of rice seedlings under salt stress. Turkish Journal of Agriculture and Forestry 2010; 34:517–527.

[43] Amirjani MR. Effect of NaCl on some physiological parameters of rice. European Journal of Biological Sciences 2010; 3:06–16.

[44] Pattanagul W, Thitisaksakul M. Effect of salinity stress on growth and carbohydrate metabolism in three rice (*Oryza sativa* L.) cultivars differing in salinity tolerance. Indian Journal of Experimental Biology 2008; 46:736–742.

[45] Sohan D, Jason R, Zajcek J. Plant-water relations of NaCl and calcium-treated treated sunflower plants. Environmental and Experimental Botany 1999; 42:105–111.

[46] Romero-Aranda R, Soria T, Cuartero S. Tomato plant-water uptake and plant-water relationships under saline growth conditions. Plant Science 2001; 160:265–272.

[47] Lutts S, Kinet JM, Bouharmont J. Effects of salt stress on growth, mineral nutrition and proline accumulation in relation to osmotic adjustment in rice (*Oryza sativa* L.) cultivars differing in salinity resistance. Plant Growth Regulation 1996; 19:201–218.

[48] Redillas MCFR, Park SH, Lee JW, Kim YC, Jeong JS, Jung H, Bang SW, Hahn TR, Kim JK. Accumulation of trehalose increases soluble sugar contents in rice plants conferring tolerance to drought and salt stress. Plant Biotechnology Reports 2012; 6:89–96.

[49] Nemati I, Moradi F, Gholizadeh S, Esmaeili MA, Bihamta MR. The effect of salinity stress on ions and soluble sugars distribution in leaves, leaf sheaths and roots of rice (*Oryza sativa* L.) seedlings. Plant Soil Environment 2011; 57(1): 26–33.

[50] Boriboonkaset T, Theerawitaya C, Pichakum A, Cha-um S, Takabe T, Kirdmanee C. Expression levels of some starch metabolism related genes in flag leaf of two contrasting rice genotypes exposed to salt stress. Australian Journal of Crop Science 2012; 6:1579–1586.

[51] Gill SS, Tuteja N. Polyamines and abiotic stress tolerance in plants. Plant Signaling & Behavior 2010; 51:26–33.

[52] Summart J, Thanonkeo P, Panichajakul S, Prathepha P, McManus MT. Effect of salt stress on growth, inorganic ion and proline accumulation in Thai aromatic rice, Khao Dawk. Mali 105, callus culture. African Journal of Biotechnology 2010; 9:145–152.

[53] Parida AK, Das AB. Salt tolerance and salinity effects on plants: a review. Ecotoxicology and Environmental Safety 2010; 60:324–349.

[54] Sumithra K, Jutur PP, Carmel BD, Reddy AR. Salinity-induced changes in two cultivars of *Vigna radiata*: responses of antioxidative and proline metabolism. Plant Growth Regulation 2006; 50:11–22.

[55] Shabala S, Demidchick V, Shabala L, Cuin TA, Smith SJ, Miller AJ, Davies JM, Newman IA. Extracellular Ca^{2+} ameliorates NaCl-induced K^+ loss from *Arabidopsis* root and leaf cells by controlling plasmamembrane K^+-permeable channels. Plant Physiology 2006; 141:1653–1665.

[56] Wang H, Zhang M, Guo R, Shi D, Liu B, Lin X, Yang C. Effects of salt stress on ion balance and nitrogen metabolism of old and young leaves in rice (*Oryza sativa* L.). BMC Plant Biology 2012; 12:194.

[57] Carillo P, Annunziata MG, Pontecorvo G, Fuggi A, Woodrow P. Salinity stress and salt tolerance. In: Shanker A (ed) Abiotic stress in plants-mechanisms and adaptations. Crotia: InTech; 2011, pp. 2–35.

[58] Zhu ZJ, Wei GQ, Li J, Qian QQ, Yu JQ. Silicon alleviates salt stress and increases antioxidant enzymes activity in leaves of salt-stressed cucumber (*Cucumis sativus* L.). Plant Science 2004; 167:527–533.

[59] Zhang Z, Liu Q, Song H, Rong X, Ismail AM. Responses of contrasting rice (*Oryza sativa* L.) genotypes to salt stress as affected by nutrient concentrations. Agricultural Sciences in China 2011; 10:195–206.

[60] Azarin KV, Alabushev AV, Usatov AV, Kostylev PI, Kolokolova NS, Usatova OA. Effects of salt stress on ion balance at vegetative stage in rice (*Oryza sativa* L.). Online Journal of Biological Sciences 2016; 16:76–81.

[61] Ahmad P, Umar S. Oxidative stress: role of antioxidants in plants. New Delhi: Studium Press; 2011.

[62] Bose J, Rodrigo-Moreno A, Shabala S. ROS homeostasis in halophyte s in the context of salinity stress tolerance. Journal of Experimental Botany 2014; 65:1241–1257.

[63] Hasanuzzaman M, Nahar K, Fujita M. Extreme temperatures, oxidative stress and antioxidant defense in plants. In: Vahdati K, Leslie C (eds) Abiotic stress—plant responses and applications in agriculture. Rijeka: InTech; 2013. pp. 169–205. doi:10.5772/54833

[64] Hernandez JA, Jimenez A, Mullineaux P, Sevilla F. Tolerance of pea (*Pisum sativum* L.) to longterm salt stress is associated with induction of antioxidant defences. Plant Cell and Environment 2000; 23(8):853–862.

[65] Tanou G, Molassiotis A, Diamantidis G. Induction of reactive oxygen species and necrotic death-like destruction in strawberry leaves by salinity. Environmental and Experimental Botany 2009; 65:270–281.

[66] Vaidyanathan H, Sivakumar P, Chakrabarty R, Thomas G. Scavenging of reactive oxygen species in NaCl-stressed rice (*Oryza sativa* L.)-differential response in salt-tolerant and sensitive varieties. Biologia Plantarum 2003; 165:1411–1418.

[67] Ghosh N, Das SP, Mandal C, Gupta S, Das K, Dey N, Adak MK. Variations of antioxidative responses in two rice cultivars with polyamine treatment under salinity stress. Physiology and Molecular Biology of Plants 2012; 18:301–313.

[68] Abdallah MMS, Abdelgawad ZA, El-Bassiouny HMS. Alleviation of the adverse effects of salinity stress using trehalose in two rice varieties. South African Journal of Botany 2016; 103:275–282.

[69] Hong C, Chao Y, Yang M, Cheng S, Cho S, Kao C. NaCl-induced expression of glutathione reductase in roots of rice (*Oryza sativa* L.) seedlings is mediated through hydrogen peroxide but not abscisic acid. Plant and Soil 2009; 320:103–115.

[70] Halliwell B, Gutteridge MJC. Free radicals in biology and medicine, 4th edn. London: Oxford University Press; 2007.

[71] Zeng L, Shannon MC. Salinity effects on seedling growth and yield components of rice. Crop Science 2000; 40:996–1003.

[72] Saleethong P, Sanitchon J, Kong-ngern K, Theerakulpisut P. Effects of exogenous spermidine (Spd) on yield, yield-related parameters and mineral composition of rice (*Oryza sativa* L. ssp. *indica*) grains under salt stress. Australian Journal of Crop Science 2013; 7:1293–1301.

[73] Ali Y, Aslam Z, Ashraf MY, Tahir GR. Effect of salinity on chlorophyll concentration, leaf area, yield and yield components of rice genotypes grown under saline environment. International Journal of Environmental Science and Technology 2004; 1:221–225.

[74] Chunthaburee S, Sanitchon J, Pattanagul W, Theerakulpisut P. Effects of salt stress after late booting stage on yield and antioxidant capacity in pigmented rice grains and alleviation of the salt-induced yield reduction by exogenous spermidine. Plant Production Science 2015; 18:32–42.

[75] Arsa IGBA, Ariffin, Aini N, Lalel HJD. Evaluation of grain yield and aroma of upland rice (*Pare Wangi* var.) as response to soil moisture and salinity. Current Agriculture Research 2016; 4:35–46.

[76] Chen H, Jiang JG. Osmotic adjustment and plant adaptation to environmental changes related to drought and salinity. Environmental Reviews 2010; 18:309–319.

[77] Teh CY, Mahmood M, Shaharuddin NA, Ho CL. In vitro rice shoot apices as simple model to study the effect of NaCl and the potential of exogenous proline and glutathione in mitigating salinity stress. Plant Growth Regulation 2015; 75:771–781.

[78] Nounjan N, Theerakulpisut P. Effects of exogenous proline and trehalose on physiological responses in rice seedlings during salt-stress and after recovery. Plant Soil and Environment 2012; 58(7): 309–315.

[79] Bhusan D, Das DK, Hossain M, Murata Y, Hoque MA. Improvement of salt tolerance in rice (*Oryza sativa* L.) by increasing antioxidant defense systems using exogenous application of proline. Australian Journal of Crop Science 2016; 10:50–56.

[80] Deivanai S, Xavier R, Vinod V, Timalata K, Lim OF. Role of exogenous proline in ameliorating salt stress at early stage in two rice cultivars. Journal of Stress Physiology and Biochemistry 2011; 7:157–174.

[81] Theerakulpisut P, Gunnula W. Exogenous sorbitol and trehalose mitigated salt stress damage in salt-sensitive but not salt-tolerant rice seedlings. Asian Journal of Crop Science 2012; 4:165–170.

[82] Ryu H, Cho Y. Plant hormones in salt stress tolerance. Journal of Plant Biology 2015; 58:147–155.

[83] Li X, Yang M, Chen H, Qu L, Chen F, Shen S. Abscisic acid pretreatment enhances salt tolerance of rice seedlings: proteomic evidence. Biochimica et Biophysica Acta 2010; 1804: 929–940.

[84] Gurmani AR, Bano A, Khan SU, Din J, Zhang JL. Alleviation of salt stress by seed treatment with abscisic acid (ABA), 6-benzylaminopurine (BA) and chlormequat chloride (CCC) optimizes ion and organic matter accumulation and increases yield of rice (*Oryza sativa* L.). Australian Journal of Crop Science 2011; 5:1278–1285.

[85] Javid MG, Sorooshzadeh A, Sanavy SAMM, Allahdadi I, Moradi F. Effects of the exogenous application of auxin and cytokinin on carbohydrate accumulation in grains of rice under salt stress. Plant Growth Regulation 2011; 65:305–313.

[86] Kim SK, Son TK, Park SY, Lee IJ, Lee BH, Kim HY, Lee SC. Influences of gibberellin and auxin on endogenous plant hormone and starch mobilization during rice seed germination under salt stress. Journal of Environmental Biology 2006; 27(2): 181–186.

[87] Sharma I, Ching E, Saini S, Bhardwaj R, Pati PK. Exogenous application of brassinosteroid offers tolerance to salinity by altering stress responses in rice variety Pusa Basmati-1. Plant Physiology and Biochemistry 2013; 69:17–26.

[88] Anuradha S, Rao SSR. Effect of brassinosteroids on salinity stress induced inhibition of seed germination and seedling growth of rice (*Oryza sativa* L.). Plant Growth Regulation 2001; 33:151–153.

[89] Kang DJ, Seo YJ, Lee JD, Ishii R, Kim KU, Shin DH, Park SK, Jang SW, Lee IJ. Jasmonic acid differentially affects growth, ion uptake and abscisic acid concentration in salt-tolerant and salt-sensitive rice cultivars. Journal of Agronomy & Crop Science 2005; 191: 273–282.

[90] Misratia KM, Ismail MR, Oad FC, Hanafi MM, Puteh A. effect of salinity and alleviating role of gibberellic acid (GA_3) for enhancement of rice yield. International Journal of Chemical, Environmental & Biological. 2013; 1(2):330–334.

[91] Bai X, Yang L, Tian M, Chen J, Shi J, Yang Y, Hu X. Nitric oxide enhances desiccation tolerance of recalcitrant *Antiaris toxicaria* seeds via protein *S*-nitrosylation and carbonylation. PLoS ONE 2011; 6:e20714.

[92] Hasanuzzaman M, Gill SS, Fujita M. Physiological role of nitric oxide in plants grown under adverse environmental conditions. In: Tuteja N, Gill SS (eds) Plant acclimation to environmental stress. New York: Springer; 2013. pp. 269–322. doi:10.1007/978-1-4614-5001-6_11

[93] Habib N, Ashraf M, Ahmad MSA. Enhancement in seed germinability of rice (*Oryza sativa* L.) by pre-sowing seed treatment with nitric oxide (NO) under salt stress. Pakistan Journal of Botany 2010; 42:4071–4078.

[94] Uchida A, Jagendorf AT, Hibino T, Takabe T. Effects of hydrogen peroxide and nitric oxide on both salt and heat stress tolerance in rice. Plant Science 2002; 163:515–523.

[95] Semida WM. Hydrogen peroxide alleviates salt-stress in two onion (*Allium cepa*L.) cultivars. American-Eurasian Journal of Agricultural and Environmental Sciences 2016; 16:294–301.

[96] Petrov VD, Van Breusegem F. Hydrogen peroxide—a central hub for information flow in plant cells. AoB Plants 2012; pls014. doi:10.1093/aobpla/pls014

[97] Wang X, Hou C, Liu J. Hydrogen peroxide is involved in the regulation of rice (*Oryza sativa* L.) tolerance to salt stress. Acta Physiologiae Plantarum 2013; 35:891.

[98] Habib N, Ashraf M. Nitric oxide regulated improvement in growth, antioxidant defense system and yield of rice plants grown under salinity. Journal of Plant Pathology & Microbiology 2016. doi:10.4172/2157-7471.C1.003

[99] Montilla-Bascon G, Rubialesa D, Altabellabc T, Prats E. Free polyamine and polyamine regulation during pre-penetration and penetration resistance events in oat against crown rust (*Puccinia coronata* f. sp. *avenae*). Plant Pathology 2016; 65:392–401.

[100] Kusano T, Berberich T, Tateda C, Takahashi Y. Polyamines: essential factors for growth and survival. Planta 2008; 228:367–381.

[101] Kubiś J, Floryszak-Wieczorek J, Arasimowicz-Jelonek M. Polyamines induce adaptive responses in water deficit stressed cucumber roots. Journal of Plant Research 2014; 127:151–158.

[102] Dobrovinskaya OR, Muñiz J, Pottosin II. Inhibition of vacuolar ion channels by poly amines. The Journal of Membrane Biology 1999; 167:127–140.

[103] Shabala S, Cuin TA, Pottosin I. Polyamines prevent NaCl-induced K^+ efflux from pea mesophyll by blocking non-selective cation channels. FEBS Letters 2007; 581:1993–1999.

[104] Zhao F, Song CP, He J, Zhu H. Polyamines improve K^+/Na^+ homeostasis in barley seedlings by regulating root ion channel activities. Plant Physiology 2007; 145:1061–1072.

[105] Ndayiragije A, Lutts S. Do exogenous polyamines have an impact on the response of a salt-sensitive rice cultivar to NaCl? Journal of Plant Physiology 2006; 163:506–516.

[106] Ndayiragije A, Lutts S. Exogenous putrescine reduces sodium and chloride accumulation in NaCl-treated calli of the salt-sensitive rice cultivar I Kong Pao. Plant Growth Regulation 2006; 48:51–63.

[107] Chattopadhayay MK, Tiwari BS, Chattopadhyay G, Bose A, Sengupta DN, Ghosh B. Protective role of exogenous polyamines on salinity-stressed rice (*Oryza sativa*) plants. Physiologia Plantarum 2002; 116:192–199.

[108] Quinet M, Alexis N, Isabella L, Beatrice L, Christine C, Dupont-G SL. Putrescine differentially influences the effects of salt stress on polyamine metabolism and ethylene synthesis in rice cultivars differing salt resistance. Journal of Experimental Botany 2010; 61:2719–2733.

[109] Saleethong P, Sanitchon J, Knog-ngern K, Theerakulpisut P. Pretreatment with spermidine reverse inhibitory effects of salt stress in two rice (*Oryza sativa* L.) cultivars differing in their tolerance. Asian Journal of Plant Science 2011; 10:245–254.

[110] Paweena S, Jirawat S, Kanlaya K, Piyada T. Effects of exogenous spermidine (Spd) on yield, yield-related parameters and mineral composition of rice (*Oryza sativa* L. ssp. indica) grains under salt stress. Australian Journal of Crop Science 2013; 7:1293–1301.

[111] Chunthaburee S, Sanitchon J, Pattanagul W, Theerakulpisut P. Alleviation of salt stress in seedlings of black glutinous rice by seed priming with spermidine and gibberellic acid. Notulae Botanicae Horti Agrobotanici Cluj-Napoca 2014; 42:405–413.

[112] Roychoudhury A, Basu S, Sengupta D. Amelioration of salinity stress by exogenously applied spermidine or spermine in three varieties of indica rice differing in their level of salt tolerance. Journal of Plant Physiology 2011; 168:317–328.

[113] Hasanuzzaman M, Nahar K, Fujita M. Silicon and selenium: two vital trace elements in conferring abiotic stress tolerance to plants. In: Parvaiz A, Rasool S (eds) Emerging

technologies and management of crop stress tolerance, vol. 1–Biological techniques. New York: Academic Press; 2014. pp. 375–420.

[114] Iqbal M, Aslam M. Effect of Zn application on rice growth under saline condition. International Journal of Agriculture and Biology 1999; 1:362–365.

[115] Mehmood EUH, Kausar R, Akram M, Shahzad SM. Is boron required to improve rice growth and yield in saline environment? Pakistan Journal of Botany 2009; 41: 1339–1350.

[116] Zayed BA, Salem AKM, El Sharkawy HM. Effect of different micronutrient treatments on rice (*Oryza sativa* L.) growth and yield under saline soil conditions. World Journal of Agricultural Sciences 2011; 7:179–184.

[117] Farooq MA, Saqib ZA, Akhtar J. Silicon-mediated oxidative stress tolerance and genetic variability in rice (*Oryza sativa* L.) grown under combined stress of salinity and boron toxicity. Turkish Journal of Agriculture and Forestry 2015; 39:718–729.

[118] Farooq MA, Saqib ZA, Akhtar J, Bakhat HF, Pasala R-K, Dietz K-J. Protective role of silicon (Si) against combined stress of salinity and boron (B) toxicity by improving antioxidant enzymes activity in rice. Silicon 2015. doi:10.1007/s12633-015-9346-z

[119] Mahdieh MN, Habibollahi MR, Amirjani MH, Abnosi M, Ghorbanpour. Exogenous silicon nutrition ameliorates salt-induced stress by improving growth and efficiency of PSII in *Oryza sativa* L. cultivars. Journal of Soil Science and Plant Nutrition 2015; 15:1050–1060.

[120] Naim A. Mitigation of salt stress in rice by exogenous application of selenium. M.S. Thesis, Department of Agronomy, Sher-e-Bangla Agricultural University, Dhaka, Bangladesh.

[121] Venkatesh J, Park SW. Role of L-ascorbate in alleviating abiotic stresses in crop plants. Botanical Studies 2014; 55:38.

[122] Wang R, Liu S, Zhou F, Ding, Hua C. Exogenous ascorbic acid and glutathione alleviate oxidative stress induced by salt stress in the chloroplasts of *Oryza sativa* L. Zeitschrift für Naturforschung C 2014; 69:226–236.

[123] Alhasnawi AN, Che Radziah CMZ, Kadhimi AA, Isahak A, Mohamad A, Yusoff WMW. Enhancement of antioxidant enzymes activities in rice callus by ascorbic acid under salinity stress. Biologia Plantarum 2016. doi:10.1007/s10535-016-0603-9

[124] Alhasnawi AN, Kadhimi AA, Isahak A, Mohamad A, Yusoff WMW, Che Radziah CMZ. Exogenous application of ascorbic acid ameliorates detrimental effects of salt stress in rice (MRQ74 and MR269) seedlings. Asian Journal of Crop Science 2015; 7:186–196.

[125] Barus WA, Rauf A, Rosmayati, Hanum C. Improvement of salt tolerance in some varieties of rice by ascorbic acid application. International Journal of Scientific & Technology Research 2015; 4:235–237.

[126] Ozfidan-Konakci C, Yildiztugay E, Kucukoduk M. Protective roles of exogenously applied gallic acid in *Oryza sativa* subjected to salt and osmotic stresses: effects on the total antioxidant capacity. Plant Growth Regulation 2015; 75(1):219–234.

[127] Sobahan MA, Arias CR, Okuma E, Shimoishi Y, Nakamura Y, Hirai Y, Mori IC, Murata Y. Exogenous proline and glycinebetaine suppress apoplastic flow to reduce Na^+ uptake in rice seedlings. Bioscience Biotechnology and Biochemistry 2009; 73:2037–2042.

[128] Rahman MS, Miyake H, Takeoka Y. Effect of exogenous glycinebetaine on growth and ultrastructure of salt-stressed rice seedlings (*Oryza sativa* L.). Plant Production Science 2002; 5:33–44.

[129] Kim YH, Khan AL, Waqas M, Shim JK, Kim DH, Lee KY, Lee IJ. Silicon application to rice root zone influenced the phyto hormonal and antioxidant responses under salinity stress. Journal of Plant Growth Regulation 2014; 33:137–149. doi:10.1007/s00344-013-9356-2

[130] Shi Y, Wanga Y, Flowers TJ, Gonga H. Silicon decreases chloride transport in rice (*Oryza sativa* L.) in saline conditions. Journal of Plant Physiology 2013; 170:847–853.

[131] Saeedipour S. Salinity tolerance of rice lines related to endogenous abscisic acid (ABA) level synthesis under stress. African Journal of Plant Science 2011; 5:628–633.

Chapter 10

Genetics and Genomics of Bacterial Blight Resistance in Rice

Yogesh Vikal and Dharminder Bhatia

Additional information is available at the end of the chapter

http://dx.doi.org/10.5772/67361

Abstract

Rice is an important food crop for half the world's population and has been in cultivation for over 10,000 years. During the last few decades, rice has evolved intricate relationships with associated pathogens and pests, bacterial blight (BB) being one of the most important among them. Utilization of resistant varieties with agricultural management practices is a more effective way to control BB. Of the 42 different resistance (*R*) genes identified to confer BB resistance, 9 have been isolated and cloned, whereas a few of the avirulence genes and a large number of candidate pathogenicity genes have been isolated from *Xanthomonas oryzae* pv. *oryzae*. The complete genome sequences of two different rice subspecies *japonica* and *indica* and three different races of BB pathogen are available. Therefore, the interaction between rice-*Xoo* could be deciphered and pave a way to study the molecular aspects of bacterial pathogenesis and host counter measures like innate immunity and R gene–mediated immunity. Although several of the type III effectors of *Xoo* have been characterized and the host targets of a few of them identified, a relatively large number of candidate effectors remain to be studied and their functional analysis may provide key for developing broad spectrum and durable resistance to BB.

Keywords: *Xanthomonas oryzae* pv. *oryzae*, *Oryza sativa*, Genome structure, *Xa* genes, mapping

1. Introduction

Rice (*Oryza sativa* L.) is a staple food for a large part of the world's human population, especially in South and Southeast Asia and tropical Latin America, making it the second most consumed cereal grain. It accounts for 35–60% of the calories consumed by more than 3 billion Asians. To meet the growing demand of nearly 5.0 billion consumers, rice-growing countries will have to produce 40% more rice by 2030 [1]. However, its production is being reduced severely by several rice diseases. Bacterial blight (BB) caused by *Xanthomonas oryzae* pv. *oryzae* (*Xoo*) is one of the major foliar diseases, causing yield losses by 20–30%. It was first reported by farmers in 1884 in the Southern region of Japan [2]. It is a vascular disease resulting in tannish grey to white lesions along the veins. The BB pathogen invades the host through natural openings in leaves, including hydathodes or wounds, and colonizes the xylem vessels. Under field conditions, symptoms are onset at the tillering stage and disease incidence increases with plant growth, which is maximum at the flowering stage. The kresek, a severe form of disease, occurs at the seedling stage resulting in a partial or total crop failure.

The incidence of BB has been widely reported in all rice-growing regions worldwide except North America [3]. The high degree of pathogenic variation in *Xoo* due to the evolution of new pathotypes often causes the breakdown of resistance. Genome structure of *Xoo* revealed the presence of effector (*avr*) genes and insertion sequences, which may be playing a major role in generating a high degree of genetic diversity and race differentiation [4]. However, with the evolution of new races, broader and more comprehensive resistance must be pursued in order to prevent super-races of *Xoo*, which can overcome all genetic sources of resistance, from becoming prevalent. Therefore, it is of immense interest to diversify the germplasm, scouting of resistance genes from wild species and pyramiding two or more effective resistance genes in developing rice cultivars with durable BB resistance to *Xoo*. For successful deployment of stable resistance (R) genes, their characterization and availability of tightly linked markers will greatly facilitate the development of new versions of cultivars. To date, more than 40 different resistance genes conferring host resistance to BB have been identified in rice so far and some of those have been characterized [5].

The rice-*Xoo* system is distinguished from *Arabidopsis—Pseudomonas syringae*, tomato—*Cladosporium fulvum*, and rice—*Magnaporthe grisea*, with respect to that about one-third of the *R* genes for *Xoo* resistance are recessive in nature [6]. Secondly, diverse types of proteins are encoded by the *R* genes such as *Xa3*/*Xa26* and *Xa21* encode leucine-rich repeat (LRR) receptor kinase-type proteins [7–9] that mediate race-specific resistance. *Xa1* encodes a nucleotide-binding LRR protein [10]. The recessive gene *xa5* encodes the γ-subunit of transcription factor IIA [11, 12]. *Xa27* encodes a novel protein [13]. The recessive gene *xa13* encodes a novel plasma membrane protein [14]. Thus, it is clear that *Xoo* has the diverse interaction mechanism within the host species. Hence, the isolation and characterization of BB resistance genes from diverse germplasm will insight depth knowledge of both pathogen and host system, which will further lead to the development of BB resistant varieties with broad spectrum and durable resistance.

2. Genetics of BB resistance

Genetic analysis of many plant-pathogen interactions has demonstrated that plants often contain a single locus that confers resistance against a complementary avirulence gene [15]. The genetic basis of host resistance to BB has been studied in depth. The genetics of resistance to bacterial blight was first carried out by Japan and IRRI, subsequently, followed by Sri Lanka, India, China, and so on. As there is diversity of *Xoo* strains in different countries, scientists found that it was difficult to characterize and distinguish the resistance genes. In order to compare the identified genes, the identical differential standard was set up [16]. About 42 BB resistance (R) genes, designated from *Xa1* to *Xa42* [5, 17, 18], conferring resistance against various strains of *Xoo*, have been identified from cultivated, mutant population, and wild rice species (**Table 1**). These genes have been mapped on six of the twelve rice chromosomes. A total of 14 recessive genes (*xa5, xa8, xa13, xa15, xa19, xa20, xa24, xa25, xa26b, xa28, xa31, xa32, xa33,* and *xa34*) in the series from *Xa1* to *Xa42* have been identified. Of these *R* genes, nine of the *R* genes have been cloned and characterized (*Xa1, Xa3/Xa26, xa5, xa13, Xa10, Xa21, Xa23, xa25,* and *Xa27*) encoding different types of proteins, suggesting multiple mechanisms of *R*-gene–mediated *Xoo* resistance [7, 8, 10, 11, 13, 14, 19–21]. Most of these genes provide complete and race-specific resistance to *Xoo* and have been used singly or in combination in rice breeding for BB resistance [22–32]. Since the bacterial races tend to evolve continually, influenced by the artificial and natural selection of genes resistant to BB, it is required to explore critically the new resistant resources to combat the evolved races.

The *Xa1* gene was identified by Sakaguchi [33] from rice cultivars Kogyoku and Java14. It conferred a high level of specific resistance to Japanese *Xoo* race 1. Since then, *Xa1* has been extensively used in Japanese rice breeding as *Xoo* race 1 is the most prevalent in Japan. Later on, a high-resolution genetic map was developed for *Xa1* using F_2 population and RFLP markers [34]. Three markers (XNpb235, XNpb264, and C600) on chromosome 4 were found to be tightly linked to *Xa1*, while another RFLP marker $U08_{750}$ was at 1.5 cM distance from *Xa1*. Screening of the YAC library with three linked markers resulted in identification of ten contiguous YAC clones. Out of these clones, Y5212 clone hybridized with all the three markers and was confirmed to possess the *Xa1* locus. Sakaguchi [33] also identified another BB resistance gene, *Xa2*, in rice cultivar Tetep. It conferred specific resistance to T7147 (Japanese *Xoo* race 2). Earlier, *Xa2* was mapped on chromosome 4, linked to *Xa1* with a recombination frequency of 2–16% [35]. He et al. [36] reported the fine mapping of *Xa2* using an F_2 population of ZZA (Zhengzhui Ai) X IRBB2 and *Xa*2 was localized between two markers spanning approximately 190 kb region. Sequence analysis of this region revealed the presence of a homologous sequence of leucine-rich repeat (LRR) kinase, which is the product of a cloned BB resistance gene, *Xa*21 [7].

The *Xa3* gene was identified in a *japonica* variety Wase Aikoku 3 [37]. It was mapped to the long arm of chromosome 11 and was found to be tightly linked to another BB resistance gene *Xa*4 [38, 39]. *Xa3* was later known to be genetically linked to *Xa26*, another BB resistance gene. For characterization of *Xa3*, the gene was fine mapped using a population segregating for only a single resistance gene and markers developed from *Xa26* gene family. Genetic analysis showed that *Xa3* co-segregated with the *Xa26* gene marker but segregated from the markers of other members of *Xa26* gene family.

DNA fingerprinting also revealed that IRBB3 carrying *Xa3* had the same copy numbers of *Xa26* gene family members as were present in Minghui 63 carrying *Xa26*. The putative coding products of *Xa3/Xa26* and the susceptible allele *xa3/xa26* shared 92% homology with difference in the LRR domains, proving that *Xa3* is the same as *Xa26* [9].

Xa gene	Resistance to Xoo race	Donor cultivar	Chromosome	Reference
Xa1	Japanese race-I	Kogyoku, IRBB1	4	[10, 33, 34]
Xa2	Japanese race-II	IRBB2	4	[33, 36]
Xa3/Xa26	Chinese, Philippine, and Japanese races	Wase Aikoku 3, Minghui 63, IRBB3	11	[9, 37]
Xa4	Philippine race-I	TKM6, IRBB4	11	[39, 40, 45]
xa5	Philippine races I, II, III	IRBB5	5	[11, 40, 46]
Xa6	Philippine race 1	Zenith	11	[47]
Xa7	Philippine races	DZ78	6	[48–51]
xa8	Philippine races	PI231128	7	[52]
xa9	Philippine races	Khao Lay Nhay and Sateng	11	[53]
Xa10	Philippine and Japanese races	Cas 209	11	[54, 55, 57]
Xa11	Japanese races IB, II, IIIA, V	IR8	3	[58, 59]
Xa12	Indonesian race V	Kogyoku, Java14	4	[60]
xa13	Philippine race 6	BJ1, IRBB13	8	[14, 61–64]
Xa14	Philippine race 5	TN1	4	[65–67]
xa15	Japanese races	M41 mutant	–	[69]
Xa16	Japanese races	Tetep	–	[70]
Xa17	Japanese races	Asominori	–	[71]
Xa18	Burmese races	IR24, Miyang23, Toyonishiki	–	[58]
xa19	Japanese races	XM5 (mutant of IR24)	–	[72]
xa20	Japanese races	XM6 (mutant of IR24)	–	[73]
Xa21	Philippine and Japanese races	*O. longistaminata*, IRBB21	11	[7, 76, 77]
Xa22	Chinese races	Zhachanglong	11	[79, 80]
Xa23	Indonesian races	*O. rufipogon* (CBB23)	11	[24]
xa24(t)	Philippine and Chinese races	DV86	2	[81, 82]
xa25/ Xa25(t) Xa25	Chinese and Philippine races	Minghui 63, HX-3 (somaclonal mutant of Minghui 63)	12	[85 86]
xa26(t)	Philippine races	Nep Bha Bong	–	[86]
Xa27	Chinese strains and Philippine race 2 to 6	*O. minuta* IRGC 101141, IRBB27	6	[26, 85]
xa28(t)	Philippine race2	Lota sail	–	[86]
Xa29(t)	Chinese races	*O. officinalis*(B5)	1	[87]
Xa30(t)	Indonesian races	*O. rufipogon* (Y238)	11	[88]

http://dx.doi.org/10.5772/67361

Xa gene	Resistance to Xoo race	Donor cultivar	Chromosome	Reference
xa31(t)	Chinese races	Zhachanglong	4	[89]
Xa32(t)	Philippine races	*Oryza australiensis* (introgression line C4064)	11	[90]
xa33(t) *Xa33(t)*	Thai races	Ba7 *O. nivara*	6	[91] [92]
Xa34(t) *Xa34(t)*	Thai races	Pin Kaset *O. brachyantha*	–	[91] [93]
Xa35(t)	Philippine races	*Oryza minuta* (Acc. No.101133)	11	[94]
Xa36(t)	Philippine races	C4059	–	[95]
Xa38	Indian Punjab races	*O. nivara* IRGC81825	–	[17, 96]
Xa39	Chinese and Philippines races	FF329	11	[97]
Xa40(t)	Korean BB races	IR65482-7-216-1-2	11	[5]
xa41(t)	Various *Xoo* strains	Rice germplasm	–	[98]
xa42	Japanese *Xoo* races	XM14, a mutant of IR24	3	[18]

Revised and updated from Ref. [147].

Table 1. Summary of resistance genes to bacterial blight in rice.

The *Xa4* and *xa5* genes were identified by Petpisit et al. [40]. The dominant gene *Xa4* confers resistance to Philippine race 1 of *Xoo* and was identified in rice cultivars TKM6, IR20, IR22, and IR72. The recessive gene *xa5* confers resistance to Philippine *Xoo* races 1, 2, 3, and 5 [41]. Yoshimura et al. [39] reported the mapping of *Xa4* and mapped an RFLP marker G181 on one flanking side of *Xa4* locus on chromosome 11. However, no marker was identified on the other flanking side of *Xa4* locus. The *Xa4* locus was further mapped between two RFLP markers, RZ536 and L457 [41]. Wang et al. [42] localized the *Xa4* locus between G181 and L1044 at a distance of 4.4 and 3.8 cM, respectively. However, using other rice molecular linkage maps as references [43, 44], Sun et al. [45] deduced that *Xa4* locus reported earlier [41, 42] was not in the same position. Using an F_2 population from IR24 X IRBB4 cross, *Xa4* gene was genetically mapped to a region less than 1 cM. A contig map was constructed consisting of six nonredundant BAC clones, spanning approximately 500 kb length. Analysis of recombination events located *Xa4* locus to one BAC "3H8" and assay of recombinants using subclones of "3H8" along with sequence analysis, further narrowed the *Xa4* locus to a 47 kb fragment.

A high-resolution genetic map of the chromosomal region harboring *xa5* was developed and mapped to a 0.5 cM interval between markers RS7 and RM611, which spanned an interval of 70 kb and contained 11 open reading frames. The potential candidate gene products of *xa5* may be basal transcription factor IIA (TFIIA), an ABC transporter, a tRNA synthase, a MAP kinase and a cysteine protease, as well as four unknown, hypothetical or putative proteins [46].

While studying the inheritance of resistance to *Xoo* isolate PXO61 in five rice cultivars, Sidhu and Khush [47] observed that the same gene conferred resistance in all the cultivars. The pattern of segregation indicated a monogenic recessive factor when plants were inoculated at booting stage, but monogenic dominant when plants were inoculated during flowering. This reversal of dominance was caused by dosage effect of the resistance gene at a specific stage of growth of the plant when exposed to the bacterial inoculum. Apparently, the gene could express itself at booting stage only when present in two dosages. Heterozygous plants with one dose of resistance allele were susceptible during booting. However, at flowering, one dose was enough for expression of resistance. The new gene identified as *Xa6* and was linked to *Xa4* with a crossover value of 26%. Later studies on the genetics of resistance to *Xoo* in 74 cultivars of *O. sativa* using PXO61 isolate from the Philippines identified an additional dominant gene, *Xa7*, that conferred resistance in DZ78 [48]. A recombination value of 8.8% between *Xa7* and G1091 located at 107.5 cM on chromosome 6 was determined [49]. Porter et al. [50] performed the AFLP analysis of a segregating, near-isogenic F_3 population of IR24 X IRBB7, which revealed one polymorphic fragment, M1, mapped to position 107.3 cM on the rice map. Sequence comparisons of resistant and susceptible lines near M1 were used to generate additional markers linked to the rice BB resistance gene *Xa7*. A sequence-tagged site (STS) named M2 was mapped proximal to M1 and farther from *Xa7*, indicating that *Xa7* lay distal to M1. Two SSRs, M3 and M4, were mapped at a distance of 0.5 and 1.8 cM from *Xa7*, respectively. The pattern of recombinants and the map distances indicated that *Xa7* was located in the region corresponding to the ends of physically mapped Clemson University Genomics Institute (CUGI) BACs 96 and 143. A complex repeat was identified in the DNA sequence from *O. sativa* cultivars 93-11 and Nipponbare that matched the end of contig 96 and an EST marker (C52865S). Amplification of the repeat revealed the presence and absence of the repeat in IR24 and IRBB7, respectively. However, no recombinants were identified between *Xa7* and the polymorphic repeat (M5). Comparison of the physical and genetic maps of rice in that region indicated that *Xa7* lay within 40 kb of M5, a distance suitable for MAS and cloning of *Xa7*. A high-resolution genetic map of the chromosomal region surrounding *Xa7* gene was constructed using SSR markers from the Gramene database (www.gramene.org/db) based on the *Xa7* gene initial mapping on chromosome 6 [51]. Primary analysis of F_2 population from the cross of IRBB7 and IR24, located *Xa7* in approximately 0.28 cM region. To walk closer to the target gene, recombinant F_2 individuals were tested using STMS markers and the gene was finally mapped to a 0.21 cM region between markers GDSSR02 and RM20593. A contig map corresponding to *Xa7* gene was constructed using reference sequence of cultivar Nipponbare through bioinformatics analysis. The target gene was assumed to span an interval of approximately 118.5 kb, containing a total of fourteen genes. Candidate gene analysis showed that the fourteen genes encoded novel domains that had no amino acid sequence similar to other cloned *Xa* (*xa*) genes.

A single recessive gene, *xa8*, conditioned resistance in rice germplasm accession PI 231129. Vikal et al. [52] mapped the gene between two consecutive SSR markers, RM21044 and RM21045 at 7.0 and 9.9 cM, respectively, on chromosome 7 and the physical distance between markers was 9.5 kb. This region harbors three intact genes, which codes for putative expressed proteins *viz.*, Zinc finger A20 and AN1 domain-containing protein (LOC_Os07g07400), oxidoreductase, 2OG-Fe oxygenase family protein (LOC_Os07g07410), and Gibberellin 20 oxidase 1-B protein (LOC_Os07g07420). All the three candidate genes are responsible for plant response to various stresses.

The BB resistance in 'Khao Lay Nhay' and 'Sateng' was recessive in nature and were allelic to each other but nonallelic to and segregated independently of *Xa3, Xa4, xa5, Xa7*, and *xa8* [53]. This new recessive gene was linked to *Xa6* with a crossover value of 5.9%. This new gene for resistance was designated as *xa9*. The *Xa10* was identified from rice cultivar Cas 209 [54, 55]. There is an extensive polymorphism on chromosome 11L between IRBB10 and IR24 and also a great distortion in *Xa10* inheritance. Rice lines that carry *Xa10* gene confer race-specific resistance to *Xoo* strains harboring avirulence (Avr) gene *avrXa10*. The *Xa10* locus was roughly integrated to a large region between RFLP markers RG103 (~83 cM) and RG1109 (~91.4 cM) on the long arm of rice chromosome 11 [56]. High-resolution mapping by Gu et al. [57] narrowed down *Xa10* to a 74 kb region between RFLP markers M491 and M419. Out of the 7 identified genes in this region, 6 are considered possible candidate genes for *Xa10*. Ogawa and Yamamoto [58] identified another BB resistance gene *Xa11*. *Xa11* confers specific resistance to strains T7156, T7147, T7133, and H75304 (Japanese *Xoo* races IB, II, IIIA, and V, respectively). *Xa11* was mapped between the loci of RM347 (2.0 cM) and KUX11 (1.0 cM) on the long arm of chromosome 3 [59]. It was reported that a single dominant gene, *Xa12*, controlled the resistance in Kogyoku and Java14 to BB race V of Indonesia [60].

The *xa13* gene was first discovered in the rice variety BJ1 and this gene specifically confers resistance to the Philippine *Xoo* race 6, the most virulent race and one not overcome by most reported *R* genes [61–63]. Sanchez et al. [63] reported the genetic and physical mapping of *xa13* gene using two F_2 populations of the cross between IR24 and IRBB13. The gene was fine mapped to a region less than 4 cM on the long arm of chromosome 8 between two RFLP markers, RG136 and R2027. Sequence analysis of the 14.8 kb fragment carrying *xa13* gene indicated only two apparently intact candidate genes—an extensin-like gene and a homologue of nodulin (MtN3) and the 5′ end of a predicted hypothetical protein [14]. The expression of the rice gene *Os8N3*, a member of the MtN3 gene family from plants and animals, was elevated upon infection by *Xoo* strain PXO99A and depended on type III effector gene PbXo-1. *Os8N3* resides near *xa13* and could not be induced in rice lines with *Xa13*. Inhibitory RNA silencing of *Os8N3* produced PX099A resistant plants [64].

Xa14 identified from Taichung Native 1 is a dominant gene against resistance to PXO112, BB race 5 of the Philippines [65]. *Xa14* was located on chromosome 4 between marker RG620 and G282, with 20.1 and 19.1 cM, respectively by Tan et al. [66]. The gene was fine mapped by using two F_2 populations. Combining recombination frequencies for the two populations together, the gene *Xa14* was mapped to 3 BAC clones spanned approximately 300 kb in length between SSR markers HZR970-8 and HZR988-1 [67].

A series of nine genes (*xa15, Xa16, Xa17, Xa18, xa19, xa20, xa26, Xa27*, and *xa28*) have been obtained from mutagenesis which had different resistance levels and resistance spectrums [58, 68–73]. A mutant rice line, XM5, was resistant to bacterial blight [72]. Allelic relationship studies with the known recessive genes, *xa5*, *xa8*, and *xa13* revealed that the gene was not allelic to them and was a new gene. It was designated as *xa19*.

Devadath [74] identified a strain of *Oryza barthii* which was resistant to all the races of *Xoo* in India. Khush et al. [75] multiplied the strain at IRRI and studied its resistance to Philippine *Xoo* races. It was found that the strain was akin to *Oryza longistaminata* which was resistant to all the six Philippine *Xoo* races [76]. F_1s of the cross of the strain with IR24 were resistant showing that the resistance was conferred by a dominant gene. The gene was different from the known BB resistance genes and was designated as *Xa21*(t). Genetic and physical analysis of *Xa21* localized it in an 8.3 cM interval on chromosome 11 and the physical size of the region containing *Xa21* gene was estimated to be about 800 kb [77].

Genomic DNA gel-blot analysis revealed that *Xa21* belonged to a multigene family containing at least eight members. Most of these members were mapped to a single locus on chromosome 11, which is linked to at least nine other major resistance genes and one QTL for resistance [7, 77]. Wang et al. [78] studied the inheritance patterns and resistance spectra of transgenic plants carrying six *Xa21* gene family members and observed that one member, designated *Xa21D*, conferred only partial resistance. Analysis of *Xa21D* sequence showed that *Xa21D* transcript terminated shortly after the stop codon introduced by retrotransposon. Sequence comparison of *Xa21* and *Xa21D* provided evidence of adaptive selection. *Xa21D* encodes a receptor-like protein carrying LRR motifs in the presumed extracellular domain. Both functional and evolutionary evidence indicated that *Xa21D* LRR domain controlled race-specific pathogen interaction.

Another BB resistance gene, *Xa22* was identified from landrace Zhachanglong (ZCL) of Yunnan province in Southwest China. It showed high level of resistance to 16 of the 17 BB strains tested [79]. The gene was nonallelic to *Xa1*, *Xa2*, *Xa4*, and *Xa14* but was found to be linked to *Xa4*. *Xa22* gene was mapped on short arm of chromosome 11, but according to new molecular map developed by Harushima et al. [43], it was present on the long arm. *Xa22* provides resistance to a broad spectrum of *Xoo* isolates. The gene was localized to a small 100 kb fragment delimited by R1506 and a subclone from M3H8 BAC clone by a combination of genetic recombination analysis and physical mapping [80].

An interspecific cross was made between RBB16 and Jiagang30 (JG30), and the F_1 plants showing highest resistance to BB were anther cultured to obtain doubled haploids for resistance. These anther-cultured progenies were inoculated with three strains of *Xoo*, and lines showing high resistance and improved plant type in H_2 generation were advanced to H_4 and collectively designated as WBB1. Zhang et al. [24] crossed WBB1 with two susceptible Chinese varieties, JG30 (*indica*) and 02428 (*japonica*), and IRBB21 (*Xa21*). Inoculation results with PXO99 at seedling stage suggested that a single dominant gene from *Oryza rufipogon* conferred resistance and the gene was nonallelic but probably linked to *Xa21*. PCR analysis of WBB1 with *Xa21* specific primers also indicated that the gene was different from *Xa21*. The new gene, designated as *Xa23*(t), was mapped within a 0.4 cM region between markers Lj138 and A83B4, and the corresponding physical distance between these markers is 49.8 kb. Six *Xa23* candidate genes have been annotated, including four candidate genes encoding hypothetical proteins, and the other two encoding a putative ADP-ribosylation factor protein and a putative PPR protein. Nine varieties (AC19-1-1, Aus 274, Chinsurah Boro II, Kalimekri 77-5, Tepa I, Long grain, Aus 295, DV85, and DV86) were crossed with BJ1 (carrying *xa13*) to confirm whether the resistance was due to *xa13* or a new gene. The results indicated that the recessive gene identified in DV86 for resistance to race 6 was nonallelic to and independent of *xa13*. DV85, DV86, and Aus 295 had the same recessive resistance gene and it was designated as *xa24*(t) [81].

The *xa24* resistance gene was fine mapped to a 71 kb DNA fragment in the long arm of chromosome 2 using PCR-based markers. It mediates resistance to the Philippine *Xoo* races 4, 6, and 10, and Chinese *Xoo* strains Zhe 173, JL691, and KS-1-21. The analysis of F_2 mapping population, developed from the cross between a resistance line DV86 (*Oryza sativa* ssp. *indica*) carrying the target gene and a sensitive line IR24 (*O. sativa* ssp. *indica*), first mapped the gene to terminal region of long arm of chromosome 2 at 8.0 and 0.9 cM from RM482 and RM138, respectively. Further, fine mapping placed the *xa24* gene between RM14222 and RM 14226 at a distance of 0.07 cM from both markers [82].

Xa25(t), a new dominant resistance gene, was identified in Minghui 63, a restorer line for a number of rice hybrids cultivated widely in China [83]. The gene conferred resistance to Philippine race 9 (PXO330) at both seedling and adult plant stages. Gao et al. [84] identified another dominant gene in a somaclonal mutant HX-3 providing resistance to BB strain Zhe173 and was also designated as *Xa25*(t). Bulked segregant analysis carried out on the DH population located the gene at the terminal region of the long arm of chromosome 4 between the two SSR markers RM6748 and RM1153 at a map distance of 9.3 and 3.0 cM, respectively.

In order to broaden the germplasm pool for breeding of disease resistance to *Xoo*, a wide hybridization project to transfer *R* genes from wild species *Oryza minuta* was initiated at IRRI in the late 1980s. A novel resistance (*R*) locus was identified from a progeny of interspecific hybrids of *O. sativa* cv. IR31917-45-3-2 and *O. minuta* acc. 101141 and was tentatively designated as *Xa27*(t) [85]. The genetic analysis of 21 rice cultivars for BB resistance divided the cultivars into two groups (group 1 and group 2) based on their resistance to Philippine *Xoo* races 1 and 2. Group 1 was resistant to race 1 and group 2 to race 2. When crossed with TN1 and evaluated for resistance with races 1 and 2, all cultivars showed monogenic inheritance of resistance. Allelic relationships of the genes were investigated by crossing these cultivars with different testers having single genes for resistance. Out of all the cultivars, three had new, undescribed genes. Nep Bha Bong had a recessive gene for moderate resistance to races 1, 2, and 3 and resistance to race 5 and was also designated as *xa26*(t). Arai Raj had a dominant gene for resistance to race 2 which segregated independent of *Xa11* and was designated *Xa27*(t). Lota Sail had a recessive gene for resistance to race 2 which segregated independently of *Xa10* and was designated as *xa28*(t) [86].

Gu et al. [26] reported the fine genetic mapping of the *Xa27*(t) gene by performing disease evaluation of an *Xa27*(t) NIL, IRBB27, with 35 *Xoo* strains from 11 countries. Resistance of *Xa27*(t) gene was developmentally regulated in IRBB27 and showed semidominant or dosage effect in the cultivar CO39 genetic background. Three markers, M336, M1081, and M1059, were closely linked to *Xa27*(t). The *Xa27*(t) locus was confirmed by chromosome landing of

M1081 and M1059 markers on rice genome. Markers derived from the genomic sequence of *O. sativa* cv. Nipponbare finally located the *Xa27*(t) gene within a genetic interval of 0.052 cM on the long arm of chromosome 6 between markers M964 and M1197, and co-segregated with M631, M1230, and M449.

Another dominant *Xa* gene was identified from a rice line 'B5' derived from *Oryza officinalis* through introgression. It proved to be highly resistant to brown plant hopper, white-backed plant hopper and bacterial blight. The resistance gene identified in B5 was designated as *Xa29*(t) [87]. Bulked segregant analysis of RILs from a cross between 'B5' and 'Minghui63' located the resistance gene within a 1.3 cM region flanked by RFLP markers C904 and R596 on chromosome 1.

A new rice BB resistance germplasm (Y238) from the wild rice species *O. rufipogon* was identified and designated as *Xa30*(t) [88]. The gene was mapped on the long arm of rice chromosome 11. Linkage analysis revealed that four molecular markers RM1341, V88, C189, and 03STS were located on the same side of *Xa30*(t), with genetic distances of 11.4, 11.4, 4.4, and 2.0 cM to the candidate gene, respectively. Wang et al. [89] identified *Xa31*(t) gene for BB resistance in Zhachanglong (ZCL), a regional rice variety from Yunnan province in southwest China, which has a high level of resistance to a broad spectrum of *Xoo* isolates. Genetic linkage analysis and fine mapping localized *Xa31*(t) within a genetic distance of 0.2 cM between two RFLP markers G235 and C600 on the end of long arm of chromosome 4. The flanking markers were used to screen the MH63 BAC library and the *Xa31*(t) locus was limited to one BAC clone with a length of about 100 kb.

Another novel BB resistance gene from a wild rice (*Oryza australiensis*) introgression line, C4064, was found to be resistant to *Xoo* strains PXO61, PXO71, PXO99, PXO145, PXO280, PXO339, and KX085 but susceptible to PXO8 and PXO79. The gene was tentatively designated as *Xa32*(t). The chromosomal position of the gene was located using BSA with SSR and EST markers between two markers at the end side of long arm and four markers on centromere side. Finally, the gene was mapped within a 2.0 cM interval flanked by two SSR markers RM2064 and RM6293 on the long arm of chromosome 11 [90]. Another BB resistance gene, *xa33*(t) was identified from rice cultivar 'Ba7'. The gene was localized on long arm of chromosome 6, where two other dominant genes (*Xa7* and *Xa27*) have been reported. RM 20590 was identified as the closest linked marker. Although *Xa7* and *xa33* shared common markers, both showed different gene actions and were not growth stage-dependent genes [91]. Natarajkumar et al. [92] placed a new BB resistance gene, also designated as *Xa33*(t) from *Oryza nivara*, on chromosome 7 flanked by the markers RM21004 and RM21177 at a genetic distance of 2.0 and 4.5 cM, respectively.

Pin Kaset (PK), a Thai rice cultivar, had a high level of resistance to BB. The resistance gene in PK was identified using BC_2F_2 plants from a cross between Ba7 and PK. Phenotypic evaluation of TB0304 and genotypic analysis with SSR markers revealed that the gene was linked to RM224 on chromosome 11 [91]. The new gene was tentatively designated *Xa34*(t). Although *Xa34*(t) and other BB resistance genes are located in the same region and shared common linked markers, no evidence was obtained whether they shared the same genomic sequence or were tightly linked to each other. Also, Ram et al. [93] screened 2 introgression lines (IR65483-118-25-31-7-1-5-B and IR65483-141-2-4-4-2-5-B) derived from IR56 X *O. brachyantha*, against a virulent isolate DXO44 and 21 different virulent isolates of BB from 11 states of India. The ILs showed resistance to 16 isolates, moderate resistance to one and susceptibility to 4 isolates, as against the differential reaction of IRBB lines to different isolates. This indicated that the introgressed gene is different from those in IRBB lines. It was designated as *Xa34*(t). Crosses of ILs with IR56 and susceptibility check BPT5204 gave all resistant F_1s and 3:1 ratio in F_2 revealing a single dominant gene in both cases. Allelic test was done by crossing ILs with IRBB21 which showed that *Xa34*(t) (resistance gene from *O. brachyantha*) and *Xa21* were nonallelic and inherited independently.

A new rice bacterial blight resistance gene *Xa35*(t) from the wild rice species *Oryza minuta* (Acc. No. IRGC101133) was identified and transferred into IR24 [94]. Through genetic analysis and identification of resistance spectrum, *Xa35*(t) showed a high level of resistance to PXO61, PXO112, and PXO339, but was susceptible to PXO86 and PXO99. The *Xa35*(t) locus was mapped to a 1.80 cM region and this locus was co-segregated with marker RM144, and was 0.7 cM from marker RM6293 on one side and 1.1 cM from marker RM7654 on the other side of the rice chromosome 11. Miao et al. [95] identified that the rice germplasm C4059 harbored a bacterial blight resistance gene and designated it as *Xa36*(t). The gene *Xa36*(t) was mapped on the long arm of rice chromosome 11 encompassing 4.5 cM region flanked by RM224 and RM2136.

An accession of *O. nivara* (IRGC 81825) was identified to be resistant to all the seven *Xoo* pathotypes prevalent in North India [96]. The F_2 population derived from a cross between PR114 and *O. nivara* acc. 81825 segregated in a 3:1 ratio of resistant: susceptible plants, indicating that the resistance was provided by a single dominant gene. The inheritance and mapping studies using F_2, BC_2F_2, BC_3F_1, and BC_3F_2 progenies of the cross *O. sativa* cv. PR114 × *O. nivara* acc. 81825 mapped the gene on chromosome 4L, spanning an approximate 38.4 kb region [17]. Since none of the known *Xa* genes mapped on chromosome 4L were effective against the *Xoo* pathotypes tested, the gene identified and transferred from *O. nivara* was considered novel and designated as *Xa38*.

A rice introgression line (IL), FF329, identified from a BC_1F_4 population derived from the cross between donor PSBRC66 (P66) and recipient Huang-Hua-Zhan (HHZ), exhibited a typical hypersensitive response (HR) with all 21 representative *Xoo* strains (14 Philippines races and seven Chinese pathotypes) [97]. By contrast, the parents were highly susceptible to 10 of the tested strains and resistant or moderately susceptible to 11 strains, but without HR symptoms. As FF329 showed broad-spectrum BB resistance, the gene identified was novel and was designated *Xa39*. Two SSR markers, RM21 and RM206, located on rice chromosome 11, were linked to the target gene by bulked segregant analysis of the F_2 population derived from the HHZ/FF329 cross. The fine mapping of *Xa39* locus placed the gene in the region of 97 · 4 kb interval flanked by markers RM26985 and DM13.

The *japonica* advanced backcross breeding lines derived from the *indica* line IR65482-7-216-1-2 in the background of cultivar Junam was resistant to all Korean BB races. Kim et al. [5]

using two F_2 populations derived from the crosses between 11325 (IR83261-3-7-23-6-2-1-1-2-1-2)/Anmi and 11325/Ilpum indicated that resistance (R) was controlled by a new resistance gene designated as *Xa40(t)*, which was co-segregated with the markers RM27320 and ID55. Thus, based on the physical map of *japonica* rice Nipponbare, the *Xa40(t)* gene was defined by RM27320 and ID55.WA18-5 located on the BAC clone OSJNBa0036K13 spanning approximately 80 kb region on chromosome 11.

Hutin et al. [98] screened a germplasm of 169 rice accessions for polymorphism in the promoter of the major bacterial blight susceptibility "S" gene OsSWEET14, which encodes a sugar transporter targeted by numerous strains of *Xoo*. They identified a single allele with a deletion of 18 bp overlapping with the binding sites targeted by several TAL effectors known to activate the gene. Further, they showed that this allele, which was designated as *xa41*(t), confers resistance against half of the tested *Xoo* strains, representative of various geographic origins and genetic lineages, highlighting the selective pressure on the pathogen to accommodate OsSWEET14 polymorphism. Analysis of *xa41*(t) demonstrated that the resistance through TAL effector-dependent loss of S-gene expression can be greatly fostered upon knowledge-based molecular screening of a large collection of host plants.

A new mutant named 'XM14' was obtained by treating IR24, which was resistant to all Japanese *Xoo* races. The gene identified in XM14 was designated as *xa42* [18]. The F_2 population from XM14 × IR24 clearly showed 1 resistant:3 susceptible segregation, suggesting control of resistance by a recessive gene. The *xa42* is located around the centromeric region of rice chromosome 3 in the chromosomal region encompassed by KGC3_16.1 and RM15189.

3. Determinants of pathogenicity

Xanthomonas oryzae pv. *oryzae* (*Xoo*) is a member of the c-subdivision of the Gram-negative proteobacteria. It continues to grow in the vascular system until the xylem vessels are clogged with bacterial cells and extracellular polysaccharide (EPS or xanthan). There are several races of *Xoo*, all of which secrete race-specific effectors into the xylem to trigger individualized response and cause infection. The bacteria also release factors which bind and activate transcription of genes that activate resistance response, known as resistance genes (*R* genes) [99]. The factors that activate *Xoo* resistance genes are known as avirulence factors that determine host specificity via gene-for-gene interactions, reducing the virulence of the pathogen as they are recognized by the host. Since each race of *Xoo* produces unique virulence and avirulence factors, *R* genes have evolved to provide resistance to individual races of *Xoo*. It has been known that the interactions of *Xoo* with plants are determined by hypersensitive response and pathogenicity (hrp) genes, which are required for pathogenicity in susceptible host plants and for the hypersensitive response (HR) in resistant and nonhost plants [3, 100]. A considerable effort has been made to identify genes involved in the pathogenesis of *Xoo* and to understand the roles of the gene products in the disease process. The knowledge of pathogenesis genes has become more clear and distinct from the genome sequence of *Xoo* strains KACC10331, MAFF311018, and PXO99A [4, 101, 102].

3.1. Avirulence genes

Members of the *AvrBs3/PthA* family of transcription activator-like effectors play a major role in the virulence of *X. oryzae* pv. *oryzae*. The *avr* gene family consists of a repeat sequence in the central domain of the encoded protein with each repeat of 34 amino acids. The number of repeats in different gene families varies, but the amino acids in each repeat are conserved except amino acids at positions 12 and 13, which are referred to as variable regions. The arrangements of the variable regions of all characterized family members appear to be a critical feature for the race specificity of the proteins. The carboxyl terminus of *AvrXa7* and *AvrXa10* contains a domain that is structurally similar to the acidic activation domain of many eukaryotic transcription factors, in addition to three nuclear localization signal (NLS) sequences as with *AvrBs3* family members. It has been postulated that, two *raxP* and *raxQ* genes of *Xoo* are required for the activation of *AvrXa21*. Both *raxP* and *raxQ* resides in a genomic cluster of sulfur assimilation genes, which encodes an ATP sulfurylase and adenosine-5′-phosphosulfate (APS) kinase. These enzymes function together to produce activated forms of sulfate, APS, and 3′-phosphoadenosine-5′-phosphosulfate (PAPS). It has also been observed that the transcription activation domain and the NLS sequences in *AvrXa7* are required for the virulence activity, suggesting that the *avr* gene products act as virulence factors that enter the host nucleus and directly affect host gene transcription [103]. Adaptation of an avirulent bacterial race to host varieties containing a single dominant *R* gene often results from the loss of function of the corresponding *avr* gene. However, the fitness penalty associated with loss of the *avr* gene may prevent disease epidemics [104].

3.2. Type III effectors

The interaction of many Gram-negative plant and animal pathogenic bacteria with their hosts depends on a conserved type III protein secretion system (TTSS). The TTSS is encoded by *hrp* genes for eliciting either HR on nonhost/resistant host plants or pathogenesis on susceptible hosts. The *hrp* gene-encoded TTSSs are responsible for translocation of avirulence proteins into the host plant cells. In xanthomonads, the *hrp* gene cluster comprises six operons (*hrpA* to *hrpF*) and is positively regulated by *HrpG* and *HrpX* [105–107], repressed in nutrient-rich media, but induced in nutrient-limited media and inside the host [108–111]. *HrpG*, which is predicted to be a member of the *OmpR* response regulator family of two-component signal transduction systems, regulates the expression of *hrpX*, encoding an *AraC*-type transcriptional activator, which then activates the expression of other *hrp* operons. The highly conserved *hrp* genes named *hrc* encode the proteins of the apparatus of the type III secretion system, and are critical for pathogenicity and the initiation of disease [100, 112, 113]. A *hrp* gene cluster was identified in the *Xoo* genome harboring 26 genes inclusive of *hpa2* and *hrpF*. The *Xoo hrp PAI* (31.3 kb) was larger due to the presence of four transposase genes (about 6 kb) located between *hpaB* and *hrpF* genes than its counterparts of *Xac* (25.6 kb) and *Xcc* (23.1 kb). Strong amino acid identity was observed between several orthologous *hrp* genes of

Xoo and *Xanthomonas axonopodis* pv. *citri* (*Xac*), as *hpaF*, *hpaP*, *hrpD5*, and *hpaA* had identity of 74, 76, 79, and 82% respectively, whereas *hrpF* (68%), *hpa1* (65%), *hrpB5* (66%), and *hrpB7* (65%) exhibited relatively low similarity. The products of *hrpF* and *hpa1* are predicted to be exposed or secreted components of the type III secretion system, which is responsible for contributing to their diversity due to distinct selective pressures in the different hosts. It has been demonstrated that the transcriptional regulator for *hrp* (trh) and *phoP* genes in *Xoo* positively regulate expression of *hrpG*, but not known whether the regulation is direct or indirect [114, 115]. The *trh* gene encodes a putative transcriptional regulator [114] and *phoP* encodes a putative response regulator of two-component regulatory systems [115]. It has not been reported whether *trh* and *phoP* influence the expression of other *XrvA* targets such as *gum* and *rpf*. It will be of interest to further study the regulation of *hrpG* expression by deciphering the functional relationship (if any) between *xrvA*, *trh*, and *phoP* to better understand *hrp* regulatory mechanisms. A homolog to *hrpW*, a proposed pectate lyase, was not readily apparent in the *Xoo* genome but several candidate pectate lyase genes were identified that could function similarly to *hrpW*.

A review classified all known and candidate TTSS effectors from strains of *Xanthomonas* spp. into 39 groups based on sequence and structural differences, and similarities [116]. A class of T3 effectors called transcription activator-like effectors (TALE) was first identified due to their relatedness to the T3 effector *AvrBs3* from *Xanthomonas campestris* pv. *vesicatoria* [117]. Some of these TALEs have avirulence activity and three different TAL effectors *avrXa7*, *avrXa10*, and *avrXa27* from *Xoo* have been cloned [13, 118]. All TALEs have nuclear localization signals (NLS) and an activation domain rich in acidic amino acids (AAD) at their C terminus. Each TALE also contains a central region of multiple 34- to 35-amino acid direct repeats that are nearly identical except the 12th and 13th amino acid residues (so-called repeat variable diaminoacids, or RVD) [119]. The combination of repeat number and composition of RVDs of individual TALEs determine the specificity of the targeted genes [119, 120]. Five additional T3 effectors from *Xoo* have known contributions to virulence under the appropriate conditions. These genes are *pthXo1*, *pthXo2*, *pthXo3*, *pthXo6*, and *pthXo7*. *PthXo1* is the major TAL effector in many strains, including the common laboratory strain PXO99A, and induces the expression of host gene *Os8N3*, a member of *nodulin 3* (*N3*) gene family and encodes a predicted membrane protein [64]. In addition to *PthXo1*, two other TAL effectors of PXO99A also contribute to virulence by inducing the expression of two different host genes [121]. *PthXo6* and *PthXo7* elevate the transcription of host genes *OsTFX1* and *OsTFIIAγ1*, respectively. *OsTFX1* is a member of bZIP family of transcription factors, which are involved in the regulation of many developmental and physiological processes. Another TAL effector gene, named *pthXo8*, has been identified in PXO99A with quantitative effects similar to *pthXo6*. Preliminary evidence indicates that the effector is involved in manipulation of the small RNA pathways of the host.

By using custom-engineered TALEs to investigate the functionality of host target genes involved in *Xoo*/rice interaction, it has been demonstrated that *xa13* can be induced by *Xoo* and confer disease susceptibility, lending further evidence for *Os8N3* (*OsSWEET11* or *Xa13*) as an S gene. This approach facilitated to identify another *SWEET* gene (*OsSWEET12*) that can act as an S gene, provided that the *Xoo* pathogen contains a corresponding TALE. Further studies revealed that the *xa27* allele can be activated and triggers resistance to the bacterium expressing a corresponding dTALE. Rice genome contains at least 21 *SWEET* (or *N3*) genes with a phylogenetic clade that harbors two known S genes (*OsSWEET11* and *OsSWEET14*) and three additional uncharacterized *SWEETs* [122, 123] generated seven artificial or designer TAL effectors based on the promoter sequences of six loci in rice and transformed them into *Xoo*. Gene expression analyses indicated all dTALEs were active in trans-activating target genes in rice achieving a 100% success rate in engineering active dTALEs. Gene activation of three alleles (*xa27*, *xa13*, and *OsSWEET12*) led to phenotypic changes in disease resistance or susceptibility in response to *Xoo* infection, suggesting the feasibility of this approach to the gene functional analysis.

The genomic sequences are available for three strains of *Xoo* [4, 101, 102]. A comparison of the repetitive regions of PXO99A and MAFF311018 indicates a high degree of rearrangements and shuffling of the genes at all of the loci, to the point where only three genes of 17 in MAFF311018 are identical [96]. The genome structure of *Xoo* MAFF 311018 was characterized by large numbers of effector (*avr*) genes of the *avrBs3/pth* family and insertion sequences (ISs). The high degree of genetic diversity and race differentiation characteristic of this pathogen is due to the presence of mobile elements, which leads to genome inversions and rearrangements. The large numbers of TAL effector genes in these species may reflect the evolutionary "investment" the strains have in utilizing the TAL effectors for virulence. The maintenance of high gene numbers may even be exacerbated by rice breeding and *R* gene deployment by farmers over the millennia. It could be inferred that *Xoo* targets different host genes to alter the host physiology and different TAL effectors to have qualitatively different effects on host susceptibility.

3.3. Type II secretion system

The bacterial type II secretion system mediates a two-step process. The proteins that are secreted through this system carry a secretion signal at their N termini and are transported into the periplasmic space through the inner membrane by either the general secretion pathway (GSP) or the twin arginine pathway (TWP) [124, 125]. Transport across the outer membrane is facilitated by the proteins of main terminal branch (MTB) of general secretion pathway (GSP). The *Xoo* genome encodes a single type II secretion system in contrast to two different type II systems in *Xac* and *Xcc* [126]. Mutations in the gene cluster *xps*, which is required for a functional type II system, result in strains defective in virulence [127]. Type II secreted proteins are mainly toxins and enzymes that target different components of the host cell, and some of these enzymes, including xylanase, cellulase, cysteine protease, cellobiosidase, and lipases, have been characterized as contributors to *Xoo* virulence. Experimental evidences indicate that rice plants perceive some type II secreted proteins and respond by hypersensitive responses (HRs), and these responses are suppressed by type III secreted effectors [125]. Other factors that contribute to the virulence of *Xoo* include EPS, the type II general secretion system and its secreted proteins, and regulation involving genes within the *rpf* cluster [3, 128–132]. The *rpfB*, *rpfC*, *rpfF*, and *rpfG* genes (regulation of pathogenicity factor) of *Xoo* have been shown to affect virulence in rice, EPS production, xylanase production, and motility [130–132]. The virulence of *Xanthomonas* also depends upon cell-to-cell signaling mediated by diffusible signal factor (DSF). Two of the genes in this cluster *rpfB* and *rpfF* are involved in the synthesis of DSF. Knockout studies on *rpfF* indicate the role of DSF in the regulation of levels of extracellular enzymes and EPS [133]. The *RpfC/RpfG* two-component system couples the DSF sensing to intracellular regulatory networks through a second messenger, cyclic *diGMP*, and a global regulator, *Clp*. Protein-protein interaction between the DSF synthase *RpfF* and the sensor *RpfC* may act as a posttranslational mechanism to modulate the biosynthesis of DSF [134, 135]. Recently, an *X. oryzae* pv. *oryzae* flagellar operon region has been isolated from *Xoo*, which contains four ORFs. One of the ORFs, *flhF*, encodes a putative GTP-binding protein which is involved in chemotaxis. Mutation in *flhF* resulted in weak chemotaxis but did not show reduced virulence if inoculated on rice leaves with a scissors-clipping method, suggesting that chemotaxis is not required for virulence once the bacterial cells enter rice leaves [136]. EPS synthesis is directed by genes at multiple chromosomal loci; one of the loci is called the *gum* cluster [137]. The *Xoo gum* cluster in strain KACC10331 is composed of 14 ORFs arranged in a tandem array, expressed from a promoter located upstream of *gumB*, but internal promoters can also be found upstream of *gumG*, *gumH*, and *gumM*, respectively [101, 138]. Two different *Xoo* mutants, one with a transposon insertion in *gum G* and another with a spontaneous mutation due to the insertion of endogenous IS element in *gum M*, were incapable of EPS production and less virulent. In *Xoo*, EPS synthesis has been found to be controlled by the *rpfC* gene, which is part of a two-component system. Strains carrying a mutation in *rpfC* have greatly reduced EPS production and virulence but still attain maximum population levels in rice plants [130] indicating that EPS is a virulence determinant in *X. oryzae* pv. *oryzae*.

Recently, screening of a transposon mutant library of a Korean *Xoo* strain, KACC10331, in rice also showed that Tn5 insertion in the *xrvA* gene (*XOO2744*) led to reduced virulence; however, the mutant was not characterized in further detail [139]. The deduced protein encoded by *xrvA* possesses an H-NS domain. H-NS and H-NS–like proteins are modular proteins associated with the bacterial nucleoid. The discovery of *xrvA* as a regulator of virulence factor synthesis came from work aimed at identification of genes involved in EPS production of *Xoo*. Disruption of *xrvA* led to a significant reduction in virulence, a delay in HR elicitation, a decrease in EPS and DSF production, and an increase in glycogen accumulation.

3.4. Type I secretion systems

Type I secretion systems of Gram-negative bacteria are secretion systems that transport proteins directly to the extracellular environment from the bacterial cytoplasm through inner and outer bacterial membranes. Three highly conserved components of type I secretion systems are an ABC transporter, which forms a channel across the inner membrane, a membrane fusion protein (MFP), and an outer membrane protein called To1C [140]. PXO99A contains a type I secretion system that is involved in triggering resistance in rice cultivars which carry the *Xa21* resistance gene. Transposon-induced mutations in PXO99A and subsequent screening for mutants that lost *Xa21*-mediated avirulence activity identified eight genes *viz. raxA*, *raxB*, *raxC*, *raxST*, *raxQ*, *raxP*, *raxH*, and *raxR* [141]. *AvrXa21* activity requires the presence of *raxA*, *raxB*, and *raxC* genes, and these three genes encode the MFP, ABC transporter, and outer membrane protein. The *AvrXa21* pathogen-associated molecule is involved in quorum sensing and the expression of *raxST* is regulated by a two-component regulatory system encoded by *raxH* and *raxR* [142] that responds to *Xoo* cell population density, and may be conserved in most *Xanthomonas* spp. The *raxST* encodes a sulphotransferase enzyme, while *raxQ* and *raxP* are involved in the production of the sulfuryl donor phosphoadenosine phosphor sulfate (PAPS). The elicitor of *Xa21* immunity, *Ax21*, was characterized as a 194 amino acid sulfated protein, which is secreted into the extracellular environment [143]. The N-terminal 17 amino acid peptide of *Ax21* is sulfated at a tyrosine residue and is sufficient to trigger *Xa21*-mediated resistance.

4. Molecular mechanism of BB resistance in rice

Out of 42 *Xa* genes identified so far, nine genes (*Xa*1, *Xa3*/*Xa26*, *xa5*, *Xa10 xa13*, *Xa21*, *Xa23*, *xa25*, and *Xa27*) have been characterized at molecular level and these encode various types of proteins (**Table 2**). Based on these studies, the molecular mechanisms of BB resistance in rice seem to be largely different from the mechanisms of resistance to rice blast, although the mechanisms of rice disease resistance remain largely to be elucidated. Most of the characterized BB resistance genes are different from the most common R protein, nucleotide-binding site-leucine-rich repeat (NBS-LRR) protein [144]. Interestingly, except for the fact that Xa21 and Xa26 encode for similar receptor-like proteins, the products of the other genes are unique and not found in other plant species [145]. These features suggest that molecular mechanism of rice-*Xoo* system is more complicated and a unique pathosystem to study the interactions between hosts and pathogens. The molecular mechanism of BB resistance gene has also been discussed in other book chapters [146, 147]. The nine characterized BB resistance genes fall into six different classes of proteins and thus may give a wide scenario of understanding at molecular level.

4.1. BB resistance conferred by LRR receptor kinase protein

The LRR receptor kinase class of BB resistance is conferred by *Xa21* and *Xa3*/*Xa26* genes. The *Xa21* was the first rice BB resistance gene characterized [7] and was one of the most intensively studied genes at molecular level. It was originally identified in the wild species *O. longistaminata*, and isolated by map-based cloning strategy from the IRBB21, a near-isogenic line in the background of IR24. This gene confers a race-specific resistance to *Xoo*, and is the most widespread BB resistance gene in the rice cultivated area, thereby providing broad-spectrum resistance. The receptor kinase-like protein encoded by *Xa21* consists of putative extracellular domain containing LRR, a single pass transmembrane domain, and an intracellular domain containing serine/threonine kinase. This protein is unique in carrying the receptor domain LRR with hypothetical function in pathogen recognition, and the kinase domain that functions in subsequent signal transduction as compared to other cloned plant resistance genes [148]. The *Xa21*-mediated resistance is not expressed in the early developmental stages and gradually increases from the seedling stage to later stages,

S. No.	Gene	Encoded protein	Reference
1	*Xa1*	NBS-LRR	[10]
2	*Xa3/Xa26*	Leucine-rich repeat receptor-like kinase (LRR-RLK)	[8]
3	*xa5*	TFIIA Transcription factor	[11, 12]
4	*Xa10*	Executor R protein, encodes 126 AA, with four potential transmembrane helices	[20]
5	*xa13*	MtN3/saliva	[14]
6	*xa21*	Receptor-like kinase	[7]
7	*Xa23*	Executor R protein, encodes 113 AA, with four potential transmembrane helices	[21]
8	*xa25*	MtN3/saliva	[19]
9	*Xa27*	Apoplast (rice unique gene)	[13]

Table 2. The BB resistance genes characterized at molecular level.

with 100% resistance at the adult stage [149]. The gradual increase in expression of *Xa21* gene during rice development is associated with development-controlled *Xa21*-mediated resistance [150]. Ectopic expression of *Xa21* gene can generate rice plants with a high level of resistance to *Xoo* at both seedling and adult stages [150, 151]. The *Ax21* (avir*Xa21* protein; as called activator of *Xa21*) protein was secreted by *Xoo* through its type I secretion system switch on the *Xa21* gene present in the host [145]. A sulfated 17-amino acid synthetic peptide derived from the N-terminal region of *Ax21* is sufficient for its initiation. The *Ax21* is highly conserved in many *Xanthomonas* species including number of pathogens of plants and human across microbial genus. Thus, *Ax21* is considered a pathogen-associated molecular pattern, and thus, *Xa21* can be classified both as a plant pattern recognition receptor (PRR) and an R protein [145].

The *Xa*21 locus is a multigene family and several key components of *Xa21* protein such as E3 ubiquitin ligase/XB3, WRKY62/XB10, protein phosphatase 2C (PP2C)/XB15, ATPase/XB24, and Bip3 (also known as glucose-regulated protein), involved in defense signaling pathway have been identified. The E3 ubiquitin ligase interacts with the kinase domain of *Xa21* protein and acts as a substrate for the *Xa21* serine and threonine kinase activity, which is necessary for full accumulation of the *Xa*21 protein, thus *Xa21*-mediated immunity [152]. The *Xa21* protein binds to WRKY62 with catalytic activity of its juxtamembrane motif and serine/threonine kinase domain. The WRKY62 and WRKY76 (another WRKY transcription factor) function as a negative regulator of *Xa21*-mediated resistance [153, 154]. The PP2C component interacts with juxtamembrane motif and kinase domain of *Xa21* protein, and can dephosphorylate autophosphorylated *Xa21*, thus acting as negative regulator of *Xa21*-mediated resistance [155]. The XB24, an *Xa21* binding protein, was isolated using yeast two-hybrid screening and belongs to a class of ATPases. The ATPase/XB24 can enhance autophosphorylation of *Xa21* protein by its physical association *in vivo* with the juxtamembrane motif and kinase domain, and downregulates the *Xa2*-mediated resistance *Xa21* protein by its enzymatic activity, because the activation of *Xa21* following interaction with pathogen-associated molecular pattern *Ax21* requires the dissociation of XB24 from *Xa21* or removal of autophosphorylation. The rice plants with reduced level of XB24 expression showed enhanced level of *Xa21*-mediated resistance, whereas *Xa21*-mediated resistance overexpressing the XB24/ATPase in rice plants showed lower level of *Xa21*-meditated resistance, as XB24 in this case cannot readily dissociate from *Xa21* and further binding of Ax21 with *Xa21* leads to conformational change in *Xa21* protein. The conformational change in *Xa21* protein exposes it to degradation in endogenous proteases leading to lower resistance 21 protein *a21* [156]. The endoplasmic reticulum chaperone Bip3 can interact with *Xa21* protein *in vivo*. Rice plants overexpressing Bip3 have decreased *Xa21* protein accumulation and inhibited *Xa21* protein processing, which results in compromised *Xa21*-mediated resistance [157].

The *Xa3*/Xa26 locus was isolated as *Xa26* from an *indica* rice cultivar Minghui 63 [8] and found to be similar to previously identified gene *Xa3*, therefore renamed as *Xa3/Xa26* [9]. It encodes a plasma membrane-localized LRR receptor kinase-type protein with an extracellular LRR domain, a transmembrane motif, and a cytoplasmic kinase domain [8]. It also confers a broad-spectrum resistance relative to *Xa21* and has been widely deployed in rice cultivars in China [158–160]. Two different alleles of *Xa3/Xa26* have been identified from "CC" genome of *O. officinalis* and *O. minuta* that can mediate a similar spectrum of resistance against *Xoo* [160]. This indicated the early origin and relatively conserved function of *Xa3/Xa*26 locus during evolution that can provide durable resistance against *Xoo* [160]. The MRKa gene, an ortholog of *Xa3/Xa*26 gene family in rice cultivar Mingui 63, can mediate partial resistance to *Xoo* when it is overexpressed [161]. The kinase domain is important for complete function of *Xa3/Xa*26 for resistance, as rice plants carrying truncated *Xa3/Xa*26 gene without kinase domain exhibits lower level of resistance. The kinase domain of MRKa protein can partially restore the function of the truncated *Xa3/Xa*26 gene in *Xoo* resistance, suggesting the partially conserved function of the orthologs of this family. This hypothesis is also supported by another study that NRKe gene from rice cultivar Nipponbare gets transcriptionally activated in response to raised temperature [162]. The kinase domain of *Xa3/Xa26* protein can replace the function of the kinase domain of NRKe protein in response to temperature change.

The genetic background and development stage of rice plant affect the *Xa3/Xa*26-mediated resistance. The higher level of resistance was observed in *japonica* as compared to *indica* background. The *Xa3/Xa26* expression gradually increases from seedling stage to adult stage and pathogen infection also differentially effects its expression in plants with different genetic backgrounds, *Xoo* strains at both seedling and adult stages, but has full resistance to other *Xoo* strains at adult stage [8, 163, 164]. Further resistance mediated by *Xa3/Xa26* increases with increase in its expression and its constitutive expression provides broad-spectrum resistance at both seedling and adult stage without affecting the agronomic performance [159].

Several members functioning downstream in the *Xa3/Xa26*-initiated defense signaling pathway have been identified, where the downstream function of these components can mediate a broad-spectrum resistance as compared to *Xa3/Xa*26 protein. Two WRKY-type transcription factors namely WRKY13 and WRKY45-2 positively regulate rice resistance to *Xoo*, of which *M. oryzae* (causal organism of rice blast), with WRKY13 putatively functions upstream of WRKY45-2 in rice-*Xoo* interaction [165–167]. Two genes namely OsDR8 encoding a protein involved in thiamine biosynthesis and C3H12, a CCCH-type zinc finger nucleic acid-binding protein, also act as positive regulator of rice resistance to *Xoo* in *Xa3*/*Xa*26-initiated defense pathway [152, 168]. Of these, C3H12 functions upstream of WRKY45-2. A rice tribe-specific gene OsDR10 that probably functions upstream of WRKY13, negatively regulates resistance to *Xoo* in *Xa3*/*Xa26*-mediated resistance pathway [169].

Both *Xa*21 and *Xa3*/*Xa26* belong to multigene family; encode same type of proteins and have 53% sequence similarity [7, 8]. The only structural difference between two genes is the number of LRR, where *Xa26* encodes 26 LRR, whereas *Xa21* encodes 23 LRR [7]. However, the respective LRR domains of *Xa3*/*Xa26* and *Xa21* are the important determinants of race-specific recognition during rice-*Xoo* interactions as evidenced from experiment on domain swapping analyses, but a juxtamembrane motif of *Xa3*/*Xa*26 also seems to contribute in resistance specificity [150]. The kinase domain of *Xa3*/*Xa*26 can partially replace the function of the kinase domain of *Xa21*, or vice versa, in *Xoo* resistance, suggesting the partially conserved nature of this domain in defense signaling pathway [150]. Both *Xa3*/*Xa26* and *Xa21* genes are dose-dependent and their expression progressively increases with developmental stage. However, *Xa3*/*Xa26* has a higher expression level in the *japonica* background than in an *indica* background [150, 164]. Further study may also be required to determine, whether *Xa3*/*Xa26* is also a PRR in addition to being an R protein.

4.2. BB resistance conferred by MtN3/saliva class protein

The MtN3 (a homologue of nodulin protein) class of BB resistance is conferred by *xa*13, a fully recessive gene for BB resistance. The gene encodes a novel protein that has no sequence similarity with any known *R* proteins, but it shows 50% sequence identity and 68% sequence similarity to the product of a nodulin MtN3 gene in legumes [14]. The *xa*13 and its dominant allele *Xa13*, which is also named *Os8N3* and OsSWEET11 [64, 122], encode identical polytopic plasma membrane proteins of the MtN3/saliva family proteins, but have crucial sequence differences in their promoter regions. Promoter swapping analysis confirms that *Xa13* (*Os8N3* and OsSWEET11) is a susceptibility gene, which is induced by direct binding of the transcription activator-like (TAL) effector Pth*Xo*1 of *Xoo* strain PXO99 to the cis element, the UPT Pth*Xo*l box, on the *Xa13* promoter [64, 170, 171]. The *Xoo* strain PXO99 that secretes Pth*Xo*1 cannot induce recessive *xa13* due to the mutation of the UPT Pth*Xo*1 box in the *xa13* promoter. Further, PXO99 is more sensitive to copper (an essential micronutrient of plants, is also an important element for a number of pesticides in agriculture) than other *Xoo* strains. The *Xa13* protein interacts with two copper transporter-type proteins, COPT1 and COPT5, to promote removal of copper from xylem vessels, where *Xoo* multiplies and spreads to cause disease [172]. As PXO99 cannot induce recessive *xa13*, the copper levels in rice plants carrying the recessive *xa*13 gene can inhibit *Xoo* growth and thus plants show resistant reaction. Besides functioning against *Xoo* resistance, product of *xa13* gene also has an essential role in pollen development. This became evident from suppressing the function of either the dominant or recessive allele of *xa13* in rice transgenic plants, where *xa13* not only enhanced the resistance but also caused male sterility [173]. The study showing the link between two unrelated biological processes further demands the detailed studies in the future on the functional overlap between pathogen-induced defense signaling and plant development.

The recessive *xa25* gene, also named *Xa25* (t), encodes a plasma membrane protein of the MtN3/saliva family similar to *xa13* and confers race-specific resistance to Philippine *Xoo* strain PXO339 [19, 83]. The protein encoded by *xa25* and its dominant allele differ in eight amino acids. The *Xoo* strain PXO339 can induce the expression of dominant *Xa25*, but not recessive *xa25*. The differences in proteins and expression pattern of *xa25* and its dominant allele *Xa25* in rice-PXO339 interaction suggest that the dominant *Xa25* may be a race-specific susceptible gene, whereas the recessive *xa25* has evolved as the mutant that cannot be induced by rice-*Xoo* interaction—similar to the recessive *xa13*. The developmental stage of rice plant also influences the *xa25*-mediated resistance. The *xa25* gene regulated by its native promoter, when transferred in the rice plants homozygous for *Xa25*, behaves in recessive nature. The rice plants homozygous for *xa25* showed resistant reaction on inoculation with PXO39. However, the rice plants heterozygous at the *xa25* locus were susceptible to PXO339 at the seedling stage, but became resistant to PXO339 at the adult stage. The dominance reversal characteristic of *xa25* may be because of suppression of PXO339-induced activation of dominant *Xa*25 at the adult stage [19]. The rice MtN3/saliva family contains more than 20 paralogs. Some MtN3/saliva proteins from different species act as glucose transporter [122, 174]. The *Xa13*, besides functioning as a susceptibility gene in race-specific rice-*Xoo* interaction, also acts as sucrose transporter. Similar role needs to be elucidated for *Xa25* gene along with rice-*Xoo* interaction.

4.3. BB resistance conferred by TAL effector-dependent class protein

Three BB resistant genes *Xa27*, *Xa10*, and *Xa23* act as transcription activator-like (TAL) effector-dependent *R* genes [13, 20, 21]. The *Xa27* mediates race-specific resistance to diverse strains of *Xoo*, including Chinese and Philippine *Xoo* races [13]. This gene encodes an apoplast protein of 113 amino acids that has no distinguishable sequence similarity to proteins from organisms other than rice [175]. The resistant and susceptible alleles of *Xa27* encode an identical protein, but they differ from each other only in the promoter region [13]. The *Xa27*-mediated race-specific resistance to *Xoo* depends upon *AvrXa27*, a TAL effector of *Xoo* that induces resistance reaction by binding to the UPTAvr*Xa*27 box of the *Xa27* promoter [176]. However, the recessive MR gene *xa5* can attenuate the *Xa*27-mediated resistance in rice, which suggests that *Xoo* TAL effector could not use protein encoded by the recessive *xa5* as a transcription machinery for activation of *Xa27* [177]. The secondary cell wall thickening in vascular bundle elements is obviously associated with *Xa*27-mediated resistance [13]. According to the molecular models of plant innate immunity, pathogen-associated molecular pattern-triggered immunity (PTI) is induced by plasma membrane-localized plant pattern recognition receptors (PRRs) and effector-triggered immunity (ETI) is initiated by NBS-LRR–type R proteins present in the cytoplasm [178, 179]. The subcellular localization and sequence specificity of *Xa27* protein suggest that it does not follow either PTI or ETI procedure [180], though its typical biochemical function in rice-*Xoo* interaction still remains to be elucidated.

Xa10, another TAL effector-dependent *R* gene for BB resistance contains a binding element for the TAL effector AvrXa10 (EBEAvrXa10) in its promoter, and AvrXa10 specifically induces *Xa10* expression [20]. The *Xa10*-encoded protein consists of 126 amino acid residues, which is predicted to have four potential transmembrane helices (M1–M4), among which, M2 contains charged amino acid residues. *Xa10* is an inducible protein that triggers programmed cell death by disruption of the endoplasmic reticulum (ER) and cellular Ca^{2+} homeostasis. The *Xa10* functions not only in rice but also in other species such as *Nicotiana benthamiana*, and in animal kingdom such as mammalian HeLa cells [20]. The *Xa10* protein localizes as hexamers in the ER and is associated with ER Ca^{2+} depletion in plant and HeLa cells. It shows hypersensitive response with faster cell death and shorter lesions in incompatible reactions. The variants of *Xa10* that abolish programmed cell death and ER Ca^{2+} depletion in *N. benthamiana* and HeLa cells, also abolish disease resistance in rice [20]. In one of the experiment, the modified *Xa10* gene designated as *XA10^{E5}* was transferred into Nipponbare and 93-11 backgrounds to generate broad-spectrum resistance to most of the BB pathogens. The *XA10^{E5}* contains five tandemly arranged effector binding elements (EBEs), each responding specifically to a corresponding virulent or avirulent TAL effector. The transgenic lines with *XA10^{E5}* showed resistance to 27 of the 28 *Xoo* strains collected from 11 countries [181].

Xa23, another TAL effector associated executor *R* gene, was isolated from *O. rufipogon* using map-based cloning and TAL effector-based technology [21]. It encodes 113 amino acid long proteins (same number of amino acids as in *Xa27*) with 64% similarity only to *Xa10* protein, but nonsignificant similarity at genomic level. In addition, the predicted three transmembrane helices of *Xa23* largely overlap with transmembrane helices (M_2–M_4) of *Xa10* indicating that Xa23 could be a homolog of *Xa10* [21]. A paralog of *Xa23* has been found to exist in Nipponbare genome with unknown function. The AvrXa23 is responsible for activation of *Xa23* transcription by a TAL effector present in all examined *Xoo* isolates [21]. As compared to *Xa10, Xa23* is a broad-spectrum *R* gene and provides strong resistance to all natural *Xoo* strains tested so far possibly due to presence of avrXa23 in natural *Xoo* strains [182]. The susceptible *xa23* allele *Xa23* but differs in promoter region by lacking the TALE binding element (EBE) for AvrXa23. *Xa23* can trigger a strong hypersensitive response in rice, tobacco, and tomato [21].

4.4. BB resistance conferred by NBS-LRR class proteins

Xa1 is the only NBS-LRR–type BB resistance gene identified so far and is an inducible gene that means its expression is induced by wounding and pathogen infection [10]. However, out of number of uncharacterized BB resistance genes, some might be providing resistance through NBS-LRR class of proteins. One such gene is *Xa38*, identified from *O. nivara* seems to be of NBS-LRR type based on the putative candidate gene sequence analysis in the targeted bacterial artificial clone (BAC) [183]. Nucleotide-binding site-leucine-rich repeat (NBS-LRR) genes play important roles in plant disease resistance. They are generally believed to be responsible for the recognition of effectors delivered by pathogens during infection and the induction of downstream disease resistance reactions [184]. Plant NBS-LRRs can be divided into several types based on the variable N-terminus. In rice, most NBS-LRRs belong to the CC-NBS-LRR (CNL) subclass, in which there is a coiled-coil domain at the N-terminus [185, 186]. In the genome of rice (*O. sativa* cv. Nipponbare), more than 400 NBS-LRR genes have been identified [19], but although these genes are believed to be defense-related, the functions of most of them are unknown in rice genome. Recently, a novel NBS-LRR gene was isolated from rice and designated as *Oryza sativa* Rp1-like 1 (OsRP1L1) based on similarity to Rp1 locus of maize, an intensively studied NBS-LRR gene providing resistance to common leaf rust in maize [187]. Overexpression of OsRP1L1 moderately elevated the resistance of plants to *Xoo* strains PXO86 and PXO341 in a susceptible *japonica* cultivar [188]. Similarly, a large number of genes out of 400 NBS-LRR in the rice genome can be isolated that could be putative candidates for BB resistance.

4.5. BB resistance conferred by other class of protein

The recessive *xa5* that belongs to this class encodes a typical gamma subunit of transcription factor IIA (TFIIAγ), which is one of general transcription factors required for transcription by RNA polymerase II [11]. It mediates specific resistance to Japanese races and Philippine races 1, 2, 3, and 5 by restriction of bacterial movement, but not multiplication [11, 189]. This gene was cloned by a map-based cloning approach combined with allele sequence analysis [11], and further complementation testing confirmed this gene [12]. TFIIAγ is involved in the recruitment of the basal transcription machinery by eukaryotic transcription factors. The *xa5* allele contains a missense mutation that does not seem to influence its function in the recruitment of the basal transcription machinery [12]. It is speculated that *Xoo* TAL effectors usurp parts of plant basal transcription machinery to regulate rice gene expression; the missense mutation of *xa5* allele does not compromise its general function in transcription, but it may evade TAL virulence functions [119, 177]. The *Xoo avrXa5* is an avirulence gene, which encodes a TAL-type protein, corresponding to *xa5* [190]. The *xa5* showed a constitutive expression pattern in different tissues, and the resistance of *xa5* is not dose-dependent [11, 12].

5. Conclusion

Exploration, identification, and utilization of new resistant germplasms in rice breeding are the strategical steps to control the bacterial blight disease of rice. The *Xa21* gene has been successfully introgressed into several elite rice varieties and hybrid rice parental lines all over the world either singly or in combination with other major resistance genes such as *Xa*4, *xa*5, and *xa*13 [27–32, 191]. Traditionally, single recessive genes are overlooked in pyramiding plans since they have to be present in the homozygous condition, which is more difficult to achieve than a heterozygous- or homozygous-dominant genotype. But, in the last decades, rice genomic research has generated a wealth of information about gene function. These advances are now accessible for rice improvement, and have been applied in MAS and genetic engineering in breeding programs. Gene silencing also paves a way to utilize these genes more efficiently. Artificial microRNA (amiRNA) technology has been developed to silence the dominant allele of *xa13*, allowing the recessive allele to be unmasked, thereby expressing the resistant phenotype, mimicking a homozygous state. This silencing conferred a higher degree of resistance to the rice line without affecting other essential traits, such as fertility [192]. Hummel et al. [99] concluded that genetic engineering of the *R* gene promoter can also effectively promote resistance towards *Xoo*. Many *R* genes are turned on in the presence of TALE released from the pathogen by binding to a specific effector binding element (EBE) and initiating the resistance response. So, by fusing new EBEs to the promoter of *Xa27*, gene transcription could be initiated by bacterial races other than the races that normally activate *Xa27*. Therefore, the *Xa27* gene will respond to a greater spectrum of pathogens, triggering resistance. Also, the addition of the EBEs resulted in resistance against a pathovar of *Xanthamonas* which is the causative agent of bacterial leaf streak, for which no *R* genes have yet been identified. This method creates a single gene with the expanded spectrum of traditionally pyramided lines but is less time-consuming and laborious than creating multigenic lines. Several EBEs could be added to a single promoter, greatly increasing the number of pathogens which initiate resistance. Thus, there are numerous factors impacting the resistance granted by different genes, and a complete understanding of the host-pathogen interaction dynamic would result in a better equipped scientific community.

Author details

Yogesh Vikal[1]* and Dharminder Bhatia[2]

*Address all correspondence to: yvikal-soab@pau.edu

1 School of Agricultural Biotechnology, Punjab Agricultural University, Ludhiana, India

2 Department of Plant Breeding & Genetics, Punjab Agricultural University, Ludhiana, India

References

[1] Khush GS. What will it take to feed 5.0 billion rice consumers in 2030? Plant Molecular Biology. 2005;**59**:1–6.

[2] Devadath S. Bacterial blight of paddy. In: Singh US, Mukhopadhyay AN, Kumar J, Chaube HS, editors. Plant Diseases of International Importance: Diseases of Cereals and Pulses. Prentice-Hall, Inc., Englewood Cliffs, NJ, USA; 1992. pp. 158–85.

[3] Nino-Liu DO, Ronald PC, Bogdanove AJ. *Xanthomonas oryzae* pathovars: model pathogens of a model crop. Molecular Plant Pathology. 2006;**7**:303–24.

[4] Ochiai H, Inoue Y, Takeya M, Sasaki A, Kaku H. Genome sequence of *Xanthomonas oryzae* pv. *oryzae* suggests contribution of large numbers of effector genes and insertion sequences to its race diversity. Japan Agricultural Research Quarterly. 2005;**39**:275–87.

[5] Kim SM, Suh JP, Qin Y, Noh TH, Reinke RF, Jena KK. Identification and fine-mapping of a new resistance gene, *Xa40,* conferring resistance to bacterial blight races in rice (*Oryza sativa* L.). Theoretical and Applied Genetics. 2015;**128**:1933–43.

[6] Zhang Q. Genetics of quality resistance and identification of major resistance genes to rice bacterial blight. In: Zhang Q, editors. Genetics and Improvement of Resistance to Bacterial Blight in Rice. Science Press, Beijing; 2007. pp. 130–77.

[7] Song WY, Wang GL, Chen L, Kim HS, Holsten T, Wang B, Zhai W, Zhu LH, Fauquet C, Ronald PC. The rice disease resistance gene, *Xa*-21, encodes a receptor kinase-like protein. Science. 1995;**270**:1804–6.

[8] Sun X, Cao Y, Yang Z, Xu C, Li X, Wang S, Zhang Q. *Xa26,* a gene conferring resistance to *Xanthomonas oryzae* pv. *oryzae* in rice, encodes an LRR receptor kinase-like protein. The Plant Journal. 2004;**37**:517–27.

[9] Xiang Y, Cao Y, Xu C, Li X, Wang S. *Xa3,* conferring resistance for rice bacterial blight and encoding a receptor kinase-like protein, is the same as *Xa*26. Theoretical and Applied Genetics. 2006;**113**:1347–55.

[10] Yoshimura S, Yamanouchi U, Katayose Y, Toki S, Wang Z, Kono I, Kurata N, Yano M, Iwata N, Sasaki T. Expression of *Xa1,* a bacterial blight-resistance gene in rice, is induced by bacterial inoculation. Proceedings of National Academy of Sciences, USA. 1998;**95**:1663–8.

[11] Iyer AS, McCouch SR. The rice bacterial blight resistance gene *xa5* encodes a novel form of disease resistance. Molecular Plant-Microbe Interactions. 2004;**17**:1348–54.

[12] Jiang GH, Xia ZH, Zhou YL, Wan J, Li DY, Chen RS, Zhai W X, Zhu LH. Testifying the rice bacterial blight resistance gene xa5 by genetic complementation and further analyzing *xa5* (*Xa5*) in comparison with its homolog TFIIA_1. Molecular Genetics and Genomics. 2006;**275**:354–66.

[13] Gu K, Yang B, Tian D, Wu L, Wang D, Sreekala C, Yang F, Chu Z, Wang GL, White FF, Yin Z. R gene expression induced by a type-III effector triggers disease resistance in rice. Nature. 2005;**435**:1122–5.

[14] Chu Z, Fu B, Yang H, Xu C, Li Z, Sanchez A, Park YJ, Bennetzen JL, Zhang Q, Wang S. Targeting *xa13,* a recessive gene for bacterial blight resistance in rice. Theoretical and Applied Genetics. 2006;**112**:455–61.

[15] Flor HH. Current status of the gene-for-gene concept. Annual Review of Phytopathology. 1971;**9**:275–96.

[16] Ogawa T. Methods and strategy for monitoring race distribution and identification of resistance genes to bacterial leaf blight (*Xanthomonas campestris* pv. *oryzae*) in rice. Japanese Agricultural Research Q. 1993; 27:71–80.

[17] Cheema KK, Grewal NK, Vikal Y, Sharma R, Lore JS, Das A, Bhatia D, Mahajan R, Gupta V, Bharaj TS, Singh K. A novel bacterial blight resistance gene from *Oryza nivara* mapped

to 38 Kb region on chromosome 4L and transferred to *Oryza sativa* L. Genetics Research Cambridge. 2008;**90**:397–407.

[18] Busungu C, Taura S, Sakagami JI, Ichitani K. Identification and linkage analysis of a new rice bacterial blight resistance gene from XM14, a mutant line from IR24. Breeding Science. 2016;**66**:636–645.

[19] Liu Q, Yuan M, Zhou Y, Li X, Xiao J, Wang S. A paralog of the MtN3/saliva family recessively confers race-specific resistance to *Xanthomonas oryzae* in rice. The Plant Cell Environment. 2011;**34**:1958–69.

[20] Tian D, Wang J, Zheng X, Gu K, Qiu C, Yang X, Zhou Z, Goh M, Luo Y, Murata-Hori M, White FF, Yin Z. The rice TAL effector-dependent resistance protein *Xa10* triggers cell death and calcium depletion in the endoplasmic reticulum. The Plant Cell. 2014;**26**:497–515.

[21] Wang C, Zhang X, Fan Y, Gao Y, Zhu Q, Zheng C, Qin T, Li Y, Che J, Zhang M, Yang B, Liu Y, Zhao K. *Xa23* is an executor R protein and confers broad-spectrum disease resistance in rice. Molecular Plant. 2014;**8**:290–302.

[22] Kinoshita T. Report of the committee on gene symbolization, nomenclature and linkage groups. Rice Genetics Newsletter. 1995;**12**:9–153.

[23] Huang N, Angeles ER, Domingo J, Magpantay G, Singh S, Zhang G, Kumaravadivel N, Benett J Khush GS. Pyramiding of bacterial blight resistance genes in rice: marker-aided selection using RFLP and PCR. Theoretical and Applied Genetics. 1997;**95**:313–20.

[24] Zhang Q, Lin SC, Zhao BY, Wang CL, Yang WC, Zhou YI, Li DY, Chen CB, Zhu LH. Identification and tagging a new gene for resistance to bacterial blight (*Xanthomonas oryzae* pv. *oryzae*) from *O. rufipogon*. Rice Genetics Newsletter. 1998;**15**:138.

[25] Gao DY, Xu ZG, Chen ZY, Sun LH, Sun QM, Lu F, Hu BS, Liu YF, Tang LH. Identification of a new gene for resistance to bacterial blight in a somaclonal mutant HX-3 (*indica*). Rice Genetics Newsletter. 2001;**18**:66–8.

[26] Gu K, Tian D, Yang F, Wu L, Sreekala C, Wang D, Wang GL, Yin Z. High-resolution genetic mapping of *Xa27*(t), A new bacterial blight resistance gene in rice, *Oryza sativa* L. Theoretical and Applied Genetics. 2004;**108**:800–7.

[27] Singh S, Sidhu JS, Huang N, Vikal Y, Li Z, Brar DS, Dhaliwal HS, Khush GS. Pyramiding three bacterial blight resistance genes (*xa5*, *xa13* and *Xa21*) using marker-assisted selection into *indica* rice cultivar PR106. Theoretical and Applied Genetics. 2001;**102**:1011–5.

[28] Joseph M, Gopalakrishnan S, Sharma RK, Singh VP, Singh AK, Singh NK, Mohapatra T. Combining bacterial blight resistance and basmati quality characteristics by phenotypic and molecular marker assisted selection in rice. Molecular Breeding. 2004;**13**:377–87.

[29] Sundaram RM, Vishnupriya MR, Biradar SK, Laha GS, Reddy AG, Rani NS, Sarma NP, Sonti RV. Marker assisted introgression of bacterial blight resistance in Samba Mahsuri, an elite *indica* rice variety. Euphytica. 2008;**160**:411–22.

[30] Sundaram RM, Vishnupriya MR, Laha GS, Rani NS, Rao PS, Balachandran SM, Reddy GA, Sarma NP, Sonti RV. Introduction of bacterial blight resistance into Triguna, a high yielding, mid-early duration rice variety by molecular marker assisted breeding. Biotechnology Journal. 2009;**4**:400–7.

[31] Perumalsamy S, Bharani M, Sudha M, Nagarajan P, Arul L, Saraswathi R, Balasubramanian P, Ramalingam J. Functional marker-assisted selection for bacterial leaf blight resistance genes in rice (*Oryza sativa* L.). Plant Breeding. 2010;**129**:400–6.

[32] Pandey MK, Rani NS, Sundaram RM, Laha GS, Madhav MS, Srinivasa Rao K, Sudharshan I, Hari Y, Varaprasad GS, Subba Rao LV, Suneetha K, Sivaranjani AKP, Viraktamath BC. Improvement of two traditional Basmati rice varieties for bacterial blight resistance and plant stature through morphological and marker-assisted selection. Molecular Breeding. 2013;**31**(1):239–46.

[33] Sakaguchi S. Linkage studies on the resistance to bacterial leaf blight, *Xanthomonas oryzae* (Uyeda et Ishiyama) Dowson, in rice (in Japanese. English summary). Bulletin of National Institute of Agricultural Science Series. 1967;**D16**:1–18.

[34] Yoshimura S, Umehara Y, Kurata N, Nagamura Y, Sasaki T, Minobe Y, Iwata N. Identification of a YAC Clone Carrying the *Xa1* allele, a bacterial blight resistance gene in rice. Theoretical and Applied Genetics. 1996;**93**:117–22.

[35] Yoshimura S, Kurata N, Sasaki T, Yoshimura A. Genetic analysis of bacterial blight resistance genes in rice by using molecular markers. Recent Advances in Breeding. 1994;**36**:33–6.

[36] He Q, Li D, Zhu Y, Tan M, Zhang D, Lin X. Fine mapping of *Xa2*, a bacterial blight resistance gene in rice. Molecular Breeding. 2006;**17**:1–6.

[37] Ezuka A, Horino O, Toriyama K, Shinoda H, Morinaka T. Inheritance of resistance of rice variety Wase Aikoku 3 to *Xanthomonas oryzae*. Bulletin of Tokai-Kinki National Agricultural Experimental Station. 1975;**28**:124–30.

[38] Yoshimura S, Nelson R, Yoshimura A, Mew TW, Iwata N. RFLP mapping of the bacterial blight resistance genes *Xa3* and *Xa4*. Rice Genetics Newsletter. 1992;**9**:136–8.

[39] Yoshimura S, Yoshimura A, Iwata N, McCouch S, Abenes M, Baraoidan M, Mew TW, Nelson RJ. Tagging and combining bacterial blight resistance genes in rice using RAPD and RFLP markers. Molecular Breeding. 1995;**1**:375–87.

[40] Petpisit V, Khush GS, Kauffman HE. Inheritance of resistance to bacterial blight in rice. Crop Science. 1977;**17**:551–4.

[41] Li ZK, Luo LJ, Mei HW, Paterson AH, Zhao XH, Zhang DB, Wang YP, Yu XQ, Zhu L, Stansel JW, Ying CS. A "defeated" rice resistance gene acts as a QTL against a virulent strain of *Xanthomonas Oryzae* pv. *oryzae*. Molecular & General Genetics. 1999;**261**:58–63.

[42] Wang W, Zhai W, Luo M, Jiang G, Chen X, Li X, Wing R A and Zhu L. Chromosome landing at the bacterial blight resistance gene *Xa4* locus using a deep coverage rice BAC library. Molecular Genetics & Genomics. 2001;**265**:118–25.

[43] Harushima Y, Yano M, Shomura A, Sato M, Shimano T, Kuboki Y, Yamamoto T, Lin SY, Antonio BA, Parco A, Kajiya H, Huang N, Yamamoto K, Nagamura Y, Kurata N, Khush GS and Sasaki T. A high-density rice genetic linkage map with 2275 markers using a single F2 population. Genetics. 1998;**148**:479–94.

[44] Wang C. Fine and Physical Mapping of Bacterial Blight Resistance Genes *Xa22*(t) and *Xa24*(t) in Rice (in Chinese). PhD thesis. Huazhong agricultural University, Wuhan, China; 1999.

[45] Sun X, Yang Z, Wang S and Zhang Q. Identification of a 47 kb DNA fragment containing *Xa4*, a locus for bacterial blight resistance in rice. Theoretical and Applied Genetics. 2003;**106**:683–7.

[46] Blair MW, Garris AJ, Iyer AS, Chapman B, Kresovich S, McCouch SR. High resolution genetic mapping and candidate gene identification at the xa5 locus for bacterial blight resistance in rice (*Oryza sativa* L.). Theoretical and Applied Genetics. 2003;**107**:62–73.

[47] Sidhu GS, Khush GS. Dominance reversal of a bacterial blight resistance gene in some rice cultivars. Phytopathology. 1978;**68**:461–3.

[48] Sidhu GS, Khush GS, Mew TW. Genetic analysis of bacterial blight resistance to seventy-four cultivars of rice *Oryza sativa* L. Theoretical and Applied Genetics. 1978;**53**:105–11.

[49] Kaji R, Ogawa T. Identification of the located chromosome of the resistance gene, *Xa-7*, to bacterial leaf blight in rice [in Japanese]. Breeding Science. 1995;**45**:79.

[50] Porter BW, Chittoor JM, Yano M, Sasaki T, White FF. Development and Mapping of markers linked to the rice bacterial blight resistance gene *Xa7*. Crop Science. 2003;**43**:1484–92.

[51] Chen S, Huang ZH, Zeng LX, Yang JY, Liu QG and Zhu XY. High-resolution mapping and gene prediction of *Xanthomonas oryzae* pv. *oryzae* resistance gene *Xa7*. Molecular Breeding. 2008;**22**:433–41.

[52] Vikal Y, Chawla H, Sharma R, Lore JS, Singh K. Mapping of bacterial blight resistance gene *xa8* in rice (*Oryza sativa* L.) Indian Journal of Genetics and Plant Breeding. 2014;74(4 Suppl.):589–95.

[53] Singh RJ, Khush GS, Mew TW. A new gene for resistance to bacterial blight in rice. Crop Science. 1983;**23**:558–60.

[54] Mew TW, Vera Cruz CM, Reyes RC. Interaction of *Xanthomonas campestris* pv. *oryzae* and a resistant rice cultivar. Phytopathology. 1982;**72**:786–9.

[55] Yoshimura A, Mew TW, Khush GS, Moura T. Inheritance of resistance to bacterial blight in rice cultivar Cas 209. Phytopathology. 1983;**73**:1409–12.

[56] Ramalingam J, Vera Cruz CM, Kukreja K, Chittoor JM, Wu JL, Lee SW, Baraoidan M, George ML, Cohen MB, Hulbert SH, Leach JE, Leung H. Candidate defense genes from rice, barley, and maize and their association with qualitative and quantitative resistance in rice. Molecular Plant-Microbe Interactions. 2003;**16**:14–24.

[57] Gu K, Sangha JS, Li Y, Yin Z. High resolution genetic mapping of bacterial blight resistance gene *Xa10*. Theoretical and Applied Genetics. 2008;**116**:155–63.

[58] Ogawa T, Yamamoto T. Inheritance of resistance to bacterial blight in rice. In: Rice Genetics. Proceedings of International Rice Genetics Symposium. IRRI, Manila, Philippines; 1986. pp. 471–80.

[59] Goto T, Matsumoto T, Furuya N, Tsuchiya K, Yoshimura A. Mapping of bacterial blight resistance gene *Xa11* on rice chromosome 3. Japanese Agricultural Research Quarterly. 2009;**43**:221–5.

[60] Ogawa T, Morinaka T, Fujii K, Kimura T. Inheritance of resistance of rice varieties of Kogyoku and Java14 to bacterial group V of *Xanthomonas oryzae*. Annals of Phytopathological Society of Japan. 1978;**44**:137–41.

[61] Ogawa T, Lin L, Tabien RE, Khush GS. A new recessive gene for resistance to bacterial blight of rice. Rice Genetics Newsletter. 1987;**4**:98–100.

[62] Zhang G, Angeles ER, Abenes MLP, Khush GS, Huang N. RAPD and RFLP mapping of the bacterial blight resistance gene *xa13* in rice. Theoretical and Applied Genetics. 1996;**93**:65–70.

[63] Sanchez AC, Ilag LL, Yang D, Brar DS, Ausubel F, Khush GS, Yano M, Sasaki T, Li Z, Huang N Genetic and physical mapping of *xa13*, a recessive gene bacterial blight resistance gene in rice. Theoretical and Applied Genetics. 1999;**98**:1022–8.

[64] Yang B, Sugio A, White FF. Os8N3 is a host disease-susceptibility gene for bacterial blight of rice. Proceedings of National Academy of Sciences, USA. 2006;**103**:10503–8.

[65] Taura S, Ogawa T, Tabien RE, Khush GS, Yoshimura A, Omura T. The specific reaction of Taichung Native 1 to Philippine races of bacterial blight and inheritance of resistance to race 5 (Pxo112). Rice Genetics Newsletter. 1987;**4**:101–2.

[66] Tan Z, Zhang Q, Zhu L, Wang C. RFLP mapping of a rice bacterial blight resistance gene *Xa-14*. Hereditas. 1998;20:30–3.

[67] Si-Yuan B, Ming-Pu T, Xing-Hua L. Genetic mapping of a bacterial blight resistance gene *Xa14* in rice. Acta Agronomica Sinica. 2010;**36**:422–7.

[68] Ogawa, T. Monitoring race distribution and identification of genes for resistance to bacterial leaf blight. In: Khush GS, editor. Rice Genetics III. Proceedings of the 3rd International Rice Genetics Symposium. International Rice Research Institute, Manila, Philippines; 1996. pp. 456–9.

[69] Nakai H, Nakamura K, Kuwahara S, Saito M. Genetic studies of an induced rice mutant resistant to multiple races of bacterial leaf blight. Rice Genetics Newsletter. 1998;**5**:101–3.

[70] Noda, T. and A. Ohuchi. A new pathogenic race of *Xanthomonas campestris* pv. *oryzae* and inheritance of resistance of differential rice variety, Tetep to it. Annals of the Phytopathological Society of Japan. 1989;**55**:201–7

[71] Ogawa, T., H. Kaku and T. Yamamoto, 1989. Resistance gene of rice cultivar, Asaminori to bacterial blight of rice. Japanese Journal of Breeding. **39** (Suppl. 1):196–7.

[72] Taura S, Ogawa T, Yoshimura A, Ikeda R, Omura T. Identification of a recessive resistance gene in induced mutant line XM5 of rice to bacterial blight. Japanese Journal of Breeding 1991;4:427–32.

[73] Taura S, Ogawa T, Yoshimura A, Ikeda R, Iwata N. Identification of a recessive resistance gene to rice bacterial blight of mutant line XM6, *Oryza sativa* L. Japanese Journal of Breeding. 1992;**42**(1):7–13.

[74] Devadath S. A strain of *Oryza barthii*, an African wild rice immune to bacterial blight of rice. Current Science. 1983;**52**:27–8.

[75] Khush GS, Mackill DJ, Sidhu GS. 1989. Breeding rice for resistance to bacterial blight. In: Bacterial Blight of Rice. International Rice Research Institute, Manila, Philippines;1989. pp. 207–17.

[76] Khush G S, Bacalangco E and Ogawa T. A new gene for resistance to bacterial blight from *Oryza longistaminata*. Rice Genetics Newsletter. 1990;**7**:121–2.

[77] Ronald PC, Albano B, Tabien R, Abenes L, Wu KS, McCouch S, Tanksley SD. Genetic and physical analysis of the rice bacterial blight disease resistance locus, *Xa21*. Molecular and General Genetics.1992;**236**:113–20.

[78] Wang GL, Ruan DL, Song WY, Sideris S, Chen L, Pi LY, Zhang S, Zhang Z, Fauquet C, Gaut B S, Whalen MC, Ronald PC. *Xa21D* encodes a receptor-like molecule with a leucine-rich repeat domain that determines race-specific recognition and is subject to adaptive evolution. The Plant Cell. 1998;**10**:769–75.

[79] Lin XH, Zhang DP, Xie YF, Gao HP, Zhang Q. Identification and mapping of a new gene for bacterial blight resistance in rice based on RFLP markers. Phytopathology. 1996;**86**:1156–9.

[80] Wang C, Tan M, Xu X, Wen G, Zhang D, Lin X. Localizing the bacterial blight resistance gene, *Xa22*(t), to a 100-kilobase bacterial artificial chromosome. Phytopathology. 2003;**93**:1258–62.

[81] Mir GN, Khush GS. Genetics of resistance to bacterial blight in rice cultivar DV86. Crop Research. 1990;**3**:194–8.

[82] Wu X, Li X, Xu C, Wang S. Fine genetic mapping of xa24, a recessive gene for resistance against *Xanthomonas oryzae* pv. *oryzae* in rice. Theoretical and Applied Genetics. 2008;**106**:1467–72.

[83] Chen H, Wang S, Zhang Q. New gene for bacterial blight resistance in rice located on chromosome 12 identified from Minghui 63, an elite restorer line. Phytopathology. 2002;**92**:750–4.

[84] Gao D Y, Liu MA, Zhou AH, Cheng Y, Xiang YH, Sun LH, Zhai WX. Molecular mapping of a bacterial blight resistance gene *Xa*-25 in rice. Journal of Genetics and Genomics = Yi Chuan Xue Bao. 2005;**32**:183–8.

[85] Amante-Bordeos A, Sitch LA, Nelson R, Dalmacio RD, Oliva NP, Aswidinnoor H, Leung H Transfer of bacterial blight and blast resistance from the tetraploid wild rice *Oryza minuta* to cultivated rice, *Oryza sativa*. Theoretical and Applied Genetics. 1992;**84**:345–54.

[86] Lee KS, Rasabandith S, Angeles ER, Khush GS. Inheritance of resistance to bacterial blight in 21 cultivars of rice. Phytopathology. 2003;**93**:147–52.

[87] Tan GX, Ren X, Weng QM, Shi ZY, Zhu LL, He GC. Mapping of a new resistance gene to bacterial blight in rice line introgressed from *O. officinalis*. Journal of Genetics and Genomics = Yi Chuan Xue Bao. 2004;**31**:724–9.

[88] Jin XW, Wang CL, Yang Q, Jiang QX, Fan YL, Liu GC, Zhao KJ. Breeding of near-isogenic line CBB30 and molecular mapping of *Xa30*(t), a new resistance gene to bacterial blight in rice. Scientia Agricultura Sinica. 2007;**40**:1094–100.

[89] Wang CT, Wen GS, Lin XH, Liu XQ, Zhang DP. Identifiation and fine mapping of the new bacterial blight resistance gene, *Xa31(t)*, in rice. European Journal Plant Pathology. 2009; **23**:235–40.

[90] Zheng CK, Wang CL, Yu YJ, Liang YT and Zhao KJ. Identification and molecular mapping of *Xa32*(t), a novel resistance gene for bacterial blight (*Xanthomonas oryzae* pv. *oryzae*) in rice. Acta Agronomica Sinica. 2009;**35**:1173–80.

[91] Korinsak S, Sriprakhon S, Sirithanya P, Jairin J, Korinsak S, Vanavichit A, Toojinda T. Identification of microsatellite markers (SSR) linked to a new bacterial blight resistance gene xa33(t) in rice cultivar 'Ba7'. Maejo International Journal of Science and Technology. 2009;**3**:235–47.

[92] Natarajkumar P, Sujatha K, Laha GS, Viraktamath BC, Reddy CS, Mishra B, Balachandran SM, Ram T, Srinivasarao K, Hari Y, Sundaram RM. Identification of a dominant bacterial blight resistance gene from *Oryza nivara* and its molecular mapping. Rice Genetics Newsletter. 2010;**25**:54–6.

[93] Ram T, Laha GS, Gautam SK, Ram Deen, Madhav MS, Brar DS, Viraktamath BC. Identification of a new gene introgressed from *Oryza brachyantha* with broad-spectrum resistance to bacterial blight of rice in India. Rice Genetics Newsletter. 2010;**25**:57.

[94] Guo SB, Zhang DP, Lin XH. Identification and mapping of a novel bacterial blight resistance gene *Xa35*(t) originated from *Oryza minuta*. Scientia Agricultura Sinica. 2010;**43**(13).

[95] Miao LL, Wang CL, Zheng CK, Che JY, Gao Y, Wen YC, Li GQ, Zhao KJ. Molecular mapping of a new gene for resistance to rice bacterial blight. Scientia Agricultura Sinica. 2010;**43**(15):3051–8.

[96] Kaur R, Grewal N, Das A, Vikal Y, Singh J, Bharaj TS, Sidhu JS, Singh K. Inheritance of bacterial blight resistance in two accessions of wild rice, *Oryza nivara*. Rice Genetics Newsletter. 2006;**22**:78–82.

[97] Zhang F, Zhuoa DL, Zhang F, Huang LY, Wang WS, Xu JL, Vera Cruz C, Li ZK, Zhou YL. *Xa39*, a novel dominant gene conferring broad-spectrum resistance to *Xanthomonas oryzae* pv. *oryzae* in rice. Plant Pathology 2014;**64**:568–75.

[98] Hutin M, Sabot F, Ghesquière A, Koebnik R, Szurek B. A knowledge-based molecular screen uncovers a broad-spectrum OsSWEET14 resistance allele to bacterial blight from wild rice. The Plant Journal. 2015;**84**:694–703.

[99] Hummel, AW, Doyle, EL, Bognadove AJ. Addition of transcription activator-like effector binding sites to a pathogen strain-specific rice bacterial blight resistance gene makes it effective against additional strains and against bacterial leaf streak. New Phytologist. 2012;**195**:883–93.

[100] Alfano JR, Collmer A. The type III (Hrp) secretion pathway of plant pathogenic bacteria: trafficking harpins, Avr proteins, and death. Journal of Bacteriology. 1997;**179**:5655–62.

[101] Lee BM, Park YJ, Park DS, Kang HW, Kim JG, Song ES, Park IC, Yoon UH, Hahn JH et al. The genome sequence of *Xanthomonas oryzae* pathovar *oryzae* KACC10331, the bacterial blight pathogen of rice. Nucleic Acids Research. 2005:**33**:577–86.

[102] Salzberg SL, Sommer DD, Schatz MC, Phillippy AM, Rabinowicz PD, Tsuge S, Furutani A, Ochiai H, Delcher AL et al. Genome sequence and rapid evolution of the rice pathogen *Xanthomonas oryzae* pv. *oryzae* PXO99A. BMC Genomics. 2008;**9**:204.

[103] Yang B, Zhu WG, Johnson WB, White FF. The virulence factor AvrXa7 of *Xanthomonas oryzae* pv. *oryzae* is a type III secretion pathway-dependent nuclear-localized double-stranded DNA-binding protein. Proceedings of National Academy of Sciences, USA. 2000;**97**:9807–12.

[104] Shen Y, Sharma P, Goes da Silva F, Ronald P. The *Xanthomonas oryzae* pv. *oryzae* raxP and raxQ genes encode an ATP sulphurylase and adenosine-5′-phosphosulphate kinase that are required for AvrXa21 avirulence activity. Molecular Microbiology. 2002;**44**:37–48.

[105] Bonas U, Schulete R, Fenselau S, Minsavage GV, Staskawicz BJ, Stall RE. Isolation of a gene cluster from *Xanthomonas campestris* pv. *vesicatoria* that determines pathogenicity and the hypersensitive response on pepper and tomato. Molecular Plant Microbe Interactions. 1991;**4**:81–8.

[106] Wengelnik K, Van den Ackerveken G, Bonas U. HrpG, a key hrp regulatory protein of *Xanthomonas campestris* pv. *vesicatoria* is homologous to two-component response regulators. Molecular Plant Microbe Interactions 1996;**9**:704–12.

[107] Wengelnik K, Bonas U. HrpXv, an AraC-type regulator, activates expression of five of the six loci in the hrp cluster of *Xanthomonas campestris* pv. *vesicatoria*. Journal of Bacteriology. 1996;**178**:3462–9.

[108] Arlat M, Gough CL, Barber C, Boucher C, Daniels MJ. *Xanthomonas campestris* contains a cluster of hrp genes related to the larger hrp cluster of *Pseudomonas solanacearum*. Molecular Plant Microbe Interactions. 1991;**4**:593–601.

[109] Schulte R, Bonas U. A *Xanthomonas* pathogenicity locus is induced by sucrose and sulfur-containing amino acids. The Plant Cell. 1992;**4**:79–86.

[110] Schulte R, Bonas U. Expression of the *Xanthomonas campestris* pv. *vesicatoria* hrp gene cluster, which determines pathogenicity and hypersensitivity on pepper and tomato, is plant inducible. Journal of Bacteriology. 1992;**174**:815–23.

[111] Tsuge S, Furutani A, Fukunaka R, Oku T, Tsuno K, Ochiai H, Inoue Y, Kaku H, Kubo Y. Expression of Xanthomonas oryzae pv. oryzae hrp genes in XOM2, a novel synthetic medium. Journal of General Plant Pathology. 2002;**68**:363–71.

[112] Bogdanove A J, Beer SV, Bonas U, Boucher CA, Collmer A, Coplin DL, Cornelis GR, Huang HC, Hutcheson SW et al. Unified nomenclature for broadly conserved hrp genes of phytopathogenic bacteria. Molecular Microbiology. 1996;**20**:681–3.

[113] Lahaye T, Bonas U. Molecular secrets of bacterial type III effector proteins. Trends in Plant Science. 2001;**6**:479–85.

[114] Tsuge S, Nakayama T, Terashima S, Ochiai H, Furutani A, Oku T, Tsuno K, Kubo Y, Kaku H. Gene involved in transcriptional activation of the hrp regulatory gene hrpG in *Xanthomonas oryzae* pv. *oryzae*. Journal of Bacteriology. 2006;**188**:4158–62.

[115] Lee SW, Jeong KS, Han SW, Lee SE, Phee BK, Hahn TR, Ronald P. The *Xanthomonas oryzae* pv. *oryzae* PhoPQ two-component system is required for AvrXA21 activity, hrpG expression, and virulence. Journal of Bacteriology. 2008;**190**:2183–97.

[116] White FF, Potnis N, Jones JB and Kloebnik R. The type III effectors of *Xanthomonas*. Molecular Plant Pathology. 2009;10:749–66.

[117] Bonas U, Stall RE, Staskawicz B. Genetic and structural characterization of the avirulence gene avrBs3 from *Xanthomonas campestris* pv. *vesicatoria*. Molecular and General Genetics. 1989;**218**:127–36.

[118] Hopkins CM, White FF, Choi SH, Guo A, Leach JE. Identification of a family of avirulence genes from *Xanthomonas oryzae* pv. *oryzae*. Molecular Plant Microbe Interactions. 1992;**5**:451–9.

[119] Boch J, Bonas U. *Xanthomonas* AvrBs3 family-type III effectors: discovery and function. Annual Review of Phytopathology. 2010;**48**(1):419–36.

[120] Bogdanove AJ, Schornack S, Lahaye T. TAL effectors: finding plant genes for disease and defense. Current Opinion in Plant Biology. 2010;**13**:394–401.

[121] Sugio A, Yang B, Zhu T, White FF. Two type III effector genes of *Xanthomonas oryzae* pv. *oryzae* control the induction of the host genes OsTFIIAg1 and OsTFX1 during bacterial blight of rice. Proceedings of National Academy of Sciences, USA. 2007;**104**:10720–5.

[122] Chen LQ, Hou BH, Lalonde S, Takanaga T, Hartung ML, Qu XQ, Guo WJ, Kim JG, Underwood W, Chaudhuri B, Chermak D, Antony G, White FF, Somerville SC, Mudgett MB, Frommer WB. Sugar transporters for intercellular exchange and nutrition of pathogens. Nature. 2010;**468**(7323):527–32.

[123] Li T, Huang S, Zhou J, Yang B. Designer TAL Effectors Induce Disease Susceptibility and Resistance to *Xanthomonas oryzae* pv. *oryzae* in Rice. Molecular Plant. 2013;**6**:781–9.

[124] Voulhoux R, Ball G, Ize B, Vasil ML, Lazdunski A, Wu LF, Filloux A. Involvement of the twin-arginine translocation system in protein secretion via the type II pathway. The EMBO Journal. 2001;**20**:6735–41.

[125] Jha G, Rajeshwari R, Sonti RV. Bacterial type two secretion system secreted proteins: double-edged swords for plant pathogens. Molecular Plant Microbe Interactions. 2005;**18**:891–8.

[126] da Silva ACR, Ferro JA, Reinach FC, Farah CS, Furlan LR et al. Comparison of the genomes of two *Xanthomonas* pathogens with differing host specificities. Nature. 2002;**417**:459–63.

[127] Ray SK, Rajeshwari R, Sonti RV. Mutants of *Xanthomonas oryzae* pv. *oryzae* deficient in general secretory pathway are virulence deficient and unable to secrete xylanase. Molecular Plant-Microbe Interactions. 2000;**4**:394–401.

[128] Dharmapuri S, Sonti RV. A transposon insertion in gumG homologue of *Xanthomonas oryzae* pv. *oryzae* causes loss of extracellular polysaccharide production and virulence. FEMS Microbiology Letters. 1999;**179**:53–9.

[129] Dharmapuri S, Yashitola J, Vishnupriya MR, Sonti RV. Novel genomic locus with atypical G+ C content that is required for extracellular polysaccharide production and virulence in *Xanthomonas oryzae* pv. *oryzae*. Molecular Plant Microbe Interactions. 2001;**14**:1335–9.

[130] Tang JL, Feng JX, Li QQ, Wen HX, Zhou DL, Wilson TJ, Dow JM, Ma QS, Daniels MJ. Cloning and characterization of the rpfC gene of *Xanthomonas oryzae* pv. *oryzae*: involvement in exopolysaccharide production and virulence to rice. Molecular Plant Microbe Interactions. 1996;**9**:664–6.

[131] Chatterjee S, Sonti RV. rpfF mutants of *Xanthomonas oryzae* pv. *oryzae* are deficient for virulence and growth under low iron conditions. Molecular Plant Microbe Interactions. 2002;**15**:463–71.

[132] Jeong KS, Lee SE, Han JW, Yang SU, Lee BM, Noh TH, Cha JS. Virulence reduction and differing regulation of virulence genes in rpf mutants of *Xanthomonas oryzae* pv. *oryzae*. The Plant Pathology Journal 2008;**24**:143–51.

[133] Wang LH, He Y, Gao Y, Wu JE, Dong YH, He C, Wang SX, Weng LX, Xu JL et al. A bacterial cell-cell communication signal with cross-kingdom structural analogues. Molecular Microbiology. 2004;**51**:903–12.

[134] Dow M. Diversification of the function of cell-to-cell signalling in regulation of virulence within plant pathogenic *Xanthomonas*. Science Signaling. 2008;**1**:pe23.

[135] He Y-W, Zhang L-H. Quorum sensing and virulence regulation in *Xanthomonas campestris*. FEMS Microbiology Reviews. 2008;**32**:842–57.

[136] Shen Y, Chern M, Silva FG, Ronald P. Isolation of a *Xanthamonas oryzae* pv. *oryzae* flagellar operon region and molecular characterization of flhF. Molecular Plant-Microbe Interactions. 2001;**14**:204–13.

[137] Tseng YH, Choy KT, Hung CH, Lin NT, Liu JY, Lou CH et al. Chromosome map of *Xanthomonas campestris* pv. *campestris* 17 with locations of genes involved in xanthan gum synthesis and yellow pigmentation. Journal of Bacteriology. 1999;**181**:117–25.

[138] Lee CK, Lee BM, Cho JY. Identification of new internal promoters of the *Xanthomonas oryzae* pathovar *oryzae* gum gene cluster. Biotechnology Letters 2008;**30**:521–7.

[139] Wang JC, So BH, Kim JH, Park YJ, Lee BM, Kang HW. Genome-wide identification of pathogenicity genes in *Xanthomonas oryzae* pv. *oryzae* by transposon mutagenesis. Plant Pathology. 2008;57:136–1145.

[140] Holland IB, Schmitt L, Young J. Type 1 protein secretion in bacteria, the ABC-transporter dependent pathway. Molecular Membrane Biology. 2005;**22**:29–39.

[141] da Silva FG, Shen Y, Dardick C, Burdman S, Yadav RC, de Leon AL, et al. Bacterial genes involved in type I secretion and sulfation are required to elicit the rice *Xa21*-mediated innate immune response. Molecular Plant Microbe Interactions. 2004;**17**:593–601.

[142] Burdman S, Shen Y, Lee SW, Xue Q, Ronald P. RaxH/RaxR: a two component regulatory system in *Xanthomonas oryzae* pv. *oryzae* required for AvrXa21 activity. Molecular Plant Microbe Interactions. 2004;**17**:602–12.

[143] Lee SW, Han SW, Sririyanum M, Park CJ, Seo YS, Ronald PC. A type I–secreted, sulfated peptide triggers *XA21*-mediated innate immunity. Science. 2009;**326**:850–3.

[144] Liu J, Wang X, Mitchell T, Hu Y, Liu X, Dai L, Wang GL. Recent progress and understanding of the molecular mechanisms of the rice–*Magnaporthe oryzae* interaction. Molecular Plant Pathology. 2010;**11**(3):419–27.

[145] Dai L, Liu X, Xiao Y, Wang G. Recent advances in cloning and characterization of disease resistance genes in rice. Journal of Integrative Plant Biology. 2007;**49**:112–9.

[146] Li H, Wang S. Disease resistance. In: Zhang Q, Wing RA, editors. Genetics and Genomics of Rice Plant Genetics and Genomics. Springer Science + Business Media, New York; 2013. pp. 161–75.

[147] Kou Y, Wang S. Bacterial blight resistance in rice. In: Varshney RK, Tuberosa R, editors. Translational Genomics for Crop Breeding, Volume 1; Biotic Stresses. John Wiley and Sons, Inc., Iowa, USA; 2013. pp. 11–30.

[148] Wang GL, Leung L. Molecular biology of host-pathogen interactions in rice diseases. In: Shimamoto K, editor. Molecular Biology of Rice. Springer-Verlag, Tokyo, Japan; 1998. pp. 201–32.

[149] Century KS, Lagman RA, Adkisson M, Morlan J, Tobias R, Schwartz K, Smith A, Love J, Ronald PC, Whalen MC. Short communication: developmental control of *Xa21*-mediated disease resistance in rice. The Plant Journal. 1999;**20**(2):231–6.

[150] Zhao J, Fu J, Li X, Xu C, Wang S. Dissection of the factors affecting development-controlled and race-specific disease resistance conferred by leucine-rich repeat receptor kinase-type R genes in rice. Theoretical and Applied Genetics. 2009;**119**(2):231–9.

[151] Park CJ, Lee SW, Chern M, Sharma R, Canlas PE, Song MY, Jeon JS, Ronald PC. Ectopic expression of rice Xa21 overcomes developmentally controlled resistance to *Xanthomonas oryzae* pv. *oryzae*. Plant Science. 2010;**179**(5):466–71.

[152] Wang G, Ding X, Yuan M, Qiu D, Li X, Xu C, Wang S. Dual function of rice OsDR8 gene in disease resistance and thiamine accumulation. Plant Molecular Biology. 2006;**60**(3):437–49.

[153] Peng Y, Bartley LE, Chen X, Dardick C, Chern M, Ruan R, Canlas PE, Ronald PC. OsWRKY62 is a negative regulator of basal and *Xa21*-mediated defense against *Xanthomonas oryzae* pv. *oryzae* in rice. Molecular Plant. 2008;**1**(3):446–58.

[154] Seo YS, Chern M, Bartley LE, Han M, Jung KH, Lee I,Walia H, Richter T, Xu X, Cao P, Bai W, Ramanan R, Park CJ, Chen X, Hwang S, Jeon JS, Ronald PC. Towards establishment of a rice stress response interactome. PLoS Genetics. 2011;**7**(4):e1002020.

[155] Park CJ, Peng Y, Chen X, Dardick C, Ruan D, Bart R, Canlas PE, Ronald PC. Rice XB15, a protein phosphatase 2C, negatively regulates cell death and *XA21*-mediated innate immunity. PLoS Biology. 2008;**6**(9):e231.

[156] Chen X, Chern M, Canlas PE, Ruan D, Jiang C, Ronald PC. An ATPase promotes autophosphorylation of the pattern recognition receptor XA21 and inhibits XA21-mediated immunity. Proceedings of National Academy of Sciences, USA. 2010;**107**(17):8029–34.

[157] Park CJ, Bart R, Chern M, Canlas PE, Bai W, Ronald PC. Overexpression of the endoplasmic reticulum chaperone BiP3 regulates *XA21*-mediated innate immunity in rice. PLoS One. 2010;**5**(2):e9262.

[158] Xu Z, Sun Q, Liu F, Chen Z, Hu B, Guo Y, Liu Y, Liu H. Race monitoring of rice bacterial blight (*Xanthomonas oryzae* pv. *oryzae*) in China. Chinese Journal of Rice Science. 2004;**18**(5):469–72.

[159] Gao J, Zhao J, Xu C, Li X, Wang S. Development of rice germplasms conferring high-level and broad spectrum resistance to *Xanthomonas oryzae* pv. *oryzae* at both seedling and adult stages. Molecular Plant Breeding. 2010;**8**(3):420–5.

[160] Li H, Li X, Xiao J, Wing RA, Wang S. Ortholog alleles at *Xa3/Xa26* locus confer conserved race-specific resistance against *Xanthomonas oryzae* in rice. Molecular Plant. 2012;**5**(1):281–90.

[161] Cao Y, Duan L, Li H, Sun X, Zhao Y, Xu C, Li X, Wang S. Functional analysis of *Xa3/Xa26* family members in rice resistance to *Xanthomonas oryzae* pv. *oryzae*. Theoretical and Applied Genetics. 2007;**115**(7):887–95.

[162] Zhang H, Cao Y, Zhao J, Li X, Xiao J, Wang S. A pair of orthologs of a leucine-rich repeat receptor kinase-like disease resistance gene family regulates rice response to raise temperature. BMC Plant Biology. 2011;**11**(1):160.

[163] Yang Z, Sun X, Wang S, Zhang Q. Genetic and physical mapping of a new gene for bacterial blight resistance in rice. Theoretical and Applied Genetics. 2003;**106**(8):1467–72.

[164] Cao Y, Ding X, Cai M, Zhao J, Lin Y, Li X, Xu C, Wang S. The expression pattern of a rice disease resistance gene *Xa3/Xa26* is differentially regulated by the genetic backgrounds and developmental stages that influence its function. Genetics. 2007;**177**(1):523–33.

[165] Qiu D, Xiao J, Ding X, et al. OsWRKY13 mediates rice disease resistance by regulating defense-related genes in salicylate- and jasmonate dependent signaling. Molecular Plant Microbe Interactions. 2007;**20**:492–9.

[166] Qiu D, Xiao J, Xie W, Cheng H, Li X, Wang S. Exploring transcriptional signalling mediated by OsWRKY13, a potential regulator of multiple physiological processes in rice. BMC Plant Biology. 2009;**9**:74.

[167] Tao Z, Liu H, Qiu D et al. A pair of allelic WRKY genes play opposite roles in rice-bacteria interactions. Plant Physiology. 2009;**151**:936–48.

[168] Deng H, Liu H, Li X, Xiao J, Wang S. A CCCH-type zinc finger nucleic acid-binding protein quantitatively confers resistance against rice bacterial blight disease. Plant Physiology. 2012;**158**:876–89.

[169] Xiao W, Liu H, Li Y et al. A rice gene of de novo origin negatively regulates pathogen-induced defense response. PLoS One. 2009;**4**:e4603

[170] Römer P, Recht S, Strauss T, Elsaesser J, Schornack S, Boch J, Wang S, Lahaye T. Promoter elements of rice susceptibility genes are bound and activated by specific TAL effectors from the bacterial blight pathogen, *Xanthomonas oryzae* pv. *oryzae*. New Phytologist. 2010;**87**(4):1048–57.

[171] Yuan T, Li X, Xiao J, Wang S. Characterization of *Xanthomonas oryzae*-responsive cis-acting element in the promoter of rice race-specific susceptibility gene *Xa13*. Molecular Plant. 2011;**4**(2):300–9.

[172] Yuan M, Chu Z, Li X, Xu C, Wang S. The bacterial pathogen *Xanthomonas oryzae* overcomes rice defenses by regulating host copper redistribution. The Plant Cell. 2010;**22**:3164–76.

[173] Chu Z, Yuan M, Yao J, et al. Promoter mutations of an essential gene for pollen development result in disease resistance in rice. Genes and Development. 2006;**20**:1250–5.

[174] Chen LQ, Qu XQ, Hou BH, Sosso D, Osorio S, Fernie AR, Frommer WB. Sucrose efflux mediated by SWEET proteins as a key step for phloem transport. Science. 2012;**335**(6065):207–11.

[175] Wu L, Goh ML, Sreekala C, Yin Z. *XA27* depends on an amino-terminal signal-anchor-like sequence to localize to the apoplast for resistance to *Xanthomonas oryzae* pv oryzae. Plant Physiology. 2008;**148**(3):1497–509.

[176] Boch J, Scholze H, Schornack S, Landgraf A, Hahn S, Kay S, Lahaye T, Nickstadt A, Bonas U. Breaking the code of DNA binding specificity of TAL-type III effectors. Science. 2009;**326**(5959):1509–12.

[177] Gu K, Tian D, Qiu C, Yin Z. Transcription activatorlike type III effector AvrXa27 depends on OsTFIIAgamma5 for the activation of Xa27 transcription in rice that triggers

disease resistance to *Xanthomonas oryzae* pv. *oryzae*. Molecular Plant Pathology. 2009;**10**(6):829–35.

[178] Bernoux M, Ellis JG, Dodds PN. New insights in plant immunity signaling activation. Current Opinion in Plant Biology. 2011;**14**:512–8.

[179] Luo S, Zhang Y, Hu Q, Chen J, Li K, Lu C, Liu H, Wang W, Kuang H. Dynamic nucleotide-binding site and leucine-rich repeat encoding genes in the grass family. Plant Physiology. 2012;**159**:197–210.

[180] Zhang H, Wang S. Rice versus *Xanthomonas oryzae* pv. *oryzae*: a unique pathosystem. Current Opinion in Plant Biology. 2013;**16**:188–95.

[181] Zeng X, Tian D, Gu K, Zhou Z, Yang X, Luo Y, White FF, Yin Z. Genetic engineering of Xa10 promoter for broad-spectrum and durable resistance to *Xanthomonasoryzaepv. oryzae*. Plant Biotechnology Journal. 2015. doi: 10.1111/pbi.12342.

[182] Wang CL, Qin TF, Yu HM, Zhang XP, Che JY, Gao Y, Zheng CK, Yang B, Zhao KJ. The broad bacterial blight resistance of rice line CBB23 is triggered by a novel TAL effector of *Xanthomonas oryzae pv. oryzae*. Molecular Plant Pathology. 2014;**15**:333–341.

[183] Bhasin H, Bhatia D, Raghuvanshi S, Lore JS, Sahi GK, Kaur B, Vikal Y, Singh K. New PCR-based sequence tagged site marker for bacterial blight resistance gene *Xa38* in rice. Molecular Breeding. 2012;**30**:607–11.

[184] Yue JX, Meyers BC, Chen JQ, Tian D, Yang S. Tracing the origin and evolutionary history of plant nucleotide-binding siteleucine-rich repeat (NBS-LRR) genes. New Phytologist. 2012;**193**:1049–63.

[185] Monosi B, Wisser RJ, Pennill L, Hulbert SH. Full-genome analysis of resistance gene homologues in rice. Theoretical and Applied Genetics. 2004;**109**:1434–47.

[186] Mchale L, Tan X, Koehl P, Michelmore RW. Plant NBS-LRR proteins: adaptable guards. Genome Biology. 2006;**7**:212.

[187] Collins N, Drake J, Ayliffe M, Sun Q, Ellis J, Hulbert S, Pryor T. Molecular characterization of the maize Rp1-D rust resistance haplotype and its mutants. The Plant Cell. 1999;**11**:1365–76.

[188] Wang X, Chen J, Yang Y, Zhou J, Qiu Y, Yu C, Cheng Y, Yan C, Chen J. Characterization of a novel NBS-LRR gene involved in bacterial blight resistance in rice. Plant Molecular Biology Reporter. 2013;**31**:649. doi:10.1007/s11105-012-0537-0.

[189] Iyer-Pascuzzi AS, Jiang H, Huang L, McCouch SR. Genetic and functional characterization of the rice bacterial blight disease resistance gene *xa5*. Phytopathology. 2008;**98**(3):289–95.

[190] Zou H, Zhao W, Zhang X, Han Y, Zou L, Chen G. Identification of an avirulence gene, avrxa5, from the rice pathogen *Xanthomonas oryzae* pv. *oryzae*. Science in China Series C Life Sciences. 2010;**53**(12):1440–9.

[191] Bhatia D, Sharma R, Vikal Y, Mangat GS, Mahajan R, Sharma N, Lore JS, Singh N, Bharaj TS, Singh K. Marker-assisted development of bacterial blight resistant, dwarf, and high yielding versions of two traditional basmati rice cultivars. Crop Science. 2011;**51**:759–70.

[192] Li C, Wei J, Lin Y, Chen H. Gene silencing using the recessive rice bacterial resistance gene *xa13* as a new paradigm in plant breeding. Plant Cell Reports. 2012;**21**:851–62.

Chapter 11

Upland Rice Breeding in Uganda: Initiatives and Progress

Jimmy Lamo, Pangirayi Tongoona, Moussa Sie,
Mande Semon, Geoffrey Onaga and Patrick Okori

Additional information is available at the end of the chapter

http://dx.doi.org/10.5772/66826

Abstract

Until recently, there was limited research on breeding upland rice varieties. Moreover, there is an increasing expansion of rice production from traditional irrigated production areas to rain-fed environments in the East African region, where drought problem is a serious challenge. To date, several initiatives aimed at increasing rice production have been made. Of the initiatives, promotion of upland rice production has been the most important in Uganda, but yield penalty due to drought continued to be a major drawback. This article traces progress in the upland rice breeding that started with improvement of late maturing varieties that had nonpreferred cooking qualities. Initially, introduced lines were evaluated and released. These varieties are the 'New Rice for Africa' (NERICA) that had been generated from interspecific crosses involving *Oryza glaberrima* and *Oryza sativa.* Several studies to understand the mode of gene action and modified pedigree breeding approaches for drought tolerance were conducted and used to develop new rice varieties. Up to 11 improved upland rice varieties were released and deployed in the country from 2002 to 2011 as a result of this initiative.

Keywords: drought tolerance, rice, gene action, NERICA, modified pedigree breeding

1. Introduction

Rice is an important food crop that is consumed mostly outside its major production areas in Uganda, with over 90% of production marketed to urban areas and major institutions within the country. This aspect makes rice to have a long value chain engaging several players. Rice is cultivated under rain-fed upland conditions, partly rain-fed lowland conditions and

	2006 B	2007 A	2007 A	2007 B	2008 A	2008 B	2009 A	2009 B	2010 A	2010 B	2011 A	2011 B	2012 A	2012 B
New lines bred in Uganda														
Selection of parental lines														
Making crosses and generating F2 (genetic studies)														
Evaluation of F2 and harvesting F3														
Evaluating F3 under rain-fed lowland and upland conditions (660)														
Evaluation F4 upland conditions														
Evaluation F5 in Advanced Trials														
Fixed lines introduced														
Evaluation F6 in NPT (12 lines)														
Evaluation F6 in NPT (10 lines)														
Testing for grain traits														
Organoleptic testing of the new varieties														
PVS selection of varieties														

Table 1. Timelines of [21]rice breeding activities 2006–2012.

irrigated conditions in Uganda, taking advantage of diverse ecosystems in Uganda [1]. Since the introduction of rice in 1904, Uganda had production under different agro-ecological conditions covering rain-fed upland conditions, partly rain-fed lowland conditions and irrigated conditions. Various production challenges are faced in these production areas.

New technologies are valuable for use in developing the new rice varieties globally including Africa. However, new tools can be most helpful if the existing varieties and candidate lines are properly characterized and documented. The purpose of this paper is to trace upland rice breeding efforts in Uganda and present for alignment, learning, and application in the new technology in their rice breeding programs with focus on breeding for drought tolerance and other stresses. This is critical considering that there is limited research on development of upland rice varieties suitable for production under mild drought conditions in the East African region and other similar Agro ecologies. There is also increasing expansion of rice production from traditional irrigated production areas to rain-fed environments, where drought problem is an inherent challenge. Indeed, drought emerged as a critical rice production constraint in East Africa [1, 2], particularly in Uganda [1], as promotion of upland rice was growing in the country.

Many upland rice varieties, earlier introduced in the country, were late maturing and did not have preferred cooking qualities. Later, more introduced lines were evaluated and released. These varieties had been generated through interspecific crossing involving *Oryza glaberrima* and *Oryza sativa*. These new genotypes were called the 'New Rice for Africa' (NERICA). They were resistant to major biological constraints but showed differential sensitivity to drought stress and new diseases, especially brown spot disease and narrow leaf spot disease. Besides, these varieties had nonaromatic characteristic which are the major concerns of the Uganda farmers. These factors made upland rice farmers to realize low yield mainly due to frequent drought stress. In addition, extensive use of irrigated rice come along with other limitations, namely need for environmental impact assessment and conflict on cultural values for use of the wetlands for farming. Subsequently, upland rice breeding involving adapted varieties led to new rice varieties with high resistance to biotic stresses and preferred agronomic traits [3]. However, abiotic stresses, especially drought stress remained a major constraint. Indeed, breeding for drought tolerance resistance in rice is challenging because the trait is quantitative and involves polygenes with low heritability. Modified pedigree breeding approaches were used in this breeding. In this paper, we review a trend of improvement of upland rice in Uganda covering three aspects: (1) screening of introductions for drought tolerance, (2) mode of gene action for drought tolerance, (3) evaluation of segregating lines, (4) Evaluation of promising lines, (5) variety release and status of deployment of the new generations of rice varieties in Uganda and within the African region. Detailed timelines of the activities are presented in **Table 1**.

2. Methodology

2.1. Screening of introductions for drought tolerance

A total of 191 rice introductions from major rice breeding centers were evaluated. Of the 191 materials, 77 were *O. sativa indica* comprising 45 from African Rice Centre (ARC), 15 lines from International Rice Research Institute (IRRI), 13 from Mali, three from Uganda, and one from China. Among the introductions, there were three *O. glaberrima* accessions. The remaining 111 were interspecific lines developed from *O. sativa* × *O. glaberrima* crosses, comprising 18 from ARC and 93 the International Center for Tropical Agriculture (CIAT), Colombia coded as the CT series. However, among the interspecific samples, two genotypes namely WAB 880-1-27-9-2-P1-HB and WAB 450-24-2-3-P-38-1-HB were duplicates from different repeated introductions from IRRI and WARDA-Africa Rice Center. The 93 interspecific lines from CIAT were BC_4F_1s developed from crossing CAIAPO, a tropical *O. sativa* japonica from Colombia with RAM 24 (*O. glaberrima*).

This experiment was conducted at National Crop Resources Institute (NaCRRI), at Namulonge in central Uganda, at 00°32′ N latitude and 32°53′ E longitudes with altitude of 1150 m above sea level during dry season. The soils of the place are clay loam. The period December to March is characteristically the long dry season, but mean long term annual rainfall is 1270 mm.

In order to assess drought stress during reproductive growth stage, drought stress was imposed by terminating irrigation, when about 50% of the population had reached a point where interauricular distance between the flag leaf and penultimate leaf was zero [4]. It is the period when it is about 10 days before anthesis. It is the time when the penultimate leaves were fully expanded. Rainfall during the trial period was recorded. Irrigation was 14 days later, when 30% of the available water had been lost from the soil at 20-cm depth. The available soil moisture was taken using the ECHO soil moisture tester (Decagon Devices, Inc., Pullman, Washington, USA). All the grains from each panicle were hand threshed and dried. The filled and unfilled grains were then separated using floatation methods.

2.2. Mode of gene action for drought tolerance

This study investigated the nature of inheritance of drought tolerance in crosses between interspecific and intraspecific rice genotypes using secondary traits. Two separate experiments were conducted, using *O. sativa* and fixed interspecific lines derived from *O. glaberrima* and *O. sativa* crosses.

Experiment 1: Genetic studies on drought tolerance traits

The aim of the first experiment was to investigate the inheritance of drought tolerance at reproductive growth stage. Eighteen crosses were generated from two sets of 3 × 3 parents using the North Carolina mating design II (NCD II). All the 18 F2 and the 12 parents were evaluated in a 2 × 15 alpha lattice design with two replicates under a rain-out shelter and nonstress conditions in the field.

Thirty genotypes comprising 18 F2 progenies from sets, A and B, along with the 12 parents were used in this experiment (**Table 2**). The 30 entries were established in a rain-out shelter at National Crops Resources Research Institute (NaCRRI), Namulonge. The rain-out shelter was constructed using translucent sheets for the roof and wire mesh on the sides of the structure to prevent rain water and to allow free air circulation, respectively. In the rain-out shelter, standard troughs that are 1m wide, 8m long, and 1.5m deep were filled with soil for fallow field from Namulonge. Four troughs were made and filled with the soil, referred to as strips. The

http://dx.doi.org/10.5772/66826

Experiment	Crossing set	Genotype no	Breeding line	Type	Parent type
1	SET A	18	CT 16334(2)-CA-2-M	Interspecific	Male
		105	WAB 365-B-1H1-HB	*O. sativa*	Male
		134	NERICA 9	Interspecific	Male
		138	NERICA 8	Interspecific	Female
		193	NERICA 13	Interspecific	Female
		196	IRAT 325	*O. sativa*	Female
	SET B	2	CT 16346-CA-20-M	Interspecific	Male
		9	CT 16350- CA-5-M	Interspecific	Male
		12	CT 16344-CA-9-M	Interspecific	Male
		96	Bonanca	*O. sativa*	Female
		121	WITA 2	*O. sativa*	Female
		129	CK 73	*O. sativa*	Female
2	SET C	18	CT 16334 (2)-CA-2-M	Interspecific	Female
		138	WAB 450-1-BL1-136-HB	Interspecific	Male

Table 2. Rice genotypes used for generating sets of F1 for drought tolerance.

seeds were planted in a 2 × 15 alpha lattice design. Two strips represented a replicate. The 12 parental genotypes were planted in three rows planted across the 1 m width strips, while the F2 populations were planted in six rows. The plant to plant spacing was 15 cm making plant population to be 36 for the parental lines and 72 for the F2 lines.

A second set of the 30 entries were planted in the field under optimal conditions. These conditions involved irrigating the field at 20 mm per week, during the period when there was no rain. In both trials, a 2 × 15 alpha lattice design planted in two replicates was used. Two seeds from each generation were drilled at a depth of 3 cm at spacing of 20 × 20 cm in each plot. In order to reduce border effects, 20 cm was left between plots. The 12 parental genotypes were planted in 5-row and 3-column plots, while the F2 populations were planted in 5-row and 6-column plots. Overall, there were 15 plants per replicate of the parents and 30 plants per replicate of each F2 genotype, thus the total number of plants were 30 and 60 for the parents and F2, respectively. The plants were thinned to one plant per hill. Standard cultural practices including hand planting and hand weeding were followed. The crops were fertilized with 25 kg N ha^{-1} at 20–25 days after transplanting (DAT) and the same rate at 40–45 DAT to enhance plant vigor.

Drought stress was imposed by terminating irrigation, when about 50% of the populations had attained an interauricular distance between the flag leaf and penultimate leaf of zero, that is the period about 10 days before anthesis [4, 5]. This method of identifying the stage of imposing drought was applied both in the field and in the rain-out shelter. In general, this is the time when the penultimate leaves were fully expanded. Rainfall during the trial period was recorded.

In the field experiment, irrigation was applied using sprinkler irrigation. The field was irrigated, every three days before imposing drought stress. On the day the irrigation was terminated, the field was irrigated to field capacity in the evening between 5:00 and 6:00 pm, which was resumed 14 days after its termination using sprinkler irrigation. The duration of drought stress was determined by testing the level of soil moisture daily, using the ECHO soil moisture tester (Decagon Devices, Inc Pullman, Washington USA). On the day, when 30% of the available water had been lost from the soil at 20-cm depth, irrigation was resumed. In the rain-out shelter, water was applied using hand irrigation cans but water was calculated for each strip at 140 L per week, which is equivalent to 20 mm per week.

The number of filled grains was counted per panicle at grain maturity period. Two panicles from each plant were randomly collected and record of number of filled grains was determined using floatation method described by these authors [6].

Experiment 2: Generation means analysis (GMA) for filled grains in rice

In the second experiment, the magnitude and direction of gene action for drought tolerances at reproductive stage was determined in five populations P1, P2, F1, F2, and F3 generated from a drought tolerant × susceptible cross using generation mean analysis (GMA). They are in set C (**Table 2**). The materials were planted in the dry season and drought was imposed by terminating at the stage of panicle initiation.

In this experiment, all the five populations generated from crossing; parents P1 and P2, their F1, F2 and F3 genotypes were planted following a randomized complete block design (RCBD) with two replicates. Two seeds from each generation were drilled at a depth of 3 cm at spacing of 20 × 20 cm in each experimental unit (plot) in the field at NaCRRI. The generations P1, P2, and F1 were planted in 5-row and 3-column plots, while F2 and F3 were planted in 5-row and 6-column plots. Overall, there were 15 plants per replicate of the parents, 30 plants per replicate of each F2 genotype, thus the total numbers of plants were 30 and 60 for the parents, F2, respectively. The cultural practice in experiment 1 was followed, and drought stress was imposed following procedures in experiment 1.

2.2.1. Data analysis

Data was analyzed in three parts, namely analyses of variance, residual maximum likelihood (REML), regression, and generation means. The analysis of variance was performed for different traits associated with drought tolerance in the two sets of populations, A and B, pooling for both stress and nonstress environments. Using REML, the separate sets were analyzed for each trait. The analyses of the variance components of genotypes were further partitioned into variations, due to parents and crosses.

General analyses of variance were performed for filled grains, grains per panicle, leaf area, plant height, tiller number, and panicle number of all hybrids including checks. Genetic analyses for the six parameters of experimental hybrids were then performed in GenStat [7] as a fixed effects model across two locations [8] as follows:

Generation mean analysis of the genotypes CT 16334 (2)-CA-2-M crossed with WAB 450-1-BL1-136-HB was used to determine additive, dominant, and epistatic effects following the

http://dx.doi.org/10.5772/66826

model [9]. The various generations did not have equal variances; therefore, weighted inverse of the variances was used in subsequent analysis according to these authors in Ref. [10]. Regression analysis procedures were used to find the best fit model. It is a graphical method used to compare the additive model with additive-dominance models. Any effect that was not significant at 5% level was excluded from the model. The parameters were fitted using weighted mean squares as described by Ref. [11].

A scaling test was conducted using linear combinations of various means according to Refs. [9–12] to detect the presence of nonallelic interactions that are known to bias estimates of additive and dominance components in the populations when present. However, in this case, where F3 populations are used instead of backcross populations, the additive effects estimate is for both additive effects and additive × additive interaction effects. Similarly, the dominance effect combined both dominance effects and dominance × dominance interaction effects as a single estimate [12]. This is not a major drawback considering that most breeding work exploits additive effects and dominance effects. Standard errors of generation means were computed by performing nested analysis of variance following methods used in Ref. [13].

In order to verify the number of genes involved in the transmission of traits associated with drought tolerance, Castle-Wrights formulae described in Ref. [14] was used.

2.3. Evaluation of segregating lines

2.3.1. Preliminary evaluation 1

Preliminary yield trials were conducted on station with objective of varietal screening, evaluation, and seed increase. These were F3 selections from the previous experiment. Overall 660 genotypes were selected from the F3 generation based on field performance. All the seed from each of the 660 hills of F3 genotypes were divided into three sets. One set was remnant; a second set was planted in the rain-fed low-land environment, while the third set was planted under rain-fed upland conditions, all on station within Namulonge. The planting was in November, 2010. All the seed from each hill was planted to 5-m long rows and evaluated.

The second set was planted under rain-fed low-land with ample moisture throughout the growth period of the lines. The evaluation focused on maturity period, tillering capacity, presence of foliar diseases, and physical grain characteristics. Lines that had longer maturity period than the variety NERICA-4, number of reproductive tillers less than 5 per hill, presence of foliar diseases, and grain discoloration were eliminated. Besides, infection by common pathogen namely rice blast, bacterial leaf blight, grain discoloration, and sheath rot were used to eliminate lines. The team that evaluated the materials comprised of scientists, farmers, and rice field workers.

The third set comprising all the 660 lines was planted under rain-fed upland conditions. Selection was made as previously stated for rain-fed low-land conditions. Unlike selection under rain-fed lowland conditions where a minimum of seven productive tillers was considered acceptable, in this production environment, five productive tillers was considered the minimum.

2.3.2. *Preliminary evaluation 2*

Set 1: Evaluation of 84 rain-fed lowland rice lines: a total 84 lines of F4 segregating populations were selected from the 660 F3 lines genotypes and planted in five sites namely Namulonge, Kigumba, Kibaale, Lira, and Doho. Each entry was planted in a 3 × 18 alpha lattice design at spacing of 20 × 20 cm planted in rows five plots each 5-m long. The evaluation had three main objectives. The first objective was to test the new genotypes under varying stress conditions. The major biotic and abiotic stresses targeted were drought stress, rice blast, RYMV, BLB and Leaf Streak, narrow leaf spot and brown leaf spot. The locations selected were major rice growing areas that had had the production constraints. The second objective was to assess yield of the whole set at Namulonge site. The third objective was to identify farmer preferred varieties using participatory variety selection method.

2.4. Evaluation of promising lines

2.4.1. *On-farm evaluation*

On-farm trials were conducted through participatory and multilocational testing of selected upland varieties. Selection of sites for participatory and multilocational testing considered the following: (i) key representative ecological zones, (ii) participation of stakeholders, and (iii) availability of resources to effectively conduct the exercise. Seed companies were invited and a proposed method of allowing most seed stakeholders to participate in variety evaluation was adhered to. Twenty lines that were tested in 2011B in two locations namely Namulonge and Kibaale. Subsequently, 12 lines were tested in 2012A and finally 8–10 in 2012 B.

In order to identify suitable upland rice varieties, the best rice lines from preliminary trials were submitted to advanced yield trial (AYT). These genotypes were WAB 95 B-B-40-HB (the best performing line among lines received through STRASA and two best lines selected from Upland Regional performance trial (ART3-11L1P1-B-B-2 and ART8-L15P14-1-2-1) as well as three genotypes that performed well among 600 new lines developed at Namulonge. The 2011 season II was suitable for selecting high yielding diseases resistant. For instance, WAB95-B-B-40-HB and WAB788-16-3-2-1-HB earlier selected had considerable symptoms of BLS and narrow leaf spot.

2.5. Variety release and status

In the year 2013, six best performing rice varieties were presented for release to the Variety Release Committee in Uganda. Among the traits and Characteristics that was provided as evidence of superiority to the existing rice varieties were, higher yield, preferred grain and cooking qualities, maturity, tolerance to stresses especially drought.

3. Results

3.1. Screening introductions for drought tolerance

A list of only 30 genotypes, including top 20 and bottom 10 least performing genotypes, in terms of filled grains are presented in **Table 3**. Among the top 20 genotypes, three namely

http://dx.doi.org/10.5772/66826

No.	Genotypes	Filled grains (%)
Top 20 genotypes		
112	WAB 56-50	96.3
53	CT 16333(1)-CA-18-M	91.1
34	CT 16326-CA-3-M	89.1
101	NERICA 14	88.7
108	WAB 56-39	88.6
137	NERICA 7	88.1
132	CO 39	87.8
142	VANDANA	87.6
124	NERICA 6	87.4
83	CT 16340-CA-9-M	86.7
190	NERICA 17	86.6
45	CT 16329-CA-10-M	85.5
177	WBK 35 (F3)	84.9
92	CT 16315(1)-CA-1-M	84.7
1	CT 16330(1)-CA-2-M	84.1
165	IR 64	83.8
188	NERICA 15	83.5
90	CT 16307-CA-5-M	83.5
10	CT 16353-CA-17-M	83.4
30	CT 16324-CA-10-M	83.1
Bottom 10 genotypes		
169	IR 57514-PMI 5-B-1-2	49.4
80	CT 16316-CA-2-M	49.4
106	IDSA 6	49.1
104	ITA 123 (FKR 28)	47.9
175	RAM 118	47.8
49	CT 16346-CA-11-M	47.8
32	CT 16312(1)-CA-1-M	47.3
166	IR 77298-14-1-2	45.7
155	LAC 23	43.8
65	CT 16307(1)-CA-2-M	27.9

No.	Genotypes	Filled grains (%)
Overall	**Mean**	**67**
	$LSD_{0.05}$	**1.88**
	CV%	**13.2**
	Range/LSD	**17.6**
	Variance	**18.6**

Table 3. The top 20 and bottom 10 genotypes in terms of percent filled grains.

NERICA 7, CO 39, and VANDANA were reference materials for high drought tolerance at reproductive growth stage. There were nine out of the 20 lines from the CT breeding lines and five from the NERICA generations.

3.2. The mode of gene action for drought tolerance

3.2.1. Gene action

Generalized linear analysis for different traits pooled across sets and sites are presented in **Table 4**. Results showed that both GCA and SCA effects within sets for filled grains, grains per panicle, leaf area tiller number, and number of panicles per plant were significant (P = 0.001), while only the GCA effects within sets for tiller number were significant (P = 0.001) but not the SCA effects.

The male and female mean squares were all significant (P = 0.05) for the filled grains under drought stress (DS) and nondrought stress (NDS) conditions for the A set population (**Table 5**).

Source of variation	Mean square value						
	d.f	Spikelet fertility	Grains per panicle	Leaf area	Plant height	Tiller number	Panicle number
Env[1]	1	146.5^{***}	1441.2^{***}	1335.9^{***}	109.3^{***}	1534.8^{***}	1708.8^{***}
Set	1	41.5^{***}	13.6^{***}	35.0^{***}	10.1^{***}	0.7	14.1^{***}
Set/GCA_f	4	10.1^{***}	10.5^{***}	17.3^{***}	7.3^{***}	3.2^{**}	5.1^{***}
Set/GCA_m	4	5.8^{***}	11.4^{***}	12.8^{***}	26.5^{***}	8.3^{***}	3.8^{**}
Set/SCA	8	3.6^{***}	8.5^{***}	8.3^{***}	26.5^{***}	1.2	14.6^{***}
Env × Set	1	14.9^{***}	15.4^{***}	35.4^{***}	1.9	6.3^{**}	13.2^{***}
Env × Set/GCA_f	4	14.8^{***}	8.4^{***}	14.7^{***}	1.4	4.5^{***}	0.3
Env × Set/GCA_m	4	5.6^{***}	7.9^{***}	11.9^{***}	3.7^{**}	2.5^{**}	0.1
Env × Set/SCA	8	2.8^{**}	7.2^{***}	8.3^{***}	1.8	1.6	0.5

* P < 0.05.
** P < 0.01.
*** P < 0.0010.
[1] Environment.

Table 4. Pooled mean square for filled grains and other secondary traits under drought stress and nondrought stress environments.

Source	d.f	Mean square values for sets A and B			
		Set A		Set B	
		Drought stress	Nondrought stress	Drought stress	Nondrought stress
Fertility[1]					
Male	2	3.21*	12.90*	0.51*	1.86
Female	2	4.07*	9.01*	2.58*	5.40
Male × female	4	1.56	6.84*	3.29*	0.80**
Total grains per panicle					
Male	2	4.58*	8.60***	2.39	12.28***
Female	2	4.14*	9.86***	1.38	9.51***
Male × female	4	2.64*	9.40***	0.62	5.83***
Leaf area					
Male	2	3.09*	11.00***	6.96***	15.51***
Female	2	1.40	16.21***	10.01***	16.01***
Male × female	4	4.76***	9.74**	17.74***	5.34***
Plant height					
Male	2	4.46*	4.45	26.70***	26.44***
Female	2	1.26	1.20**	7.66***	7.27***
Male × female	4	8.14***	8.14***	15.70***	14.70***
Tiller no					
Male	2	23.13*	42.99	26.70***	4.61**
Female	2	3.36	49.08	7.66***	0.75
Male × female	4	2.17***[23]	13.97	15.70***	1.67

* $P < 0.1$,
** $P < 0.05$,
*** $P < 0.001$
1 Filled grains in percentage.

Table 5. Mean squares for filled grains, total grains per panicle, leaf area, plant height, and tiller number under drought and nondrought stress.

There was significant ($P < 0.05$) mean square for male × female interaction for the filled grains under NDS for set A. In the case of the B crossing set, the male × female interaction mean squares were significant under DS and NDS conditions. In addition, the mean square of male and female were significant under DS but not under NDS. The male, female, and male × female interaction mean squares were all highly significant ($P = 0.001$) for the total number of grains per panicle under NDS conditions for the A set and significant ($P = 0.05$) under DS conditions. In the case of B crossing set, male, female, and the male × female interaction, mean squares were significant under NDS conditions, but not the case under DS conditions. The set A had highly significant ($P < 0.001$) male, female, and male × female mean squares for leaf area under NDS conditions. Mean squares for male and male × female interactions were significant for leaf area under DS conditions, but not the case of female mean square. The results of the B crossing set revealed that male, female, and the male × female interaction mean squares were all highly significant ($P < 0.001$) under the NDS and DS conditions.

The male × female interaction mean squares were all highly significant ($P = 0.001$) for the plant height under DS, and NDS conditions for the A and B populations. There was significant mean square for female effects under NDS, but not under DS conditions for the A and B populations. In the case of B crossing set, male, female, and the male × female interaction mean squares were all highly significant under both NDS and DS conditions. Mean squares for male, female, and male × female interaction mean squares were not significant for the tiller number under NDS conditions for the A populations. There was, however, significant mean square for male and the male × female interactions but not female mean squares under DS conditions. On the other hand, the B crossing set had male, female, and the male × female interaction mean squares, all highly significant, under both DS and only the male under NDS conditions.

3.2.2. Relative contribution of GCA and SCA

General combining ability (GCA) for female and male (GCA_f and GCA_m) in A populations under DS and NDS conditions are presented in **Figure 1**. The total GCA for both male and female parents (GCA_t) under DS were more than 55% for all five traits except leaf area that had 42%. All the SCA values were less than 50%. The SCA effects of tiller number, under DS, was 14% and the male × female interaction was not significant (**Table 5**). Similarly, the SCA effects for filled grains were not important under DS conditions because there were lack of significance (**Table 5**). This finding, therefore implies that the additive effects was more important than nonadditive effects for filled grains, total number of grains per panicle, plant height, and tiller number.

Under NDS, however, filled grains, leaf area, and tiller number had GCA_t more than 55%. The total number of grains per panicle and the plant height had nearly equal GCA_t, when compared with SCA. This finding implies that the additive effects are more important than nonadditive effects for filled grains, leaf area, and tiller number, while additive and nonadditive effects had nearly equal effects for total number of grains per panicle. However, the lack of significance in the male × female interactions for tiller numbers (**Table 5**) makes the importance of SCA not valid.

Results of the GCA effects for female and male (GCA_f and GCA_m) for B populations, under DS and NDS conditions, are shown in **Figure 2**. The GCA total (GCA_t) under drought was more than 55% for filled grains, total number of grains per panicle, and tiller number under DS. The GCA_t and SCA for plant height were nearly equal, while a very high SCA value of 68% was found for leaf area. All the SCA values were less than 50%; moreover, the SCA effects for tiller number were not significant (**Table 5**) and that of total number of grains per panicle were also not significant (**Table 5**). This finding implies that the additive effects were more important than nonadditive effects for filled grains, total number of grains per panicle, and tiller number, while additive and nonadditive effects had nearly equal effects for plant

http://dx.doi.org/10.5772/66826

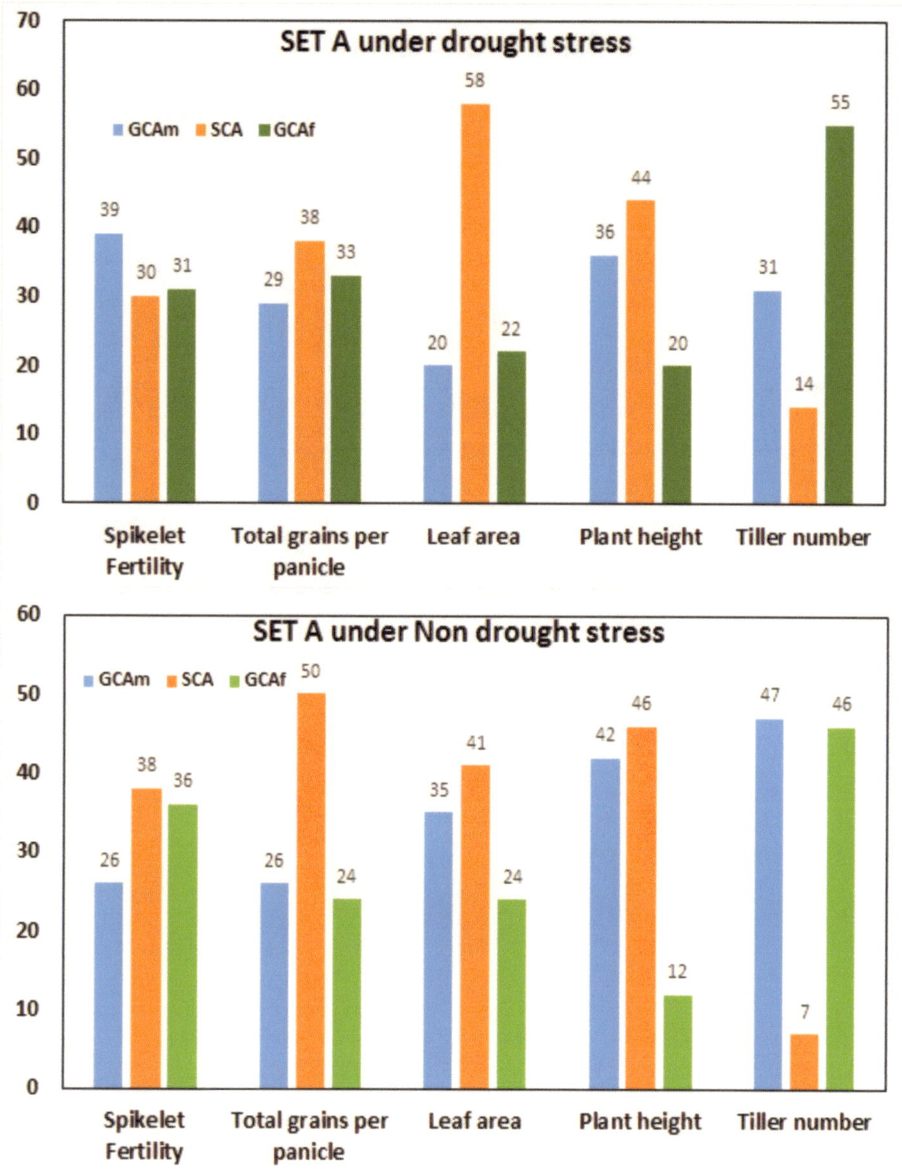

Figure 1. Relative (%) contribution of GCA and SCA effects to the cross sum of squares in set A under drought stress and nondrought stress.

height. Under NDS conditions, however, filled grains, total number of grains per panicle, leaf area, and tiller number had GCA_t more than 55%. The plant height had nearly equal GCA_t when compared with the SCA. This finding implies that the additive effects were more important than nonadditive effects for filled grains, total number of grains per panicle, leaf area and tiller number, while additive and nonadditive effects had nearly equal effects for plant height.

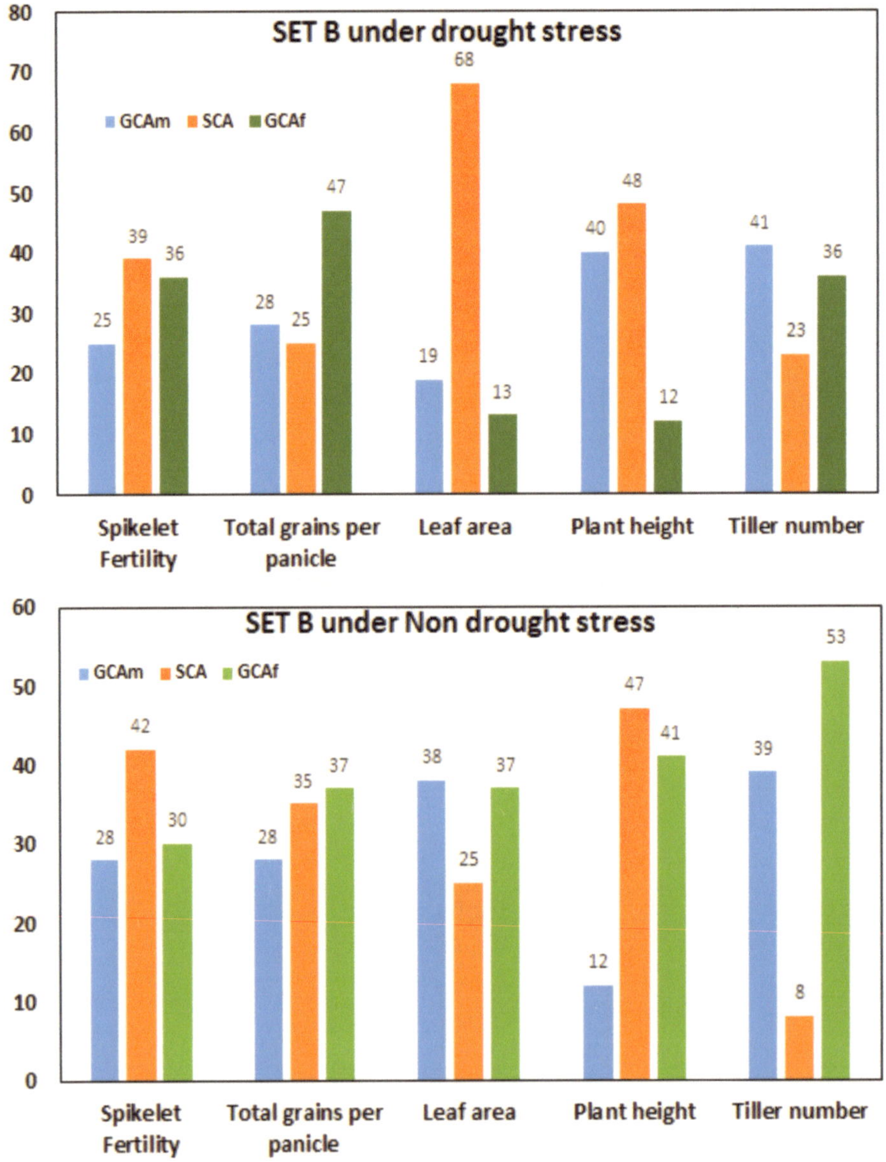

Figure 2. Relative (%) contribution of GCA and SCA effects to the cross sum of squares in set B under drought stress and nondrought stress.

3.2.3. General combining ability effects

Table 6 showed the GCA effects for filled grains for interspecific and intraspecific rice. The GCA values for filled grains were the only one presented, because other secondary traits had weak correlation with the filled grains, which is a trait associated with drought tolerance. Positive GCA effect is desirable in breeding for improved drought tolerance. Strong negative

http://dx.doi.org/10.5772/66826

	Filled grains	
	Nondrought stress	Drought stress
Female		
NERICA 8	-0.46	-3.42**
NERICA 13	-1.69	3.45**
IRAT 325	2.15*	-0.03
Bonanca	-2.80**	1.04
WITA 1	-4.98***	-6.23***
CK 73	-2.18*	5.19***
Male		
CT 16334(2)-CA-2-M	-1.29	-1.18
WAB 365-B-1H1-HB	-0.71	-2.12*
NERICA 9	2.00*	3.30**
CT 16346-CA-20-M	-2.54**	-2.26*
CT 16350- CA-5-M	5.26***	-1.16
CT 16344-CA-9-M	-2.72**	3.42**
SE	**±0.83**	**±0.97**

* Significant at 0.05 (2.15).
** Significant at 0.01 (2.98).
*** Significant at 0.001 (4.14).

Table 6. Estimates of general combining ability (GCA) effects for filled grains under drought and nondrought stress conditions.

values of GCA effects of parents show contribution of GCA towards low filled grains, while high positive values show high filled grains. Since both GCA effects and SCA effects were significant for filled grains, the individual values for both GCA and SCA effects are presented (**Tables 6** and 7). Parents CT 16350- CA-5-M, IRAT 325, and NERICA 9 had positive and significant scores of filled grains under NDS conditions. In the DS environment, CK 73 was highly positive and significant at P = 0.001, while CT 16344-CA-9-M, NERICA 9, and CT 16346-CA-20-M had positive and significant filled grain scores at P = 0.01.

3.2.4. *Specific combining ability effects*

Superior crosses were observed, with positive SCA effects (**Table 7**). Under nondrought stress conditions, crosses WITA 1 × CT 16350- CA-5-M, Bonanca × CT 16346-CA-20-M, and CK 73 × CT 16344-CA-9-M had significant filled grain score of 0.01%, 0.01%, and 0.001%, respectively. The cross WITA 1 × CT 16344-CA-9-M had significant filled grain score at 0.05 level of significance. In the drought stress conditions, the cross NERICA 8 × WAB 365-B-1H1-HB were highly significant at P = 0.001, and Bonanca × CT 16344-CA-9-M and IRAT 325 × CT 16334(2)-CA-2-M were positive and significant at 0.01.

	Filled grains	
	Nondrought stress	Drought stress
NERICA 8 × CT 16334(2)-CA-2-M	-3.46	-12.86***
NERICA 13 × CT 16334(2)-CA-2-M	1.47	4.48*
IRAT 325 × CT 16334(2)-CA-2-M	0.73	7.21**
NERICA 8 × WAB 365-B-1H1-HB	-0.59	8.53***
NERICA 13 × WAB 365-B-1H1-HB	-2.01	-4.93*
IRAT 325 × WAB 365-B-1H1-HB	1.90	-5.71*
NERICA 8 × NERICA 9	3.60	0.92
NERICA 13 × NERICA 9	-1.12	3.91
IRAT 325 × NERICA 9	-0.46	-1.52
Bonanca × CT 16346-CA-20-M	6.15**	-3.19
WITA 1 × CT 16346-CA-20-M	-6.39**	-0.88
CK 73 × CT 16346-CA-20-M	-2.32	1.86
Bonanca × CT 16350- CA-5-M	1.30	-1.74
WITA 2 × CT 16350- CA-5-M	7.12**	-1.63
CK 73 × CT 16350- CA-5-M	-3.17	2.26
Bonanca × CT 16344-CA-9-M	-10.27***	6.02**
WITA 2 × CT 16344-CA-9-M	4.24*	-3.67
CK 73 × CT 16344-CA-9-M	3.30***	1.12

* Significant at 0.05.
** Significant at 0.01.
*** Significant at 0.001.

Table 7. Estimates of specific combining ability (SCA) effects for filled grains under drought and nondrought stress conditions.

3.2.5. Summary of analysis of generation of means

The mean, variance, and mean variance of filled grains for P1, P2, F1, F2, and F3 are shown in **Table 8**. The F2 populations had the highest variance followed by F3 and F1. Scaling tests for dominance × dominance and additive × additive interactions were nonsignificant for both levels. Dominance main effects were not significant, but additive main effects were significant at P = 0.01. When the mean scores were fitted to an additive model, it fitted with $r^2 = 0.77$ (**Figure 3**). Mean filled grains score was best linear unbiased estimator (BLUE) of the traits

The estimate of the number of genes that control filled grains trait based on Castle-Wrights method was 0.9 ≈ 1 gene. Estimate of the degree of dominance in the F1 and F2 generation based on the [15] method was -3 ≈ 0 and 0.9 ≈ 1 level of dominance, respectively.

http://dx.doi.org/10.5772/66826

Descriptive summary of generations				
Generations	**d.f**	**Mean**	**Variance**	**Mean variance**
P1	29	77.80	42.92	2.59
P2	29	56.93	16.89	1.90
F1	29	74.73	80.47	2.49
F2	59	72.07	141.08	1.20
F3	59	64.60	81.87	1.08
Scaling test for filled grains				
Interactions	Scale	SE	d.f	t (Scale/SE)
dominance × dominance	-15	3.559	146	-1.184^{NS}
additive × additive	3	3.361	176	-0.893^{NS}
Components of means (three parameters)				
Gene effects	Expectation estimates	SE	t = (component/SE)	d.f
Mean	57.0	1.176	48.46**	59
Additive effects	10.5	0.707	14.85**	58
Dominance effects	0.8	36.842	0.22	147

Table 8. Summary of generations in variety 18 × 138 cross, scaling test, and components of means for filled grains score.

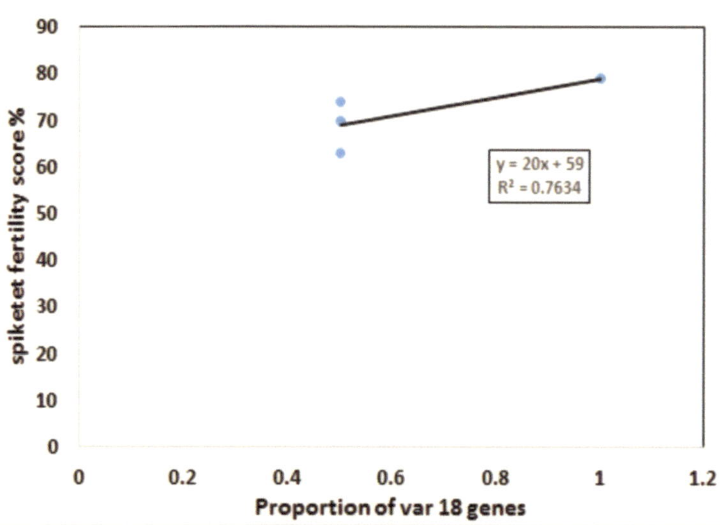

Figure 3. Proportions of genes contributing to filled grains score.

The narrow sense heritability in the generations from the cross between CT 16334 (2)-CA-2-M and WAB 450-1-BL1-136-HB using regression of F1 on mid-parents and F2 to F1 based on single seed decent are shown in **Figures 4** and **5** respectively. In the F1 to midparent regression, heritability of 60% was realized, but when F2 was regressed onto F1 means, the heritability estimate was 74%.

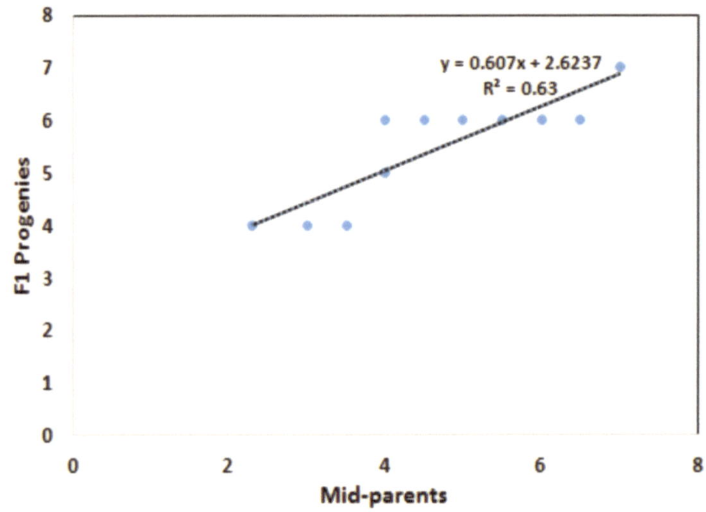

Figure 4. Regression of F1 progenies on midparents for 12 × 138 cross using filled grains.

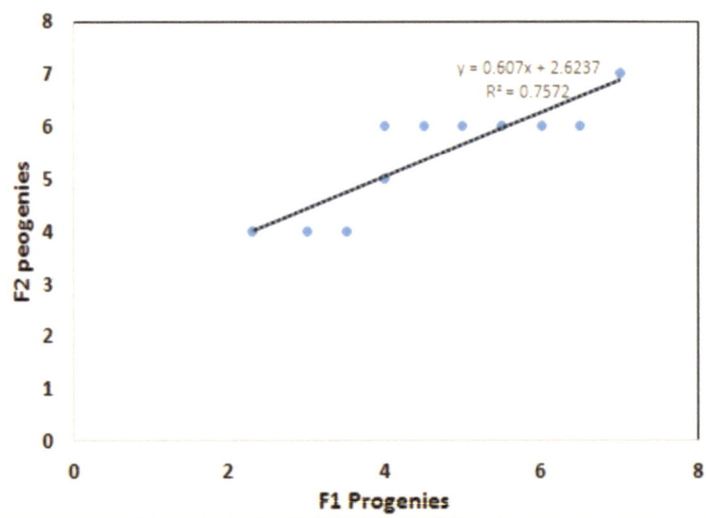

Figure 5. Regression of F2 progenies on F1 parental means for 12 × 138 cross using filled grains.

http://dx.doi.org/10.5772/66826

3.3. Evaluation of segregating lines

3.3.1. Preliminary evaluation of 660 F3

Results of evaluation of two sets of new 660 at NaCRRI are presented in this section. The first set grown under optimum moisture throughout the growth period is presented in **Table 9**. The selection pressure was 11.4% (75 out of 660 rows selected) for rain-fed lowland conditions and 9.85% (65 out of 660 rows selected) for rain-fed upland conditions. Candidate varieties CAIAPO/CT 16324-CA-9-M, WAB 450-1-BL1-136-B/WAB 450-B-136-HB, CT 16317-CA-4-M/ WAB 365-B-1H1-HB, IRAT 325/WAB 450-B-136-HB, and CT 16342-CA-25-M/CK 73 are among the lines. Overall, 84 genotypes were selected for further evaluation.

Selection under rain-fed upland conditions			**Selection under rain-fed lowland conditions**		
No	**Groups of crosses**	**Total**	**No**	**Groups of crosses**	**Total**
	One row crosses selected at F4			**One row crosses selected at F4**	
1	Bonanca × WAB 881-10-37-18-3-P1-HB		1	WAB 56-104 × CT 16324-CA-9-M	
2	IRAT 325 × WAB 450-B-136-HB		2	CT 16350- CA-5-M × WITA 2	
3	CT 16355-CA-15-M × IRAT 112		3	CT 16355-CA-15-M × IRAT 112	
4	WAB 365-B-1H1-HB × WAB 450-1-BL1-136-HB		4	CT 16317-CA-4-M × IRAT 104	
5	WAB 450-B-136-HB × IRAT 325		5	WAB 450-B-136-HB × IRAT 325	
6	CT 16344-CA-9-M × WITA 2		6	WAB 365-B-1H1-HB × IRAT 325	
7	WAB 365-B-1H1-HB × IRAT 325		7	CT 16313-CA-4-M × Caiapo	
8	WBK 35 (F3) × WAB 450-1-BL1-136-HB		8	WAB 56-104 × CT 16313-CA-4-M	
9	Bonanca × CT 16346-CA-20-M		9	CK 73 × CT 16346-CA-20-M	9
10	CT 16342-CA-25-M × IRAT 257			**Two rows crosses selected at F4**	
11	WAB 450-B-136-HB × WAB 365-B-1H1-HB		1	IRAT 13 × CT 16342-CA-25-M	
12	CT 16334(2)-CA-2-M × IRAT 325		2	CT 16346-CA-20-M × Bonanca	
13	CT 16344-CA-9-M × CK 73		3	CT 16342-CA-25-M × CK 73	
14	CT 16344-CA-9-M × Bonanca		4	WAB 365-B-1H1-HB × IRAT 325	
15	Bonanca × WAB 450-I-B-P-38-HB	15	5	IRAT 112 × WAB 365-B-1H1-HB	10
	Two rows crosses selected at F4			**Three rows crosses selected at F4**	
1	Caiapo × CT 16324-CA-9-M		1	Bonanca × WAB 881-10-37-18-3-P1-HB	
2	CT 16313-CA-4-M × WAB 56-104		2	WAB 365-B-1H1-HB × IRAT 325	6
3	IRAT 325 × WAB 365-B-1H1-HB			**Four rows crosses selected at F4**	
4	CT 16324-CA-9-M × WAB 56-104		1	Caiapo × CT 16324-CA-9-M	
5	WAB 365-B-1H1-HB × IRAT 325		2	WAB 450-1-BL1-136-HB × WAB 450-B-136-HB	
6	WAB 450-B-136-HB × IRAT 112		3	CT 16324-CA-9-M × WAB 56-104	12

Selection under rain-fed upland conditions			Selection under rain-fed lowland conditions		
No	**Groups of crosses**	**Total**	**No**	**Groups of crosses**	**Total**
	One row crosses selected at F4			**One row crosses selected at F4**	
7	CT 16334(2)-CA-2-M × IRAT 325	14		**Five rows crosses selected at F4**	
	Three rows crosses selected at F4		1	IRAT 325 × WAB 450-B-136-HB	5
1	WAB 450-1-BL1-136-HB × WAB 450-B-136-HB			**Six row crosses selected at F4**	
2	CT 16317-CA-4-M × WAB 365-B-1H1-HB		1	WAB 450-B-136-HB × WAB 365-B-1H1-HB	6
3	IRAT 112 × WAB 365-B-1H1-HB			**Seven rows crosses selected at F4**	
4	CT 16334(2)-CA-2-M × WAB 450-1-BL1-136-HB		1	WAB 365-B-1H1-HB × WAB 450-1-BL1-136-HB	
5	WAB 56-104 × CT 16313-CA-4-M		2	Bonanca × CT 16346-CA-20-M	14
6	CK 73 × CT 16350-CA-5-M			**Thirteen rows crosses selected at F4**	
7	IRAT 257 × CT 16355-CA-15-M	21	1	CT 16317-CA-4-M × WAB 365-B-1H1-HB	13
	Five rows crosses selected at F4				
1	IRAT 112 × WAB 450-B-136-HB				
2	IRAT 13 × CT 16342-CA-25-M				
3	CT 16342-CA-25-M × IRAT 13	15			
	Rows selected	**65**			**75**
	Total Hills planted	**660**			**660**
	Selection pressure	**9.85**			**11.4**

Table 9. Selection of F4 genotypes from 660 F3 genotypes.

3.3.2. Evaluation of 84 F4-F5 lines

Results of evaluation of 84 rain-fed segregating lines showed that 20 genotypes were resistant to RYMV, blast, and BLB in all the five locations, namely Namulonge, Kigumba, Kibaale, Lira, and Doho (**Table 10**). Results of yield in Lira are presented in **Table 11**. The best six genotypes in yield in descending order are P27-H14 (11,950 kg/ha), P29-H4 (9750 kg/ha), P36-H17 (9313 kg/ha), P5-H1 (9111 kg/ha), P36-H9 (8688 kg/ha), and P36-H4 (8417 kg/ha). When yield, pest and disease resistance, plant height, and panicle length was considered a total of nine lines were nominated for National Performance evaluation.

3.4. Evaluation of promising lines

Results of evaluation of nine selected lines along with two earlier selected lines and a local check is presented in **Table 12**. Six lines were selected and presented for release to the National Variety Release Committee of Uganda

No	Genotype	RYMV	Blast	BLB
1	P 22 H13 WAB 450-1-BL1-136-HB × WAB 450-B-136-HB	v	0	0
2	P 36 H1 WAB 365-B-1H1-HB × WAB 450-1-BL1-136-HB	0	0	0
3	16-16 CT 16344-CA-9-M × Bonanc	0	0	0
4	13-13 CT 16344-CA-9-M × CK 73	0	0	0
5	NERICA 4	0	0	0
6	P 25 H1 CT 16346-CA-20-M × Bonanca	0	0	0
7	P 8 H2 Caiapo × CT 16324-CA-9-M	0	0	0
8	77 WAB95-B-B-40-HB	0	0	0
9	96 WAB56-77	0	0	0
10	152 AB788-16-3-2-1-HB	0	0	0
11	P 24 H8 IRAT 13 × CT 16342-CA-25-M	0	0	0
12	P 1 H14 Bonanca × WAB 881-10-37-18-3-P1-HB	0	0	0
13	P 4 H6 CT 16350-CA-5-M × WITA 2	0	0	0
14	P 29 H1 CT 16342-CA-25-M × CK 73	0	0	0
15	P 23 H1 CT 16346-CA-20-M × WITA 2	0	0	0
16	P 45 H15 WAB 365-B-1H1-HB × IRAT 325	0	0	0
15	P 24 H9 IRAT 13 × CT 16342-CA-25-M	0	0	0
18	P 27 H10 CT 16317-CA-4-M × WAB 365-B-1H1-HB	0	0	0
19	P 5 H2 IRAT 325 × WAB 450-B-136-HB	0	0	0
20	P 29 H4 CT 16342-CA-25-M × CK 73	0	0	0

Table 10. List of 20 varieties that was resistant to RYMV, blast, and BLB in five locations: Namulonge, Kigumba, Kibaale, Lira, and Doho.

Seven genotypes namely 1. P5H2 (IRAT 325/WAB 450-B-136-HB-F6), 2. P29H4 (CT 16342-CA-25-M/CK 73-F6), 3. P8H2 (Caiapo/CT 16324-CA-9-M-F6), 4. ART3-11L1P1-B-B-2, ([WAB56-104 (WAB56-104/CG14)]/Moroberekan), 5. P27H1 (CT 16317-CA-4-M/WAB 365-B-1H1-HB-F6), 6. WAB 95 B-B-40-HB (ITA257/(IDSA6/ROK16)), and 7. P24H9 (IRAT 13/CT 16342-CA-25-M-F6), higher yields that NERICA-4 the local check. Under optimum conditions, six genotypes had higher yield than NERICA-4.

3.5. Varietal release and status of release

Breeding background, characteristics, and selected agronomic information on six varieties were presented to the variety release committee. These varieties were released based on important characteristics detailed in **Table 13**. The names proposed and accepted by the variety released committee were NamChe-1. NamChe-2, NamChe-3, NamChe-4, NamChe-5, and NamChe-6.

Rank	Acc no.	Yield	Rank	Acc no.	Yield	Rank	Acc no.	Yield	Rank	Acc no.	Yield
1	P27-H14	**11,950**	22	P35-H5	**6625**	43	P36-H16	**6075**	64	P22-H3	**5156**
2	P29-H4	**9750**	23	P59-H13	**6625**	44	P5-H14	**6025**	65	P55-H9	**5139**
3	P36-H17	**9313**	24	P59-H19	**6625**	45	P1-H14	**6000**	66	P27-H9	**5100**
4	P5-H11	**9111**	25	P27-H15	**6550**	46	P55-H2	**5975**	67	P27-H12	**5071**
5	P36-H9	**8688**	26	P36-H4	**6500**	47	P31-H3	**5950**	68	P28-H3	**5025**
6	P36-H4	**8417**	27	P55-H17	**6500**	48	P27-H10	**5900**	69	P5-H3	**4825**
7	P51-H17	**8400**	28	P5-H2	**6500**	49	P26-H17	**5700**	70	P8-H10	**4594**
8	P33-H3	**7625**	29	P50-H1	**6469**	50	P59-H9	**5700**	71	P26-H13	**4500**
9	P22-H6	**7600**	30	P1-H17	**6400**	51	P59-H8	**5675**	72	P7-H2	**4300**
10	P25-H14	**7600**	31	P22-H13	**6375**	52	P49-H3	**5625**	73	P45-H15	**4275**
11	P37-H13	**7575**	32	P59-H17	**6375**	53	P33-H6	**5583**	74	P24-H9	**4250**
12	P31-H15	**7250**	33	P26-H6	**6350**	54	P36-H1	**5583**	75	P27-H3	**4179**
13	P34-H2	**7188**	34	P27-H11	**6350**	55	P36-H8	**5500**	76	P8-H17	**3700**
14	P25-H1	**7125**	35	P38-H15	**6325**	56	P27-H1	**5464**	77	P55-H19	**3650**
15	P33-H1	**7000**	36	P7-H19	**6325**	57	P26-H18	**5450**	78	P1-H20	**3500**
16	P26-H1	**6889**	37	P55-H5	**6275**	58	P5-H4	**5429**	79	P4-H6	**3400**
17	P27-H18	**6850**	38	P22-H16	**6250**	59	P7-H14	**5357**	80	P27-H17	**3300**
18	P24-H8	**6833**	39	P59-H10	**6214**	60	P29-H1	**5333**	81	P59-H17	**3125**
19	P55-H10	**6775**	40	P35-H12	**6188**	61	P56-H19	**5325**	82	P27-H6	**2800**
20	P8-H2	**6708**	41	P55-H20	**6125**	62	P23-H1	**5300**	83	P27-H2	**2679**
21	P27-H7	**6625**	42	P58-H16	**6083**	63	P8-H15	**5194**	84	P58-H11	**2607**

NB: Yield of NERICA 4 was 5600 tons/ha

Table 11. Yield of 84 breeding lines screened at Lira.

Genotype	Arua	NaCRRI	Masindi	Soroti	Kibaale	Kanungu	Mean yield	Mean rank	Yield under optimal condition
1	2913	2653	2906	3464	2619	3284	**2973**	**7.6**	**3500**
2	2747	3026	3364	3703	2802	3673	**3219**	**4.2**	**3600**
3	4093	2954	3069	3984	3230	3561	**3482**	**4.2**	**4300**
4	2183	2538	2887	3196	2287	3187	**2713**	**9.6**	**4500**
5	3785	2351	2420	3454	2731	2949	**2948**	**9.4**	**5800**
6	3399	2364	2496	3369	2604	2974	**2868**	**9.6**	**3800**
7	4990	2214	2072	3651	3068	2773	**3128**	**9.2**	**3750**

http://dx.doi.org/10.5772/66826

Genotype	Arua	NaCRRI	Masindi	Soroti	Kibaale	Kanungu	Mean yield	Mean rank	Yield under optimal condition
8	2660	3086	3446	3726	2809	3737	**3244**	**3.2**	**4013**
9	4532	3235	3326	4305	3567	3837	**3800**	**1.4**	**4550**
10	4219	2937	3030	4003	3264	3540	**3499**	**4.4**	**3650**
11	3656	3023	3218	3928	3121	3644	**3432**	**4**	**3600**
12	3928	2080	2084	3286	2606	2666	**2775**	**11.2**	**3780**

Index[24]:
1. P5H2 (IRAT 325/WAB 450-B-136-HB-F6).
2. P29H4 (CT 16342-CA-25-M/CK 73-F6).
3. P8H2 (Caiapo/CT 16324-CA-9-M-F6).
4. ART3-11L1P1-B-B-2 ([WAB56-104/(WAB56-104/CG14)]/Moroberekan).
5. P27H1 (CT 16317-CA-4-M/WAB 365-B-1H1-HB-F6).
6. WAB 95 B-B-40-HB (ITA257/(IDSA6/ROK16)).
7. P24H9 (IRAT 13/CT 16342-CA-25-M-F6).
8. NERICA-4.
9. P29H1 (CT 16342-CA-25-M × CK 73–F6).
10. P23H1 (CT 16346-CA-20-M/WITA 2-F6).
11. ART8-L15P14-1-2-1.
12. P22 H13 (WAB 450-1-BL1-136-HB/WAB 450-B-136-HB-F6).

Table 12. Yield of 12 genotypes in six locations in the country and under optimal conditions.

Variety name	NamChe 1	NamChe 2	NamChe 3	NamChe 4	NamChe 5	NamChe 6
Year of release	**2013**	**2013**	**2013**	**2013**	**2013**	**2013**
Local name	NamChe 1	NamChe 2	NamChe 3	NamChe 4	NamChe 5	NamChe 6
Pedigree	WAB95-B-B-40-HB	NM7-8-2-B-P-11-6	NM7-29-4-B-P-80-8	ART3-11L1P1-B-B-2	NM7-27-1-B-P-77-6	NM7-5-2-B-P-79-7
Parents	ITA257/(IDSA6/ROK16)	Caiapo/CT 16324-CA-9-M-F6[25]	CT 16342-CA-25-M/CK 73	[WAB56-104/(WAB56-104/CG14)]/Moroberekan	CT 16317-CA-4-M/WAB 365-B-1H1-HB	IRAT 325/WAB 450-B-136-HB–
Test names	WAB95-40	NM7-1	NM7-8	ART3-10	NM7-6	NM7-7
Breeding center	AfricaRice, Senegal	NaCRRI, Uganda	NaCRRI, Uganda	AfricaRice, Ibadan	NaCRRI, Uganda	NaCRRI, Uganda
Characteristics						
Leaf planotype	Semi-erect	Semi-erect	Erect	Erect	Semi-erect	Erect
Culm inclination	Semi-erect	Semi-erect	Erect	Erect	Erect	Erect
Culm length (cm)	64	66	66	65	60	62
Duration from germination to harvest (days)	110	132	125	120	125	125

Variety name	NamChe 1	NamChe 2	NamChe 3	NamChe 4	NamChe 5	NamChe 6
Year of release	**2013**	**2013**	**2013**	**2013**	**2013**	**2013**
Milling percentage	66.2	68.7	72.7	72.1	71.4	70.4
Volume expansion	1.7	1.6	1.6	1.9	2	1.9
Yield (kg/ha)	3800	4300	4550	4500	5800	5000
1000 grain weight (g)	29	28	24	27	26	23
Grain length dehusked (mm)	6.3	6.7	6.3	6.5	6.4	6.4
Grain width dehusked (mm)	2.4	2.2	2.2	2.1	2.2	2.2

Table 13. Major characteristics of the released varieties.

4. Discussions

4.1. Screening introductions for drought tolerance

Results that nine out of 20 best lines were from the CT breeding work imply there was adequate variability in the selected set for selection of suitable genotypes. There were five out of 20 genotypes also indentified as drought tolerant. Also, result that three reference genotypes namely NERICA 7, CO 39, and VANDANA were suitable for identified as drought tolerant in this study implies that the method of screening was acceptable.

4.2. Genetic analysis for filled grains and other agronomic traits

The analysis of F2 crosses revealed the various components of gene action controlling various drought tolerance traits in rice. Both male GCA and female GCA effects were significant for filled grains under both DS and NDS, for the A populations and the B populations under DS. The finding that the additive effects were more important than nonadditive effects for total number of grains filled, grains in set A under NDS and B under NDS and DS, implies that additive effects control the traits in different populations and the nonadditive effects varied with populations under study. This result is contrary to the finding by Mohapatra and Mohanty [16] that filled grains was predominantly controlled by nonadditive gene effects under drought stress. However, in the study by Mohapatra and Mohanty [16], the populations were generated by crossing *O. sativa* with *O. sativa*. The mechanism of drought tolerance in *O. glaberrima*, a parent of the interspecific rice used in the current study, was reported to be different from that of *O. sativa* [17]. This could explain the apparent differences in findings of the current study when compared with that of Ref. [16]. Based on the current study, breeding methods that involve selection in the early generations are recommended. The methods include single seed decent, pedigree selection, and modified bulk methods. Studies using more populations generated from *O. sativa* and interspecific rice could confirm our finding that the importance of SCA varies with population under study.

http://dx.doi.org/10.5772/66826

Findings of this study that nonadditive effects for total number of grains per panicle was important in both set A and set B under NDS conditions, implies that breeding methods that involve late selection could improve drought tolerance under NDS using number of grains per panicle trait. The use of yield components including grains per panicle has been demonstrated to be effective in improving yield under drought stress by selecting under NDS conditions [18]. The differences between the responses under DS and NDS conditions for total number of grains per panicle could be due to fewer loci within the set B that could segregate for the trait than in A. Set B comprised of lines with more susceptibility to drought stress than those in set A.

Additive effects for number of grains per panicle were important in all the population in set A and set B, under DS and NDS. This implies that breeding methods that involve selection in the early generations especially, single seed decent, pedigree selection, and modified bulk methods could improve drought tolerance through selection of number of grains per panicle. In another study involving *O. sativa* parents that included susceptible, moderately susceptible, moderately resistant and resistant lines, number of grains per panicle was reported to be controlled by additive effects under NDS conditions [19]. Genes with additive effects were predominant in the inheritance of number of grains per panicle [16]. Both additive and nonadditive effects were nearly equal in populations in set A, under NDS. These set of populations could be used to improve drought stress using methods that involve selection in the early and late generations of the populations. These methods include modified bulk methods and repeated crossing at the segregation stage. Similarly, additive and nonadditive gene effects were significant for number of spikelets per panicle under both normal and saline conditions, and repeated crossing has successfully been used to improve salinity tolerance [20].

Findings of this study that nonadditive effects for leaf area were more important than additive effects in both set A and set B under DS conditions, suggests that late selection could improve drought tolerance. In addition, the findings that additive effects were more important than nonadditive effects for the populations in sets A and B under NDS implied that selection methods that involve early selection could be employed under NDS. In the populations in sets A and B, interspecific rice genotypes generated from *O. glaberrima* crosses were the majority of the parents. *O. glaberrima* is known to have high vegetative growth as a drought stress adaptation mechanisms [21, 17]. It is likely that these traits were transmitted to the populations under study and it is expressed more under DS than under NDS conditions.

Results of this study that additive and nonadditive effects for plant height were nearly equal with contribution for total GCA, varying between 45 and 55% for both set A and set B under DS and NDS conditions, implied that that breeding methods that involve both early and late selection could be employed in the improvement of drought tolerance using this trait. Modified bulk method of selection method could be appropriate. In another study involving *O. sativa* parents that included susceptible, moderately susceptible, moderately resistant and resistant lines, and plant height was controlled by additive effects under NDS conditions [19]. In the current study, both additive and nonadditive effects were important when the B generations were tested under DS and NDS conditions. Drought traits were controlled quantitatively.

The current study found that additive effects were the more important in the transmission of drought tolerance using tiller number as evidenced by the lack of significance for male × female interaction effects for tiller number. This finding is contrary to the work reported by other scientists that nonadditive effects were more important under drought stress conditions [22, 23]. In another study, however, expression of tiller number, under both NDS and DS situations, was found to involve nonallelic gene interactions [20].

Overall, in situations where nonadditive effects are more important, selection should be delayed until later generations. In these types of populations, repeated crossing in the segregating generations may be useful to pool all the desirable genes in one genotypes according to Ref. [24]. The modified bulk method is another useful method of improvement. However, when additive affects are more important, then a modified pedigree method that involves bulking germplasm before evaluation is appropriate. However, when both additive and nonadditive effects are important, two options can be taken depending on the objective of the breeding and the relative importance of the additive or nonadditive effects. In case, if the objective is to develop hybrid rice, as it is planned in Uganda, then pure line selection should be employed. In this approach, additive effects will be extracted because rice is autogamous [25]. In a situation, where both additive and nonadditive gene action are to be exploited, a modified bulk breeding method would hasten the rate of genetic improvement. Similar exploitation of both additive and nonadditive gene action has been conducted in the improvement of cold tolerance [26] and sodicity tolerance in rice [27].

4.3. Combining abilities filled grains under drought stress

Generally, there was no clear distinction in combining ability between *O. sativa* and interspecific rice lines under nondrought stress conditions, but the interspecific lines were better combiners under drought stress conditions. Among the *O. sativa* line, IRAT 325 was a good general combiner, while CT 16350-CA-5-M and WAB 450-B-136-HB (NERICA 9) were good combiners under nondrought stress conditions. In the drought stress condition, however, CK 73, an *O. sativa* genotype, was the best combiner for improved filled grains. Other parents with lower levels of significance were CT 16344-CA-9-M, WAB 450-B-136-HB (NERICA 9), and CT 16346-CA-20-M.

Specific combining ability analysis revealed that crosses WITA 1 × CT 16350-CA-5-M, Bonanca × CT 16346-CA-20-M, and CK 73 × CT 16344-CA-9-M were best under NDS condition. The cross CK 73 × CT 16344-CA-9-M had both parents as good combiners indicating additive × additive type of gene action. It is expected that these crosses could provide transgressive segregants that could be selected using pedigree methods [28]. The others crosses had mixed combiners, therefore additive and nonadditive gene action could be the major contributors. In such crosses, bulk breeding methods could exploit both gene actions.

4.4. Generation means for filled grains under drought stress

There were significant differences among generations for filled grains indicating the presence of sufficient genetic variability. Variability for various traits of rice has been reported [29–32]. The scaling test showed that additive genetic effects but not dominance and epistatic genetic

http://dx.doi.org/10.5772/66826

effects were important in the inheritance of filled grains. Fitting means of filled grains on the additive model showed that additive effects accounted for 77% of the genetic variation. In addition, the finding that dominance level was 0 in the F1 population showed that there were no dominance effects.

The generation means analysis confirmed that additive effects were significant in the transmission of filled grains in the populations generated. This study had no inconsistencies in detecting that additive effects were the most important genetic factor in the population under study. In addition, results where narrow sense heritability was high indicated that a high proportion of genetic components of variance can be fixed in segregating generations. Since the selection was conducted under drought stress, it is appropriate that selection for improved drought stress is conducted as early as F2 in the study location. According to Ref. [31], it is appropriate that selection for improved drought stress is conducted using heritability estimates for target traits. There is limited information on the inheritance of filled grains trait under drought stress. However, various reports indicated that additive effects were the main components that controlled the transmission of this trait under high temperature [33, 34]. A single gene pair was estimated to control filled grains under drought stress. A single gene was found to be responsible for the transmission of filled grains under high temperatures [33, 34].

4.5. Evaluation of segregating lines

Results of evaluation of two sets of new 660 genotypes showed that CT lines namely CAIAPO, CT 16324-CA-9- CT 16317-CA-4-M and CT 16342-CA-25-M had the highest number of parents that could improve the landraces. These lines were developed for drought tolerance through CIAT Colombia Breeding program.

Results of evaluation of 84 rain-fed genotypes that P27-H14 P29-H4 P36-H17, P5-H1 P36-H9 and P36-H4 were the preferred genotypes based on resistance to diseases and yield concurs with other reports (3, 23) that rice varieties with tropical Japonica have higher resistance to RYMV and other diseases.

4.6. Evaluation of promising lines

Results of evaluation of nine selected lines along with two earlier selected lines and a local check is presented in **Table 12**. Although seven varieties were more had higher yields than NERICA-4, only six were presented for release when information on milling and cooking qualities were considered. These genotypes were: **1**. P5H2 (IRAT 325/WAB 450-B-136-HB-F6), **2**. P29H4 (CT 16342-CA-25-M/CK 73-F6), **3**. P8H2 (Caiapo/CT 16324-CA-9-M-F6), **4**. ART3-11L1P1-B-B-2, ([WAB56-104/(WAB56-104/CG14)]/Moroberekan), **5**. P27H1 (CT 16317-CA-4-M/WAB 365-B-1H1-HB-F6), and **6**. WAB 95 B-B-40-HB (ITA257/(IDSA6/ROK16).

4.7. Varietal release and status of release

Breeding background, characteristics, and selected agronomic information on six varieties were presented to the variety release committee. These varieties were released based on important characteristics summarized in **Table 13**. Information from genetic studies during F2 generation guided selection of promising lines from F2 through F6. Subsequently, promising varieties were nominated for National Performance Trials and eventually released. Four new varieties were released namely, NM7-8-2-B-P-11-6 generated from CAIAPO/CT 16324-CA-9-M cross, NM7-29-4-B-P-80-8 (CT 16342-CA-25-M/CK 73), NM7-5-2- B-P-79-7 (IRAT 325/WAB 450-B-136-HB), NM7-27-1- B-P-77-6 (CT 16317-CA-4-M/WAB 365-B-1H1-HB), and NM7-5-2- B-P-79-7 (IRAT 325/WAB 450-B-136-HB). These varieties were assigned release names, where WAB95-B-B-40-HB was named NamChe-1 at the release in Uganda and ARICA 5 by the AfricaRice Breeding Task Force. ARICA acronym means advanced rice for Africa, implying that the harmonized names are to be used by all parties involved. Another variety bred by AfricaRice is NamChe-1 (ARICA-5) with designation ART3-11L1P1-B-B-2. Of the six varieties released, four were bred from Uganda with support from Alliance for Green Revolution in Africa (AGRA) and the other two were developed by AfricaRice through the AfricaWide Rice Breeding Task Force with support from Stress-tolerant rice for poor farmers in Africa and South Asia. These were NamChe-2 (NM7-8-2-B-P-11-6), NamChe 3 (NM7-29-4-B-P-80-8), NamChe 5 (NM7-27-1- B-P-77-6), and NamChe 6 (NM7-5-2- B-P-79-7). The acronym NamChe means **Nam**ulonge **Mche**re (Mchere means uncooked rice in Kiswhili rice). In 2015, over 20,000 ha was under production based on figures of direct seed sale by different producers.

5. Conclusion

This research found that there was adequate variability in the rice population studied for secondary traits for drought tolerance namely, leaf roll and filled grains. However, the filled grains were found to be more informative and therefore recommended for further studies. Of the three rice groups *O. sativa*, interspecific lines, and *O. glaberrima*, there was high similarity between *O. sativa* and interspecific lines. This similarity could make crossing easy.

The genetic studies for drought provided information on the gene action for drought tolerance at reproductive stage of crosses between interspecific and *O. sativa* genotypes. Evidence of additive, nonadditive, additive × additive, and dominance effects were found for drought stress at reproductive stage. Additive effects were the most important components that controlled filled grains in most of the populations. This suggests that breeding methods that involve selection in the early generation could therefore be helpful in improving rice for filled grains. These methods include pedigree breeding, pure line selection, mass selection, single seed decent and progeny selection. In a few crosses, however, proportion of filled grains was controlled by nonadditive effects. Methods that involve a delay in selection of genotypes would be appropriate for improving filled grains in these populations. Modified bulk methods of selection are proposed to be employed in this breeding. Tests for magnitude of the gene action for filled grains using additive-dominance model confirmed that additive gene effects were the most important and additive × additive, as well as, additive × dominance effects were not important. Genotypes *O. sativa,* namely WITA 1 (*O. sativa* indica), IRAT 325 (*O. sativa* japonica), CT 16350-CA-5-M (*O. sativa* japonica), and WAB 450-B-136-HB (NERICA

9) (interspecific) were good combiners under nondrought stress condition for filled grains. In the drought stress condition, however, CK 73, an *O. sativa* genotype, was the best combiner for improved filled grains. Specific combining ability analysis revealed that crosses WITA 2 × CT 16350- CA-5-M, Bonanca × CT 16346-CA-20-M, WITA 2 and CT 16344-CA-9-M were best under NDS condition.

Follow up of their performance in countries in the region shows that NamChe-3 (NM7-29-4-B-P-80-8) and NamChe-2 (NM7-8-2-B-P-11-6) could be mega variety and a major source of disease resistance. In 2015, over 20,000 ha was under production based on figures of direct seed sale by different producers. This is a success story demonstrating the benefit of collaboration and rigorous breeding in the development of locally adapted rice varieties.

Acknowledgements

This work was supported by the Government of Uganda, AGRA-PASS program, and AfricaRice Breeding Task Force program of AfricaRice and University of Kwa-Zulu Natal, in South Africa.

Author details

Jimmy Lamo[1*], Pangirayi Tongoona[2], Moussa Sie[3], Mande Semon[4], Geoffrey Onaga[1] and Patrick Okori[5]

*Address all correspondence to: jlamoayo@gmail.com

1 National Agricultural Research Organization (NARO), Uganda

2 West Africa Center for Crop Improvement (WACCI), University of Ghana, Legon, Ghana

3 Africa Rice Center, Madagascar Center, Madagascar

4 Africa Rice Center, Ibadan Nigeria Center, Nigeria

5 International Crops Research Institute for the Semi-Arid Tropics ICRISAT, Malawi

References

[1] Lamo J, Imanywoha J, Bigirwa G, Walusimbi M, Kyetere D, Kikafunda J, and Kalule T. 2010. First NERICA rice released in Uganda tops farmer's rankings. *International Rice Research Notes*, 35:2.

[2] ECARRN. 2005. Five Year Priority setting for the East, Central and Southern Africa Rice Research Network(EACRRN) 140 p. Mikocheni Agricultural Research Institute, Mwenge Coca Cola Road P O Box 6226, Dae salaam, Tanzania.

[3] Lamo J, Asea G, Otim M, Stella Adur, Serrumaga J, Onaga G, Tsuboi, Baboucarr M, Mande S, Moussa S, Cesear M, Okanya S, Ekebu J, and Ochen S. 2014. Release of 6 upland rice varieties in Uganda in 2014. 41 pages. NARO Annual Report 2014.

[4] Ji X, Ende WE, Schroeven L, Clerens S, Geuten K, Cheng S, and Bennett J. 2007. The rice genome encodes two vacoular invertases with fructan exohydrolase activity but lacks the related fructan biosyntheis genes of the Pooidaee. *The New Phytologist*, 173:50–163.

[5] Itoh JI, Nonomura KI, Ikeda K, Yamaki S, Inukai Y, Yamagishi H, Kitano H, and Nagato Y. 2005. Rice plant development: from zygote to Spikelet Plant Cell Physiol 2005;46:23–47. doi: 10.1093/pcp/pci501

[6] IRRI. 2002. Field evaluation hand book of the International Rice Research Institute. IRRI Publications, Manila, Philippines.

[7] Payne RW, Murray DA, Harding SA, Baird DB, and Soutar DM. 2007. GenStat for Windows, 10th Ed. Introduction, VSN International, Hemel Hempstead.

[8] Hallauer AR and Miranda F. 1988. Quantitative genetics in maize breeding, Iowa State University Press, Ames, Iowa.

[9] Hayman BI. 1958. The separation of epistatic from additive and dominance variation in generation means. *Heredity*, 12:371–390.

[10] Nigam SN, Upadhyaya S, Chandra S, Nageswara Rao RC, Wright GC, and Reddy AGS. 2001. Gene effects for specific leaf area and harvest index in three crosses of groundnuts, *Arachis hypogea. Annuals of Applied Biology*. 139(3):301–306. DOI: 10.1111/j.1744-7348.2001.tb00143.x.

[11] Rowe KE and Alexander WL. 1980. Computations for estimating the genetic parameters in joint-scaling test. *Crop Science*, 20:109–110.

[12] Jinks JL and Jones RM. 1958. Estimation of components of heterosis. *Genetics*, 43:223–234.

[13] Nunir M, Chowdhry MA, and Ahsan M. 2007. Generation mean studies in bread wheat under drought condition. *International Journal of Agriculture and Biology*, 9:282–286.

[14] Sharma JR. 1995. Statistical and biometrical techniques in plant breeding, New Age International Limited, New Delhi, 432 pp.

[15] Peter FC and Frey KJ. 1966. Genotypic correlations, dominance and heritability of quantitative characters in oats. *Crop Science*, 6:259–262.

[16] Mohapatra K and Mohanty HK. 1985. Inheritance of some quantitative characters including heterosis in rice by combining ability analysis, pp. 579–591, *In*. Swaminathan MS, ed. Rice Genetics. Proceedings of International Rice Genetics Symposium, Island Publishing House, Manila, Philippines.

[17] Fujii M, Andoh C, and Ishihara S. 2005. Drought resistance of NERICA (New Rice for Africa) compared with *Oryza sativa* L. and millet evaluated by stomatal conductance and soil content. Proceeding of the fourth International Crop Science Conference.

http://dx.doi.org/10.5772/66826

[18] Atlin G. 2003. Improving drought tolerance by selecting for yield, pp. 14–22, *In*. Fischer KS, Lafitte R, Fukai S, Atlin Q, and Hardy B, eds. Breeding rice for drought-prone environments. IRRI Publication, Los Baños, Philippines.

[19] Ahmad L, Zakri AH, Jalani BS, and Omar D. 1985. Detection of additive and nonadditive variation in rice pp. 555–564, *In*. Swaminathan MS, ed. Rice Genetics. Proceedings of International Rice Genetics Symposium, Island Publishing House, Manila, Philippines.

[20] Narayanan KK and Sree Rangasamy SR. 1990. Genetic analysis for salt tolerance in rice, pp. 167–174, *In*. Lampe KJ, ed. Rice Genetics. Proceedings of Second International Rice Genetics Symposium. Island Publishing House Manila, Philippines, International Rice Research Institute (IRRI), Manila, Philippines.

[21] Jones MP, Mande S, Aluko K (1997). Diversity and potential of *Oryza glaberrima* Steud in upland rice breeding. Breed Sci. 47:395–398.

[22] Sarker U, Biswas PS, Prasad B, and Mian MAK. 2002. Heterosis and genetic analysis in rice hybrids. *Pakistan Journal of Biological Sciences*, 5:1–5.

[23] Efisue A, Tongoona P, Derera J, Langyintuo A, Laing M, and Ubi B. 2008. Farmers perceptions on rice varieties in Sikasso region of Mali and their implications to rice breeding. *Journal of Agronomy and Crop Science*, 4:212–218.

[24] Manonmani S and Fazlullah Kahn AK. 2003. Studies on combining ability and heterosis in rice. *Madras Agricultural Journal*, 90:228–231.

[25] Chandraratna MF. 1964. Genetics and breeding of rice p. 389 Tropical Science Series, Longman, London, U.K.

[26] Acharya S. 1987. Genetic parameters and their implications in breeding cold tolerance varieties of rice (*Oryza sativa* L.). *Crop Improvement*, 14:100–103.

[27] Geetha S, Shanthi P, Jebaraj S, and Mohammed SEN. 2006. Gene action for sodicity tolerance in rice. *Indian Journal of Crop Science*, 1:201–202.

[28] Chakraborty S, Hazarika M, and Hazarika G. 1994. Combining ability analysis in rice. *Oryza* 31:281–283.

[29] Singh BN and Mackill DJ. 1990. Genetics of leaf rolling under drought stress. Rice Genetics, Proceedings of the Second International Rice Genetics Symposium II:159–166.

[30] Garrity DP and O'Toole JC. 1994. Screening rice for drought resistance at the reproductive phase. *Field Crops Research*, 39:99–110.

[31] Fukai S and M Cooper. 1995. Development of drought-resistant cultivars using physiomorphological traits in rice. *Field Crops Research*, 40:67–86.

[32] Lafitte R, Blum A, and Courtois G. 2003. Secondary traits to help identify drought-tolerant genotypes, pp. 37–48, *In*. Fischer KS, Lafitte R, Fukai S, Atlin Q, and Hardy B, eds. Breeding rice for drought-prone environments, IRRI, Los Baños, Philippines.

[33] Sun ZX, Xiong ZM, Min SK, and M SH. 1989. Identification of the temperature-sensitive male sterile rice. *Chinese Journal of Rice Science*, 3:49–55.

[34] Ali J, Siddiq EA, Zaman FU, Abraham MJ, and Ahmed IM. 1995. Identification and characterization of temperature sensitive genic male sterile sources in rice (*Oryza sativa* L.). *Indian Journal of Genetics*, 55:243–259.

Section 4

New Technologies in Rice Research

Chapter 12

Deciphering Histone Modifications in Rice by Chromatin Immunoprecipitation (ChIP): Applications to Study the Impact of Stress Imposition

Liliana J. Ferreira, Sebastião Ravasco,
Duarte D. Figueiredo, Christoph Peterhänsel,
Nelson J.M. Saibo, Ana Paula Santos and
M. Margarida Oliveira

Additional information is available at the end of the chapter

http://dx.doi.org/10.5772/66424

Abstract

The spatial organization of chromatin, the methylome, and histone modifications represents epigenetic layers that greatly intersect each other, influencing genome regulation and allowing high flexibility in stress response. Although changes in specific histone modification marks could be extensively associated with transcriptional regulation of stress-responsive genes, a link between specific epigenetic signatures and plant stress tolerance has not yet been established. This chapter includes some examples of the associations found between fluctuations in these marks and regulation of plant stress-responsive genes. Chromatin immunoprecipitation (ChIP) has been widely used to uncover the landscape of histone modifications. However, ChIP involves multiple steps and requires optimizations targeting the tissue and the plant species. Here, we detail the ChIP procedure currently used in our laboratory, for leaf tissues of young rice seedlings, to decipher the dynamic feature of specific chemical modifications of histones that may influence the expression of stress-responsive genes. We show the success achieved after introducing specific optimizations and highlight the key critical steps and trouble shootings that may occur. A thorough understanding of stress-induced fluctuations of specific histone modifications may unveil new strategies to improve plant adaptation and performance in suboptimal conditions.

Keywords: abiotic stress, chromatin, epigenetics, histone modifications, rice

INTECH
open science | open minds

1. Introduction: histone modifications and gene expression regulation under stress

Rice is a very important crop whose production is being affected due to the climate changes we have been witnessing (higher and more variable temperatures, increased soil salinity, and extreme drought or flooding) have been negatively impacting rice production and sustainability [1]. Environmental stress experiences have been implicated in extensive changes of chromatin structure (e.g., decondensation of heterochromatic domains), and also in the plasticity of epigenetic marks [2, 3]. The flexibility of chromatin structure, chromatin loops, and epigenetic marks (DNA methylation and histone modifications) all play a role in gene expression regulation under stress [4]. This chapter focuses on the ChIP (chromatin immunoprecipitation) strategy to decipher the pattern of histone modifications particularly at specific promoter regions of selected stress-responsive genes. We describe in detail the ChIP protocol we are currently applying in our laboratory for leaf tissues of young rice seedlings.

At the basic level of chromosome structure, the DNA is bound to histone proteins forming the nucleosomes which represent the basic element of chromatin, being made of DNA around histones [5]. The histone tails can be modified by chemical groups, such as acetylation, methylation, phosphorylation, or ubiquitinylation, which can affect chromatin accessibility to the transcriptional machinery [6–8]. There are over 60 different residues on histones where chemical modifications have been detected and these posttranslational modifications (PTMs) of histones are the basis of a "histone code." The "histone code" was originally proposed by Strahl and Allis [9] and postulates that specific combinations of histone variants and PTMs can influence chromatin states [9]. For example, histone hyperacetylation has been generally correlated with transcription activation, while hypoacetylation has been associated with transcriptional silencing [9]. This biological effect is commonly explained by the fact that acetylation can lead to chromatin unwinding thus, decreasing their affinity for DNA and subsequently influencing the way transcription factors access DNA [10]. On the other hand, the presence of methyl groups at lysine residues can be reflected in different meanings including gene activation or repression, based on whether the lysine residue is mono-, di-, or trimethylated and also on which lysine residue is methylated [11]. Gene expression regulation can be greatly influenced by histone modifications landscape along the gene promoter region [12–14]. The analysis of histone modifications landscape has been mainly performed by chromatin immunoprecipitation which includes the cross-linking of histones and DNA, chromatin isolation and sonication, chromatin immunoprecipitation with antibodies specific for a given histone modification, de–cross-linking of histone-DNA complexes, DNA recovery, and gene-specific real-time quantitative polymerase chain reaction (PCR). Quantification of the relative proportion of the different loci to which the PTM is associated can be achieved by quantitative PCR (ChIP-qPCR) or microarray-based techniques (ChIP-chip), depending on the amount of loci one wants to analyze [15]. Large-scale enrichment analysis can also be performed through ChIP followed by DNA sequencing (ChIP-seq). This technique allows obtaining information regarding *in vivo* analysis of the protein-binding position in the

http://dx.doi.org/10.5772/66424

genome and thus can also be used if one wants to determine the specific targets of a given transcription factor.

In plants, environmental stress responses have been associated with the plasticity of histone modification marks, which in turn have been related to alterations in the expression of genes underlying responses to distinct stress types (**Table 1**). Nevertheless, the mechanistic links behind such connections are still largely unknown. This chapter is descriptive regarding the chromatin immunoprecipitation protocol that we are currently using to decipher the plasticity of histone modifications with particular focus on selected stress-responsive genes. The ChIP protocol we present refers to leaf tissues of young rice seedlings and is based on what was previously described for maize leaves [15] and thus, in its essence, is not new. However, here we describe important optimizations that take into account intrinsic specificities of rice leaf tissues. The whole procedure, from harvesting the rice leaves to the recovery of immunoprecipitated DNA, can be carried out within 3 days. The critical

Plant	Stress	Histone modifications	Biological effects	Reference
Rice	Submergence	↑H3 acetylation	↑ ADH1 and PDC1 stress responsive genes	[16]
Rice	Drought	↑H3K4me$_3$	↑ Dehydrin genes	[17]
Rice	Salt	↑H4K20me3 ↑H3K9ac, ↑H4K5ac	↑ OsRMC (salt-responsive gene)	This work (**Figure 4**)
Arabidopsis	Drought	↑H3K9ac ↑H3K4me3	↑ RD29A, RD29B, RD20,and RAP2.4 (drought-responsive genes)	[18]
Arabidopsis	Salt	↑H3K9K14ac ↑H3K4me3 ↓H3K9me2	↓ ABI1, ABI2, RD29A, RD29B, DREB2 (abiotic stress-responsive genes like)	[19]
Arabidopsis	Salt	↓ H3K27me3	HKT1 (Salt stress-responsive gene)	[20]
Arabidopsis	Dehydration	↑H3K4me1 ↑H3K4me2 ↑H3K4me3	↑Dehydration-responsive genes	[21]
Arabidopsis	Cold	↑H3 acetylation	↑CBF1	[22]
Arabidopsis	Cold	↑H3K27me3	↑COR15A and GOLS3 (cold-responsive genes)	[23]
Soybean	Salt	↑H3K4me3 ↑H3K9ac	Activation of members of AP2/EREB, bZIP, NAC and MYB transcription factors	[24]
Maize	Cold	↓H3 acetylation ↓H4 acetylation	↑Histone deacetylases	[25]

Table 1. Connections between plasticity of histone modifications marks and biological effects due to stress imposition.

steps and trouble shootings are clearly indicated along the procedure, including important optimizations made to improve cross-linking and increase sonication efficiency for young rice leaves. Regarding the antibody selection, ,to assess quality and specificity, we routinely perform immunodetection using distinct antibodies for histone modifications in rice root tissue sections. This *in situ* approach provides information regarding the spatial organization pattern of specific histone modifications in individual cells during interphase when transcription is intensely occurring. Very briefly, the procedure includes the vibratome sectioning of root tips of 3-day-old rice seedlings, three-dimensional (3D) *in situ* immunofluorescence on preserved tissue sections followed by confocal microscopy analysis; for details on this protocol, see [3].

2. ChIP protocol for rice young leaves

2.1. Materials

Reagents	Supplies	Equipment
p-Formaldehyde 37%	Kitasato	Stirrer
Sodium butyrate	Rubber tubes	Vacuum pump
Sucrose	Sieve	Barometer
Tris	Small paintbrush	Chronometer
β-Mercaptoethanol	Liquid nitrogen	Mortar and pestle
PMSF	50 ml, 15 ml falcon tubes	Rotating mixer (2 ml, 50 ml tubes)
Glycine	Small funnel	Cold room
Protease inhibitor	Miracloth	Centrifuge
Magnesium chloride	DNA purification kit (Roche)	Sonicator (Bioruptor, Diagenode)
Triton X-100	Paper towels	Incubator (65°C)
EDTA	1.5 ml tubes	Vortex mixer
SDS		
Sodium chloride		
DTT		
Protein-A agarose		
Lithium chloride		
Sodium deoxycholate		
NP-40		

http://dx.doi.org/10.5772/66424

2.2. Buffers and solutions

Buffers	Solutions
A	10 mM sodium butyrate, 0.4 M sucrose*, 10 mM Tris (pH 8.0), 5 mM β-mercaptoethanol, 0.1 mM PMSF, 0.8% formaldehyde
B	10 mM sodium butyrate, 0.4 M sucrose, 10 mM Tris (pH 8.0), 5 mM β-mercaptoethanol, 0.1 mM PMSF, 1 M protease inhibitor**
C	10 mM sodium butyrate, 0.25 M sucrose*, 10 mM Tris (pH 8.0), 5 mM β-mercaptoethanol, 0.1 mM PMSF, 10 mM $MgCl_2$, 1 M Triton X-100, 1 M protease inhibitor*
D	10 mM sodium butyrate, 1.64 M sucrose*, 10 mM Tris (pH 8.0), 5 mM β-mercaptoethanol, 0.1 mM PMSF, 2 mM $MgCl_2$, 150 mM Triton X-100, 1 M protease inhibitor*
E	25 mM Tris (pH 8.0), 5 mM EDTA, 0.5 M SDS, 0.1 mM PMSF, 1 M protease inhibitor*
F	50 mM Tris (pH 8.0), 1 mM EDTA, 150 mM NaCl, 100 mM Triton X-100
G	62.5 mM Tris (pH 6.8), 200 mM NaCl, 2 M SDS, 10 mM DTT
Low salt	0.15 M NaCl, 0.1 M SDS, 1 M Triton X-100, 2 mM EDTA, 20 mM Tris (pH 8.0)
High salt	0.5 M NaCl, 0.1 M SDS, 1 M Triton X-100, 2 mM EDTA, 20 mM Tris (pH 8.0)
LiCl	0.25 M LiCl, 0.1 M NP-40, 24 mM sodium deoxycholate, 1 mM EDTA, 20 mM Tris (pH 8.0)
TE	10 mM Tris (pH 8.0), 1 mM EDTA

*The sucrose solution should be filtered and kept at 4°C.
**We use the protease inhibitor "complete," from Roche.

2.3. Procedure

A. Collection of plant material and tissue fixation

1. Harvest 3 g of young rice leaves (around 70 plants), cut them into pieces of approximately 1–2-cm length and put the leaves immediately, as loose as possible, in a vacuum flask with a stirrer.

   Notes:

   a. The ChIP assays were performed in 14-day-old leaves of rice (*Oryza sativa* L. ssp japonica AA, 2*n* = 24) cv. Nipponbare. Two days before applying stress, for example, salt or cold, the plants, grown in a hydroponic system inside glass tubes, were transferred to larger flasks for better manipulation and faster harvesting of leaves.

   b. Place a sponge on top of the flask to avoid losing leaves by the suction applied during vacuum.

2. Submerge leaves in 200 ml Buffer A (0.8 % p-formaldehyde). Vacuum infiltrate at room temperature for 2 min (meaning vacuum up to 50 mbar, a short release of vacuum and then repeating the cycle six to seven times), allowing penetration of fixative into leaf

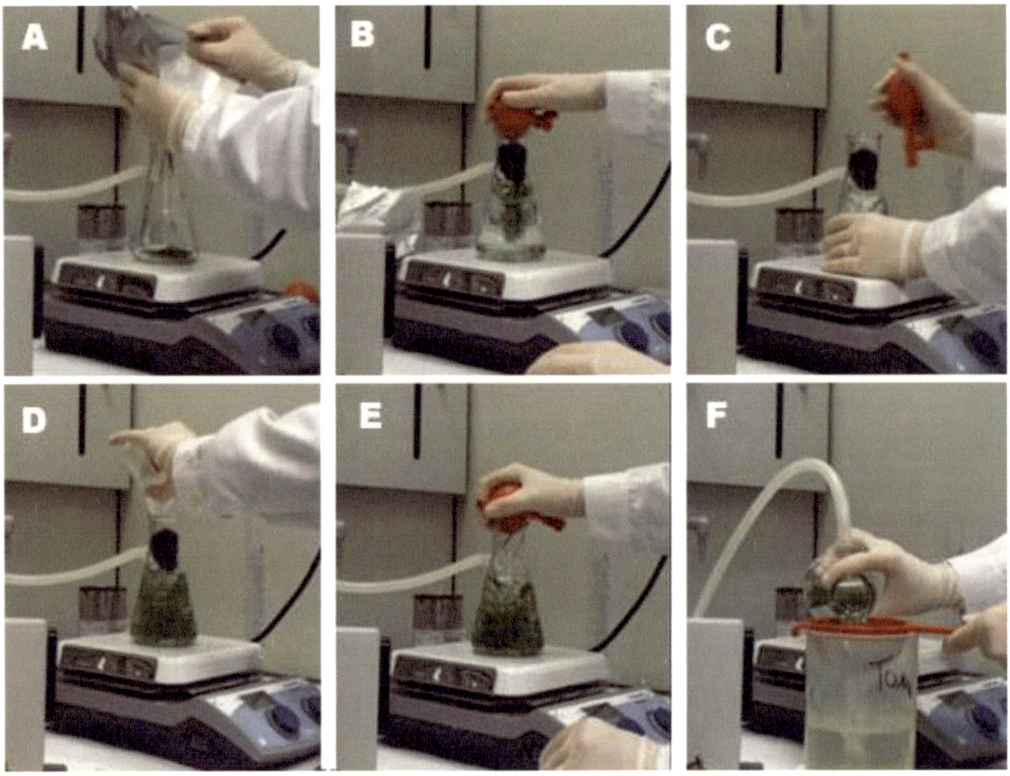

Figure 1. Schematic description of the cross-linking step in ChIP. (A) Leaf fragments with approximately 1-cm length are placed in a flask and immediately mixed with the formaldehyde solution (Buffer A). (B) The flask is covered for vacuum infiltration. (C) When pressure reaches 50 mbar, the vacuum is released and the cycle is repeated approximately 6–7 times for 2 min, followed by stirring an additional 1 min without vacuum. (D) The cross-linking reaction is stopped by adding 20 ml of 2M glycine. (E) Vacuum is again applied for 5 min with pressure release every 30 s. (F) The leaves are then washed in water (using a sieve), carefully dried between paper towels, and finally frozen in liquid nitrogen until further use.

tissues. Stir for an additional 1 min without vacuum. Key steps of this procedure are depicted in **Figure 1**.

Notes:

a. A FAIRE (formaldehyde-assisted isolation of regulatory elements) test for assessing chromatin cross-link efficiency is routinely performed in our laboratory (**Figure 2**).

b. The color of the p-formaldehyde solution should be evaluated. A yellowish color indicates some oxidation and this will negatively affect the efficiency of the cross/de–cross-linking process.

c. When reducing the pressure to 50 mbar, one should see foam at the surface of the suspension; if not the vacuum gear insulation should be checked to make sure there is no leak. Also, ensure that leaf samples are fully submerged in solution A.

Samples	CROSSLINKED CTs	[DNA]	Normaliz.	DE-CROSSLINKED CTs	[DNA]	Normaliz.
0%	16,600	1,03E-05	1,000	22,417	1,85E-07	1,000
0.8%	23,993	6,21E-08	0,006	22,253	2,07E-07	1,120
1%	24,327	4,93E-08	0,005	24,237	5,25E-08	0,284

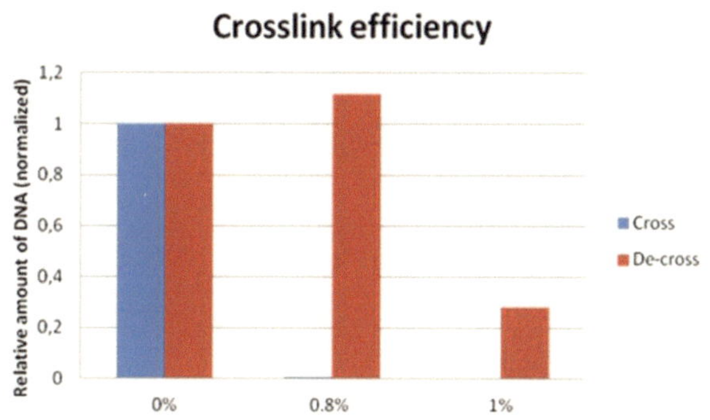

Figure 2. FAIRE assay to optimize chromatin cross-link efficiency. Two formaldehyde concentrations (0.8% and 1%) were tested against cross-linking with water (0%). Primers specific for the *OsUBC2* gene were used and their efficiency was calculated according to the formula E = 10(1/slope) (efficiency = 1.996933). This efficiency value was then used to estimate the initial concentration of DNA present in the samples, following the formula [DNA] = Efficiency-Ct. The concentrations were normalized against 0% formaldehyde values and plotted. Both formaldehyde concentrations assured a high degree of cross-linking, but DNA recovery after de–cross-link was higher under the use of 0.8% formaldehyde; thus, this formaldehyde concentration was elected as ideal for our material and assay conditions.

Caution: Formaldehyde is toxic and potentially carcinogenic, thus, particular precautions should be considered (e.g., working in the fume hood).

3. Add 20 ml glycine (2 M) to the flask, and mix vigorously to stop cross-linking. Vacuum infiltrate again during 5 min, releasing vacuum every 30 s.

4. Pour off the fixative solution and wash the leaves in a sieve with plenty of water. Dry the leaves carefully with paper towels and insert them into 50 ml tubes. Freeze in liquid nitrogen.

Pause point: Leaf tissue samples can be put in storage for several months at −80°C.

B. Isolation of nuclei and chromatin fragmentation

5. Grind the leaves, in liquid nitrogen (N_2), to a fine powder. Insert into 50 ml tubes and resuspend the ground material in 40 ml Buffer B.

   Note: Do not add the entire buffer at once. Tap the tube on the bench to get the N_2 out of the mix.

Caution: β-mercaptoethanol containing solution; work in the fume hood.

6. Incubate for 15 min at 4°C. Carefully, shake to release nuclei from cells (use a rotating mixer in a 4°C chamber).

7. Place a small funnel on top of a new 50 ml tube and filter the previous solution through it, using four layers of Miracloth.

   Note: Keep the tubes on ice.

8. Centrifuge the filtered solution for 20 min at 2880 × *g* at 4°C.

   Note: In the meantime, prepare Buffer C.

9. Gently, remove supernatant and resuspend the pellet in 1 ml Buffer C.

   Note: First, add 50 μl of Buffer C then, resuspend using a small paintbrush and then add the other 950 μl.

10. Transfer the solution to 1.5 ml Eppendorf tube and proceed to centrifugation at 12,000 × *g* for 10 min at 4°C.

    Note: In the meantime, prepare Buffer D.

11. Gently, remove supernatant and resuspend the pellet in 300 μl Buffer D.

    Note: First, add 50 μl of Buffer D, resuspend with a small paintbrush, and then add the other 250 μl.

12. Pipet 1.5 ml of Buffer D to a new 2 ml tube. Overlay this 1.5 ml of Buffer D with the resuspended pellet in the 2 ml tube.

    Note: The process should end up with a layer of extract (green) on top of the 1.5 ml Buffer D (colorless).

13. Centrifuge for 1 h at 16,000 × *g* at 4°C.

14. Remove supernatant and resuspend the chromatin pellet in 300 μl Buffer E.

    Note: Take a 30-μl aliquot for later gel analysis of "unsheared chromatin."

15. Sonicate the chromatin solution for successive cycles, 13 times, each time 30 s ON followed by 30 s OFF, selecting "LOW" power sonication, using the Bioruptor® Plus Sonicaton System, Diagenode. Water bath should be previously cooled to 4°C or less.

    Note: The shearing step is determinant on ChIP efficiency. To determine shearing efficiency, release bead-bound complexes from the sheared and unsheared samples by adding 100 μl Buffer G, vortex for 5 min, centrifuge briefly and incubate overnight at 65 °C. Purify these samples with a kit, eluting in 80 μl to get the most DNA possible out of the column, and then concentrate to ±15 μl with a speed vac. Add loading buffer and run on a 1.5% agarose gel. A DNA smear with 200–500-bp size range should be ideally obtained (see **Figure 3**).

16. Centrifuge the sonicated chromatin solution for 5 min at 16,000 × *g* at 4°C to sediment cell debris.

17. Take out the supernatant to a new tube and save again a 30 μl aliquot for gel analysis of "sheared chromatin."

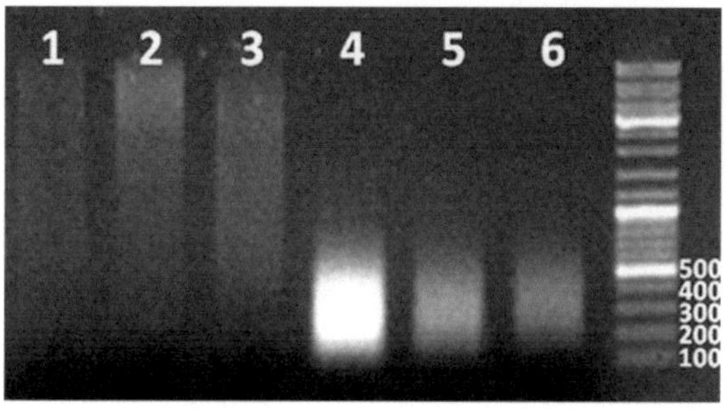

Figure 3. Chromatin fragmentation for ChIP. Unsheared (1, 2, 3) and sheared (4, 5, 6) chromatin samples. Sonication for 13 cycles, 30 s on and 30 s off, low-intensity sonication. The size of chromatin fragments was determined through agarose gel electrophoresis.

Pause point: The chromatin, once sonicated, can be kept at −80°C for a few months (not more than 3 months).

C. Preclearing

18. Use 200 μl chromatin solution and add 1.8 ml Buffer F and 30 μl protein-A agarose. Preclear the solution for 3 h at 4°C on a rotating mixer in the 4°C chamber.

19. Centrifuge at 4°C for 5 min at 500 × *g* and incubate for 5 min on ice. Collect the supernatant in a 2-ml tube and discard the beads.

D. Immunoprecipitation

20. Use the supernatant following this scheme:

Sample	Supernatant	Antibody	Protein-A Agarose
Antibody 1	400 μl	2–5 μl	30 μl
Antibody 1	400 μl	2–5 μl	30 μl
Antibody 1	400 μl	2–5 μl	30 μl
No Ab	400 μl	–	30 μl
Input	40 μl	–	–

Notes:

a. Freeze the input sample, as well as the supernatant that may be left.

b. Make sure to vortex the agarose beads prior to each use.

c. A successful ChIP assay depends on the quality of the antibody. In our laboratory, we routinely use ChIP-validated antibodies. Also, as a prior checking of the antibodies efficacy, we previously conducted in situ immunofluorescence in tissue sections and Western blotting analysis with commercial antibodies to histone modifications.

d. If the immunoprecipitation is inefficient or produces very low signals, the amount of antibody should be increased.

21. Incubate overnight at 4°C on a rotating mixer.

22. Pull down the agarose beads by centrifugation (5 min, 500 × *g*) and incubate on ice for 5 min. Collect the supernatant and add to the remainder chromatin solution frozen on step 20 and freeze it. This supernatant can be re-used to perform ChIP again. Proceed with the washing of the agarose beads.

E. Washes

23. Wash the beads using 900 μl buffer per wash followed by pelleting the beads (10 min on the rotating mixer, 4°C; spin 5 min at 500 × *g* and remove supernatant). Apply the washes in the following order: 1× low-salt wash buffer, 1× high-salt wash buffer, 1× LiCl wash buffer, 2× TE wash buffer.

    Note: The buffers should be prepared fresh. Remove TE totally after the final wash.

F. Reverse cross-linking

24. Release bead-bound complexes by adding 200 μl Buffer G, vortex for 5 min, centrifuge briefly, and incubate overnight at 65°C. Do this also with the "*input*" sample frozen on step 20.

25. Centrifuge shortly to sediment the agarose, collect supernatant (~100 μl), and purify it with a kit, eluting in 80 μl.

    Note: We use the high pure PCR product purification kit (Roche).

Pause point: ChIP samples, once purified, can be stored at −20°C for at least 1 month. Do not dilute prior to storage.

G. Quantitative PCR and data analysis

26. Using the ChIP-purified DNA, proceed to quantitative PCR using gene-specific primers.

    Notes:

    a. Real-time quantitative PCR was performed using the LightCycler 480 system (Roche). The PCR was carried out in a final volume of 20 μl containing 10 μl SYBR Green PCR Master Mix from Roche (2×), 5-μl ChIP DNA template, 1-μl primers (forward and reverse, 1 mM each, 3-μl sterile ddH_2O.

    b. qPCR conditions: one cycle at 95°C for 5 min and 45 cycles of amplification at 95°C for 10 s, 52°C for 10 s, and 72°C for 10 s. All qPCR experiments were performed on at least three biological replicates and the CT values were calculated from means of three technical replicates.

    c. Several methods exist for data normalization, namely the background subtraction [26], percentage of input (% IP) [27], fold enrichment [28], normalization relative to a control sequence [29], and normalization relative to nucleosome density [30]. In our ChIP experiments, we used the % input method, in which the pPCR signals derived from the ChIP samples were divided by the qPCR signals from the input sample. Additionally,

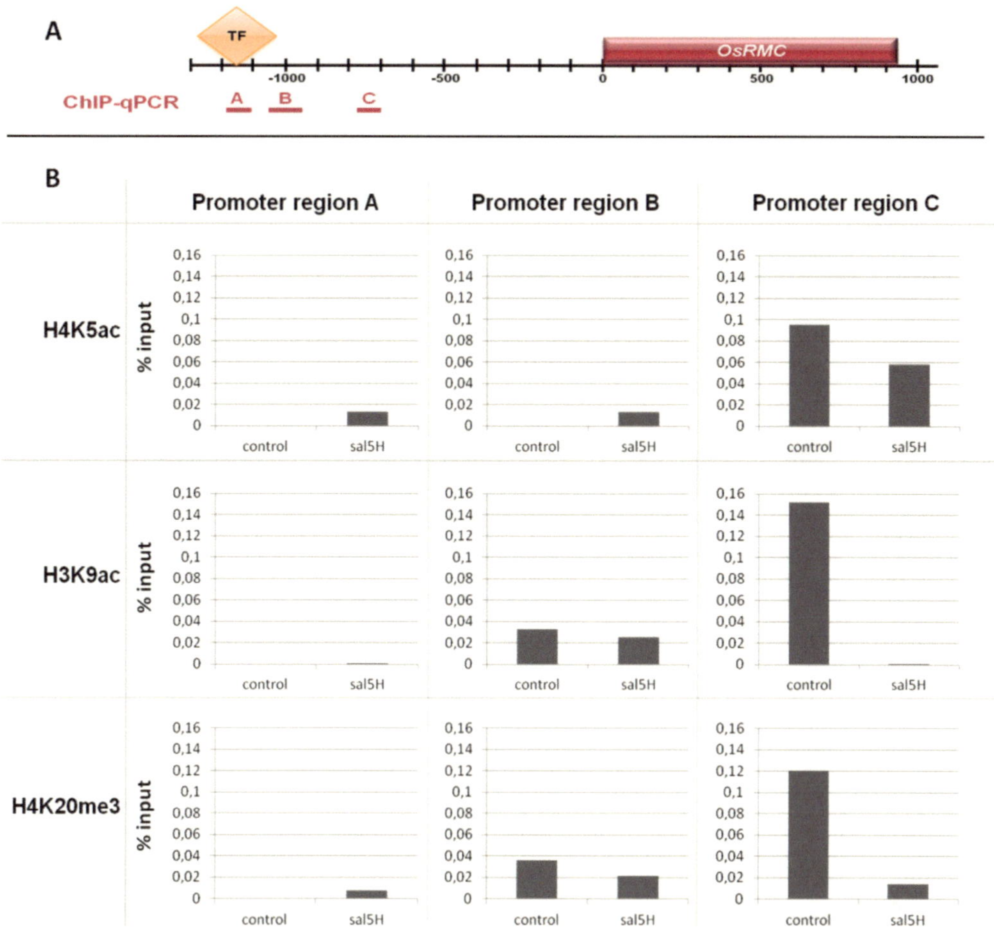

Figure 4. ChIP assay to determine the dynamics of H3K4ac, H3K9ac, and H4K20me3 marks at *OsRMC* gene promoter after salt stress. (A) Schematic representation of the *OsRMC* promoter regions analyzed: promoter region A [−1159; −1073], promoter region B [−1079; −967], and promoter region C [−773; −693]. Primer A forward TTGACGAGCAGGCATAGGTA, reverse CTGGATTGTCTCGGTGGAAT; primer B forward ATCCAGTTCGTTGCCATCTC, reverse CGGAATGAACGGTGATCCTA; primer C forward GGCACAGATATCCCC TTTGA, reverse CCGTGAGAGCCCATTTTTAC. The diamond shape indicates the binding site of the transcription factors *OsEREBP1* and *OsEREBP2* as reported by [31]. (B) The levels of histone modification marks were determined by ChIP using specific antibodies for acetylation of histone H3 at the lysine 4 and 9 (H3K4 and H3K9) and for trimethylation of histone H4 at lysine 20 (H4K20). The samples were analyzed using real-time qPCR to quantify *OsRMC* gene promoter DNA enriched in the immunoprecipitates. The distribution of specific histone modification marks was not homogeneous along distinct *OsRMC*-promoter regions. The promoter region C, the closest to ATG, presented a higher enrichment in all the histone marks analyzed as compared to the promoter regions more far away from ATG. The landscape of histone modifications was dynamics and salt stress responsive. Under control conditions, the histone marks present on the promoter region A were barely detected. However, after 5 h of salt treatment, there was an increase of the H4K5ac and H4K20me3 marks. The promoter region B, on the vicinity of the TFs-binding site, was depleted of H4K5ac in control conditions but got an enrichment on this mark under salt stress. On the contrary, the levels of H3K9ac and H3K20me3 marks decreased with salt stress. Concerning the promoter of region C, all histone marks analyzed were detected in high levels in control conditions but were drastically reduced upon salt stress. This example shows a differential enrichment of euchromatic marks dependent on the promoter region which may be interpreted from the viewpoint of gene expression regulation under stress.

the background signal evidenced by the NoAb sample subtracted to the ChIP samples, according to the formula % IP = (AB-NoAB)/input.

d. The analysis of immunoprecipitated DNA by qPCR enabled to evaluate the dynamics of specific histone modifications along specific regions of the OsRMC promoter under salt stress as exemplified in **Figure 4**.

3. Trouble shooting and future directions

The study of histone modifications has been mainly based on ChIP analyses, which is a very time-consuming process involving multiple stages. Some steps are particularly critical, such as cross-linking, sonication, and antibody immunoprecipitation, and must be previously optimized for each plant species and tissue. One critical point of the protocol concerns the formaldehyde cross-linking. With a low-efficiency cross-linking with formaldehyde, many DNA/protein interactions can be lost. On the other hand, if there is an excessive cross-linking, the DNA may not be recovered. Thus, various cross-linking times, as well as different formaldehyde concentrations, should be tested. Another possible problem is when a specific signal is observed but at very low levels (low input). That may happen because the chromatin structure itself may have been altered during the process and thus affecting the detection of specific regions. Also, the DNA-protein complexes may remain bound to the tubes during the procedure and in this case, the use of siliconized 1.5-ml tubes can help to solve this problem. A third critical point of the ChIP protocol refers to sonication efficiency since a deficient sonication can influence antigen accessibility that often results in a huge variability between experiments. In humans, some of these limitations have been overcome by combining ChIP procedure with microfluidic devices that in a semi-automated manner enables the identification of multiple marks, while requiring smaller volumes of samples and reagents and less human manipulation [32, 33]. This technology must still be extended to plants. At last, it must also be referred that ChIP studies have been mainly based on using heterogeneous cell populations which can lead to misleading results since epigenetic patterns are cell- and tissue-specific. Therefore, we anticipate a growing importance of methodologies allowing a cell-based resolution analysis of histone modifications. For such resolution, the isolation of single cells may be very important, and techniques such as droplet encapsulation, fluorescence-activated cell sorting (FACS), or microfluidic processing [34] are particularly relevant.

Acknowledgements

The authors thank Louisa Heimann for technical help on establishing the ChIP protocol. This work was supported by the FCT (Portuguese Foundation for Science and Technology) through the project "Epigenetic regulation of the rice genome under environmental stresses" '[BIA-BCM/111645/2009]' and through the R&D unit, UID/Multi/04551/2013 (GREEN-IT). NS was supported by the Programa FCT Investigator, financed by POPH (QREN). APS and LF were supported by FCT grants "[BPD/74197/2010]" and "[BD/61428/2009]," respectively.

http://dx.doi.org/10.5772/66424

Author details

Liliana J. Ferreira[1], Sebastião Ravasco[1], Duarte D. Figueiredo[1,2,3], Christoph Peterhänsel[4], Nelson J. M. Saibo[1], Ana Paula Santos[1,*] and M. Margarida Oliveira[1]

*Address all correspondence to: apsantos@itqb.unl.pt

1 Instituto de Tecnologia Química e Biológica António Xavier, Universidade Nova de Lisboa, Oeiras, Portugal

2 Department of Plant Biology, Swedish University of Agricultural Sciences, Uppsala, Sweden

3 Linnean Center of Plant Biology, Uppsala, Sweden

4 Leibniz University Hannover, Institute of Botany, Hannover, Germany

References

[1] Lobell DB, Schlenker W, Costa-Roberts J (2011) Climate trends and global crop production since 1980. Science. 29;333(6042):616–20.

[2] Santos AP, Ferreira L, Maroco J, Oliveira MM (2011) Abiotic stress and induced DNA hypomethylation cause interphase chromatin structural changes in rice rDNA loci. Cytogenet Genome Res. 132(4):297–303.

[3] Ferreira LJ, Azevedo V, Maroco J, Oliveira MM, Santos AP (2015) Salt Tolerant and Sensitive Rice Varieties Display Differential Methylome Flexibility under Salt Stress. PLoS One. 1;10(5):e0124060.

[4] Madlung A, Comai L (2004) The effect of stress on genome regulation and structure. Ann Bot. 94(4):481–95.

[5] Kornberg RD (1974) Chromatin structure: a repeating unit of histones and DNA. Science. 24;184(4139):868–71.

[6] Workman JL, Kingston RE (1998) Alteration of nucleosome structure as a mechanism of transcriptional regulation. Annu Rev Biochem. 67:545–79.

[7] Kornberg RD, Lorch Y (1999) Twenty-five years of the nucleosome, fundamental particle of the eukaryote chromosome. Cell. 98(3):285–94.

[8] Rothbart SB, Strahl BD (2014) Interpreting the language of histone and DNA modifications. Biochim Biophys Acta. 1839(8):627–43.

[9] Strahl BD, Allis CD (2000) The language of covalent histone modifications. Nature. 6;403(6765):41–5.

[10] Kouzarides T (2007) Chromatin modifications and their function. Cell. 23;128(4):693–705.

[11] Zhang X, Bernatavichute YV, Cokus S, Pellegrini M, Jacobsen SE (2009) Genome-wide analysis of mono-, di- and trimethylation of histone H3 lysine 4 in Arabidopsis thaliana. Genome Biol. 10(6):R62.

[12] Bernstein BE, Meissner A, Lander ES (2007) The Mammalian Epigenome. Cell. 23;128 (4):669–81.

[13] Berger SL (2007) The complex language of chromatin regulation during transcription. Nature. 24;447(7143):407–12.

[14] Li B, Carey M, Workman JL (2007) The role of chromatin during transcription. Cell. 23;128(4):707–19.

[15] Haring M, Offermann S, Danker T, Horst I, Peterhansel C, Stam M (2007) Chromatin immunoprecipitation: optimization, quantitative analysis and data normalization. Plant Methods. 24;3:11.

[16] Tsuji H, Saika H, Tsutsumi N, Hirai A, Nakazono M (2006) Dynamic and reversible changes in histone H3-Lys4 methylation and H3 acetylation occurring at submergence-inducible genes in rice. Plant Cell Physiol. 47(7):995–1003.

[17] Zong W, Zhong X, You J, Xiong L (2013) Genome-wide profiling of histone H3K4-tri-methylation and gene expression in rice under drought stress. Plant Mol Biol. 81(1-2):175–88.

[18] Kim JM, To TK, Ishida J, Morosawa T, Kawashima M, Matsui A, Toyoda T, Kimura H, Shinozaki K, Seki M (2008) Alterations of lysine modifications on the histone H3 N-tail under drought stress conditions in Arabidopsis thaliana. Plant Cell Physiol. 49(10):1580–8.

[19] Chen LT, Luo M, Wang YY, Wu K (2010) Involvement of Arabidopsis histone deacetylase HDA6 in ABA and salt stress response. J Exp Bot. 61(12):3345–53.

[20] Sani E, Herzyk P, Perrella G, Colot V, Amtmann A (2013) Hyperosmotic priming of Arabidopsis seedlings establishes a long-term somatic memory accompanied by specific changes of the epigenome. Genome Biol. 14(6):R59.

[21] van Dijk K, Ding Y, Malkaram S, Riethoven JJ, Liu R, Yang J, Laczko P, Chen H, Xia Y, Ladunga I, Avramova Z, Fromm M. (2010) Dynamic changes in genome-wide histone H3 lysine 4 methylation patterns in response to dehydration stress in Arabidopsis thaliana. BMC Plant Biol. 10:238.

[22] Pavangadkar K, Thomashow MF, Triezenberg SJ (2010) Histone dynamics and roles of histone acetyltransferases during cold-induced gene regulation in Arabidopsis. Plant Mol Biol. 74(1-2):183–200.

[23] Kwon CS, Lee D, Choi G, Chung WI (2009) Histone occupancy-dependent and -independent removal of H3K27 trimethylation at cold-responsive genes in Arabidopsis. Plant J. 60(1):112–21.

[24] Song Y, Ji D, Li S, Wang P, Li Q, Xiang F (2012) The dynamic changes of DNA methylation and histone modifications of salt responsive transcription factor genes in soybean. PLoS One. 7(7):e41274.

[25] Hu Y, Zhang L, Zhao L, Li J, He S, Zhou K, Yang F, Huang M, Jiang L, Li L (2011) Trichostatin A selectively suppresses the cold-induced transcription of the ZmDREB1 gene in maize. PLoS One. 6(7):e22132.

[26] Mutskov V, Felsenfeld G (2004) Silencing of transgene transcription precedes methylation of promoter DNA and histone H3 lysine 9. Embo J. 23(1): 138–149.

[27] Nagaki K, Talbert PB, Zhong CX, Dawe RK, Henikoff S, Jiang J (2003) Chromatin immunoprecipitation reveals that the 180-bp satellite repeat is the key functional DNA element of Arabidopsis thaliana centromeres. Genetics. 163(3):1221–5.

[28] Tariq M, Saze H, Probst AV, Lichota J, Habu Y, Paszkowski J (2003) Erasure of CpG methylation in Arabidopsis alters patterns of histone H3 methylation in heterochromatin. Proc Natl Acad Sci U S A. 100(15):8823–7.

[29] Mathieu O, Probst AV, Paszkowski J (2005) Distinct regulation of histone H3 methylation at lysines 27 and 9 by CpG methylation in Arabidopsis. EMBO J. 24(15):2783–91.

[30] Kristjuhan A, Svejstrup JQ (2004) Evidence for distinct mechanisms facilitating transcript elongation through chromatin in vivo. EMBO J. 23(21):4243–52.

[31] Serra TS, Figueiredo DD, Cordeiro AM, Almeida DM, Lourenço T, Abreu IA, Sebastián A, Fernandes L, Contreras-Moreira B, Oliveira MM, Saibo NJ (2013) OsRMC, a negative regulator of salt stress response in rice, is regulated by two AP2/ERF transcription factors. Plant Mol Biol. 82(4-5):439–55.

[32] Shen L, Shao NY, Liu X, Maze I, Feng J, Nestler EJ (2013) diffReps: Detecting Differential Chromatin Modification Sites from ChIP-seq Data with Biological Replicates. PLoS One. 10;8(6):e65598.

[33] Shen J, Jiang D, Fu X, Guo H, Feng B, Pang Y, Streets AM, Tanq F, Huang Y (2015) H3K4me3 epigenomic landscape derived from ChIP-Seq of 1 000 mouse early embryonic cells. Cell Res. 25(1):143–7.

[34] Clark SJ, Lee HJ, Smallwood SA, Kelsey G, Reik W (2016) Single-cell epigenomics: powerful new methods for understanding gene regulation and cell identity. Genome Biol.17:72.

Chapter 13

The Application of Genomic Approaches in Studying a Bacterial Blight-Resistant Mutant in Rice

Chang-Sheng Wang and Da-Gin Lin

Additional information is available at the end of the chapter

http://dx.doi.org/10.5772/67331

Abstract

Rice bacterial blight disease (BBD), caused by *Xanthomonas oryzae* pv. *oryzae* (*Xoo*), is one of the serious diseases in most rice production regions. In this report, we screened for resistance mutants from the mutation pool of TNG67 variety derived by sodium azide (SA) mutagenesis with phenotype investigation and assisted with fluorescent detection. SA0423 is a mutant of broad range resistance against *Xoo* for many years; the resistance was studied following the concept of central dogma. The inheritance of resistance was characterized, and three QTLs were mapped onto the genome of SA0423 using simple sequence repeat (SSR) markers and R/qtl by genomic approach. In transcriptomic approach, only one differential expression QTLs (eQTLs) were identified; two differentially expressed proteins (pQTLs) were identified and genetically characterized by proteomics after *Xoo* challenged in SA0423 mutant. To improve the bacterial blight resistance, makers are developed from QTLs, eQTLs and pQTLs to pyramid the resistance genes through marker-assisted breeding in our rice breeding programs.

Keywords: rice, bacterial blight disease (BBD), resistance, mutant, genetics, genomics, transcriptomics, proteomics, marker-assisted breeding (MAB)

1. Introduction

Rice is a staple food crop and provides more than one-fifth of the calories to humans [1]. However, rice production is often challenged by bacterial blight disease (BBD), which is one of the most destructive diseases caused by *Xanthomonas oryzae* pv. *oryzae* (*Xoo*). This disease was first found in rice by Japanese farmers in 1884. It was not a serious problem in rice production until the release of high-yielding varieties during the 1960s–1970s [2–4]. Some field observations displayed that this disease can lead up to 50% losses in rice planting areas [3, 5]. In Taiwan, BBD often occurs in the second crop season, and its annual pathogenesis area is usually more than 20,000 hectares, accounting for about 4% of the total rice production area. Because of climate change, this disease has become more and more serious recently [6, 7]. Furthermore, International Rice Research Institute (IRRI) proposed that BBD can cause up to 70% of yield loss when susceptible varieties are grown in the environments suitable for *Xoo* pathogens (http://www.knowledgebank.irri.org/index.php?option=com_zoo&task=item&item id=806&Itemid=606).

The existing prevention of BBD includes field management, fertilizer control, pesticide application and resistance varieties with the major resistance gene (*R* gene) or the pattern recognition receptor gene (*PRR* gene). In field management, appropriate spacing could prevent rice plants from the infection of pathogens. Appropriate nitrogen fertilizer application could prevent rice plants from pathogens' infection [8]. The spray of probenazole or other chemicals might prevent the infection before transplantation, but this treatment could not be applied in tropical regions [8–10]. So far, the use of resistant varieties is considered to be the most effective strategy against this disease. In recent years, there is no specific bactericide which could effectively control BBD, and chemicals application also increases production cost, plant injury and environmental pollution. On the other hand, the evolution of pathogens increases the diversity and the difficulty in the breeding program for durable or broad-spectrum resistance [11–13]. Therefore, breeding the bacterial blight-resistant varieties is urgently required to meet the demand of a safe rice production.

Previous studies demonstrated that climate change has been proposed to affect the microflora of *Xoo* in the field and even change the life cycle and evolution of *Xoo* pathogen. Large-scale and long-term cultivation of *Xa4*-mediated resistant varieties also altered the *Xoo* population. Consequently, resistant varieties carried with only *Xa4* have become susceptible to *Xoo* in Southeast and South Asia [14]. Bacterial blight is one of the serious diseases often occurring in the second crop season (August to November) in Taiwan. Our previous results also displayed that the top 20 cultivars with large-scale cultivation in Taiwan were susceptible to *Xoo* (**Table 1**). Therefore, if the bacterial blight disease is endemic, it will cause serious loss to the rice production. These results indicated again that breeding of the bacterial blight-resistant varieties is urgently required to meet the demand of the Taiwanese rice industry.

The availability of resistant sources is the major limitation in breeding. A series of near isogenic lines (NILs) harboured various *Xa* genes (IRBB NILs) that were developed on the susceptible cultivar, IR24, at the International Rice Research Institute (IRRI) [15]. The IRBB lines, often applied in the domestic resistance breeding, were introduced and inoculated with Taiwan local pathogens to test their responses in our previous work. The results indicated that only the IRBB lines carried *Xa5* or *Xa7* showed moderate resistance while all other IRBB lines carried single *Xa* gene showed susceptibility to the local pathogens (**Figure 1**) [16]. Many of the resistance genes were introduced into the susceptible varieties by marker-assisted selection (MAS) to improve bacterial blight resistance [17, 18]. However, many of these genes lose their resistance due to the fast evolution of pathogen [19]. It has been reported that durable or broad-spectrum resistance can prolong the bacterial blight resistance in rice [20, 21]. Actually, broad-spectrum and durable resistance can be accomplished by the introduction of one very resistance gene and pyramiding with two to three other resistance genes [22]. However, large-scale and long-term cultivation of resistant varieties might result in changes of pathogen race in the *Xoo* population and cause the breakdown of resistance [14, 23]. These findings indicate that exploration of new germplasms with novel resistance genes become a crucial subject in breeding resistance variety.

Planting area during 2010–2015		**Variety**	**Response for *Xoo***	
Order	**Ha**		**XM42**	**XF89-b**
1	484,063	Tai Nan No. 11	7	7
2	142,132	Taikeng No. 8	7	7
3	119,641	Taikeng No. 14	7	9
4	98,404	Taikeng No. 16	7	9
5	47,139	Taikeng No. 9	9	9
6	46,204	Taichung-Hsien No. 10	9	9
7	41,424	Taikeng No. 2	7	9
8	39,374	Kaohsiung 139	9	9
9	39,149	Taikeng No. 11	9	7
10	23,767	Taichung-Hsien No. 1	7	9
11	22,965	Taichung 192	9	7
12	19,357	Taikeng No. 4	7	7
13	19,108	Tail Nung No. 71	5	7
14	13,989	Tail Nung No. 67	9	9
15	13,867	Taikeng No. 5	7	9
16	9341	Tai Tung No. 30	9	7
17	9178	Kaohsiung 145	7	7
18	7318	Taoyuan No. 1	5	7
19	7152	Taikeng-No. 1	7	7
20	5660	Taichung-Hsien No. 17	9	9

Note: The resistance of the top 20 rice cultivars was investigated according to the Kauffman's method [66]. The lesion level can be classified by a scale of five scores, such as 1 (HR), 3 (MR), 5, 7 (MS) and 9 (HS).

Table 1. The resistance investigation of the top 20 rice cultivars grown in Taiwan.

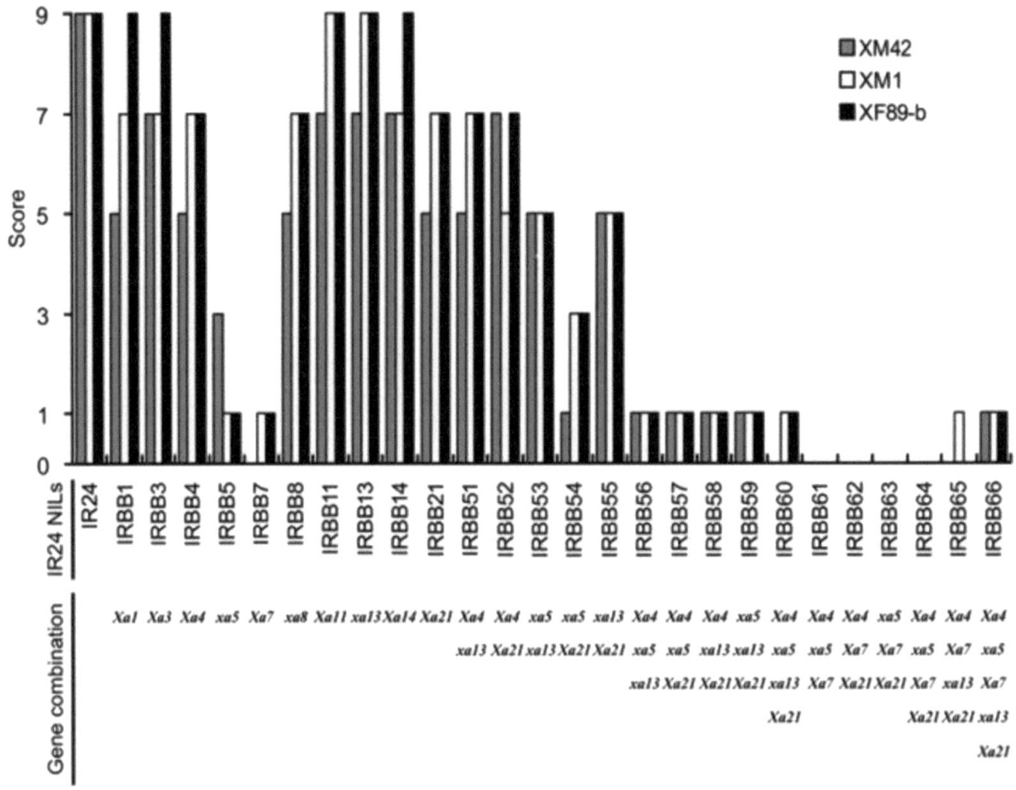

Figure 1. The resistance investigation of IRBB lines against the local pathogens in Taiwan [16].

2. Mutant screening

Sodium azide (NaN_3, SA) induced mutants can be applied to any rice breeding program at any facility, while genetically modified mutants can only be handled in the isolating facilities under the governmental regulation. A TNG67 mutant pool was developed by SA mutagenesis at the Taiwan Agricultural Research Institute (TARI) in our previous breeding program. All the mutants were screened and purified according to their morphological traits by at least 10 generations of self-crossing, selection and purification following pedigree procedures. Over 3000 pure line mutants on the same genetic background of TNG67 variety were maintained in the pool [24]. The genetic diversity of mutant lines in this pool includes disease resistance (blast, bacterial blight and sheath blight) [25], pest resistance (brown planthopper, white backed planthopper and leafroller) [26], herbicide resistance (bentazon, glufosinate and glyphosate) [27] and many agronomic traits; grain quality and morphology diversities seldom found in rice cultivars. These results suggested that the TNG67 mutant pool should have high potential in basic research as well as variety improvement [24].

To improve the bacterial blight resistance for local rice varieties, we attempted to obtain the local resistant germplasms from the selection of TNG67 mutant pool [7]. So far, at least 50 bacterial blight-resistant mutants have been selected from the mutant pool (**Figure 2**). These

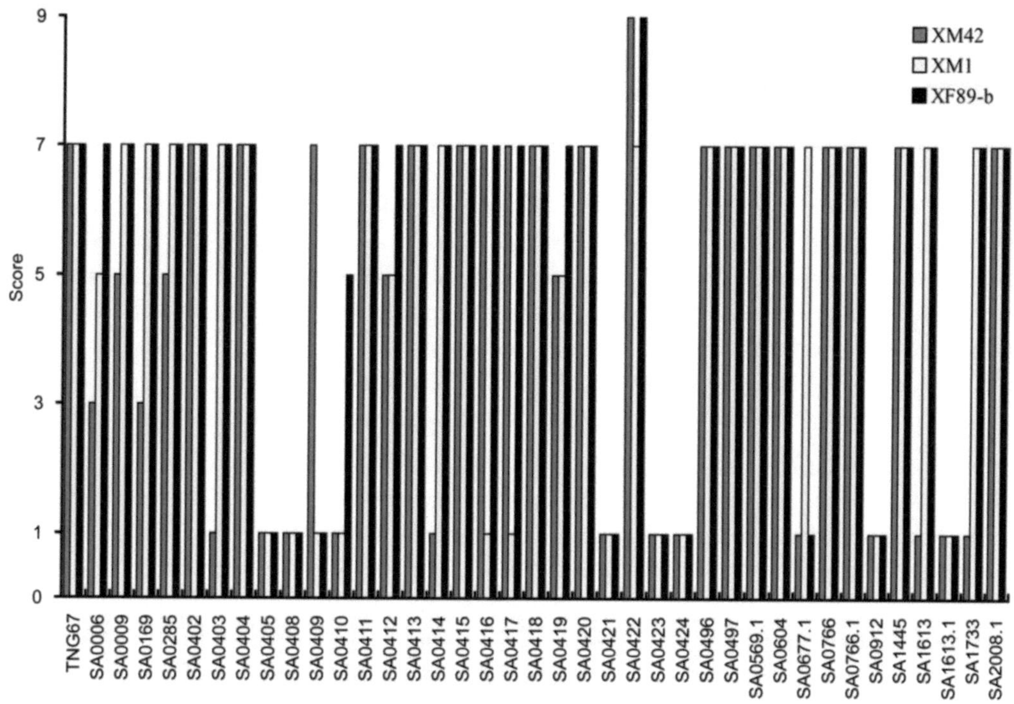

Figure 2. The screening of resistance mutants from TNG67 mutation pool by inoculation of local pathogens in Taiwan.

mutants might carry various genotypes of resistance and participate in the resistant pathway. Among them, SA0423 and SA0424 showed stable resistances against various *Xoo* pathogens for many years. The genetic analysis displayed that these two mutants might carry multiple resistant genes to confer broad-spectrum resistance and show different resistant phenotypes (data shown in the following section).

3. Genetic and mapping of resistant genes

At present, planting resistant varieties is accepted as the most efficient, reliable and economic strategy against bacterial blight. It has been proposed that the durable and broad spectrum resistance of plants was usually governed by multiple genes or quantitative trait loci (QTLs) [28]. Therefore, the discovery of novel resistance genes against *Xoo* is very important in the breeding program for disease resistance. So far, 42 resistance loci (*Xa*) for BBD have been identified and characterized [refer to: http://www.nig.ac.jp/labs/PlantGen/english/oryzabase-e/; http://www.gramene.org/; http://www. ricedata.cn/and previous reviews] [29–31]. Most of these genes were found to be controlled by dominance [32], but 14 of them, such as *xa5, xa8, xa13, xa15, xa19, xa20, xa24, xa25, xa26b, xa28, xa31, xa32, xa33* and *xa34*, were found to be regulated in a recessive manner [33, 34]. These genes distribute among 9 chromosomes of rice genome, and 16 of them are clustered on chromosome 4 (*Xa1, Xa2, Xa14, Xa31*(*t*) and *Xa38*) and chromosome 11 (*Xa3/26, Xa4, Xa10, Xa21, Xa22, Xa23, Xa30*(*t*), *xa32*(t), *Xa35*(t), *Xa36*(t) and *Xa40*), respectively. At present, *Xa1, Xa3/Xa26, xa5, Xa10, xa13, Xa21, Xa23, Xa25, Xa27* and *Xa40* have been cloned and characterized to encode six types of proteins, i.e. NBS-LRR, receptor kinase like protein, ER membrane protein, Os8N3 protein, MtN3/saliva family member and WAK3, indicating the existence of multiple mechanisms of bacterial blight resistance in rice [35–47].

Near isogenic lines (NILs) with various *Xa* genes on the background of IR24, a very susceptible cultivar, named IRBB NILs were applied as the donor parents [15]. Besides, molecular markers linked with *Xa* genes in IRBB NILs were developed through comparative mapping strategy for improving the BBD resistance of commercial cultivars [17, 18]. However, it has been reported that the plant resistance genes may breakdown due to the fast evolution of pathogen isolates [19]. Many studies suggested that large-scale and long-term cultivation of resistant varieties may result in changes of pathogen race in *Xoo* population and caused the breakdown of resistance [14, 23]. These findings indicated that exploration of new resistance genes has become an important subject for breeding resistance variety.

Among the previously selected resistant mutants, SA0423 shows a stable resistance to Taiwan local pathogens for years. Hence, their genetic properties and BBD resistance genes were characterized in our team. Except for the bacterial blight resistance, SA0423 also has thinner leaf blades, shorter plants, more erect plant type and less tiller number than its mutagenesis parent, TNG67 (**Figure 3**). A strong and stable Taiwanese epidemic pathogen, *Xoo* XF89b, has been used for genetic analysis and mapping the bacterial blight-resistance genes. Taichung Native 1 (TN1), a very susceptible *indica* rice cultivar, was used as the recipient parent. The cross TN1/SA0423 was made to generate F_1 and F_2 materials for genetic analysis and mapping of resistant genes. After pathogen infection, the lesion lengths of TN1, SA0423 and TN1/SA0423 F_1 were 17.2 ± 1.1, 1.2 ± 0.7 and 3.4 ± 0.9 cm, respectively, indicating that

Figure 3. Morphology of TNG67, SA0423 and their disease responses at 28 days after inoculation (DAI) with Taiwanese *Xanthomonas oryzae* pv. *oryzae* XF89b.

http://dx.doi.org/10.5772/67331

the BBD resistance of SA0423 is partial dominance (**Figure 4**). The lesion lengths of the TN1/SA0423 F_2 population showed a continuous distribution (**Figure 5**) and indicated that the disease resistance of SA0423 is controlled by multiple genes or quantitative trait loci (QTLs).

A linkage map covering 12 chromosomes with an average distance of 11.2 cM was constructed and applied to map the resistance of SA0423 using 361 TN1/SA0423 F_2 individuals [48]. QTL analysis was performed using the R program language platform (version 3.1.0; http://www.r-progect.org/) with an add-on package, qtl [46, 47]. Three QTLs are detected on chromosomes 11, 8 and 6 and account for 21.1, 11 and 9.6% of the observed phenotypic variance, respectively (**Table 2** and **Figure 6**). Three QTLs are localized to 6, 7 and 14 cM intervals, respectively; they contribute to approximately 47% of the total phenotypic variation (resistance) and no epistatic effect could be detected among them [48].

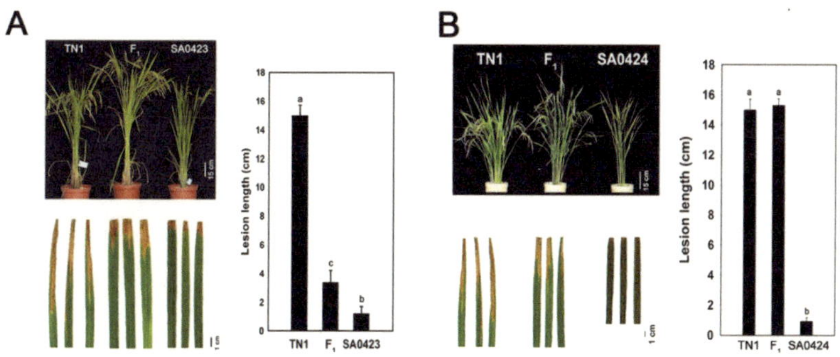

Figure 4. Morphology of TN1, SA0423 and their F_1 individual, and the disease response at 28 days after inoculation (DAI) against Taiwanese *Xanthomonas oryzae* pv. *oryzae* XF89b (A, upper panel). The lower panel of (B) shows the morphology of leaf lesion at 28 DAI; left panel shows the leaf lesion (cm) investigated at 28 DAI. Error bar is the standard error of mean ($n = 3$). Means with the same letter are not significantly different at 5% level by LSD test.

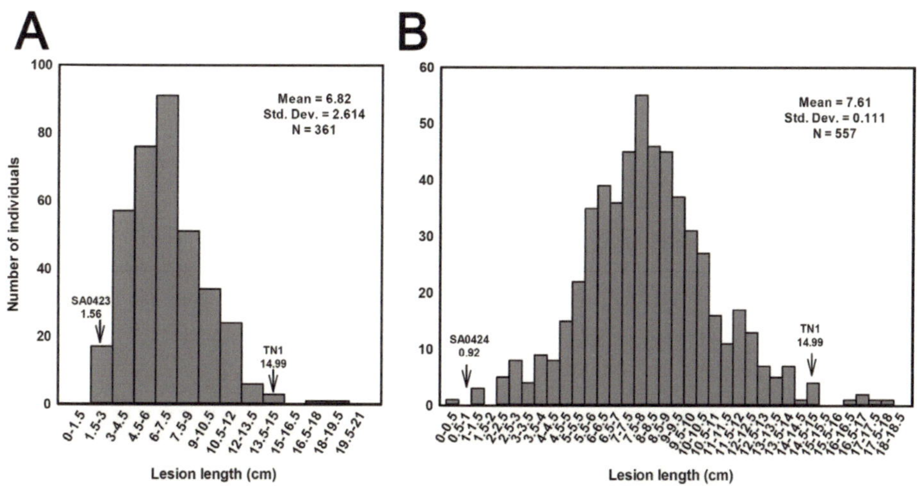

Figure 5. Distribution of lesion length (cm) after inoculation with Taiwanese *Xanthomonas oryzae* pv. *oryzae* XF89b in an F_2 population from the cross, TN1/SA0423.

QTL	Chr.	QTL (Confidence interval) (cM)	LOD	Phenotyping variance (%)	Additive effect	Dominance effect
qBBR11.1 (Q1)	11	124 (121–127)	26.60	21.10	–1.64	–0.44
qBBR08.1 (Q2)	8	39 (34–41)	15.04	11.04	–1.20	–0.82
qBBR06.1 (Q3)	6	120 (111–125)	13.20	9.58	–1.13	0.79

Note: QTLs are labelled according to the principles of previous publications [67, 68].

Table 2. Putative QTLs were identified from the F_2 population of TN1/SA0423.

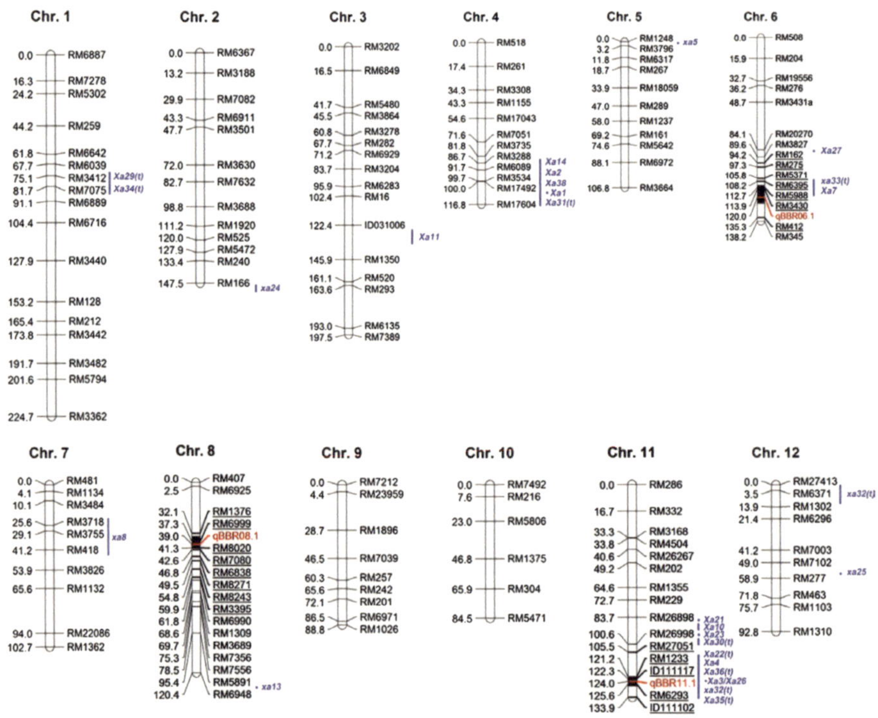

Figure 6. The linkage mapping of SSR/InDEL markers and SA0423 resistance QTLs in the F_2 population of TN1/SA0423. The markers and genetic distances (cM) are labelled to the right and left of the chromosome, respectively. The QTLs are coloured with red, and other published genes and QTLs associated with BBR are labelled as blue dots and lines, respectively.

4. Transcriptomic studies

According to QTL analysis, all the three identified QTLs contribute to 47% of the resistance indicating that other resistance genes may exist in SA0423 [48]. Therefore, the transcriptomes of TNG67 and SA0423 were determined by microarray technologies to explore the bacterial-resistant genes in SA0423.

For a precise and non-destructive investigation in the infection process of bacterial blight pathogen after inoculation, a *Xanthomonas fluorescent* expression plasmid, pRBBZsGFP, was constructed with a strong fluorescent gene *ZsGFP* and the pBBR1MCS vector for simultaneous detection of bacterial blight pathogen infection and the gene expression [49]. Pathogens infection with $XF89b_{ZsGFP}$ was conducted on the dark-treated albino seedlings of TNG67 rice variety; the multiplication and colonization of $XF89b_{ZsGFP}$ could be detected in 0.5 hour after inoculation, and the maximum fluorescence was observed on the same leaf in 1 hour after inoculation (**Figure 7**). However, the fluorescence was reduced in the following time course indicating that the multiplication and colonization of $XF89b_{ZsGFP}$ might be suppressed by the endogenous immune system of rice. At 7 DAI (days after inoculation), the stronger fluorescence was observed again on the same leaf and extended continuously to the leaf base, suggesting that the rice immune system was broken down by the $XF89b_{ZsGFP}$.

After the infection of *Xoo* XF89b, RNA samples prepared from the leaves of TNG67 and SA0423 collected at 0, 0.5, 1, 2 and 6 hours, respectively, were applied in the transcriptomic analysis with Agilent Oligo Microarray (60K, custom-made, Agilent Technologies) [50]. The results demonstrated that 2727, 3585 and 18,432 differentially displayed transcripts were identified in SA0423, TNG67 and in both, respectively. Among them, 58 genes involved in SA0423 resistance were further conducted by bioinformatics strategies [refer to: http://www.nig.ac.jp/labs/PlantGen/english/oryzabase-e/; http://www.gramene.org/; http://www.ricedata.cn/and previous reviews] [29–31] as well as "plant-pathogen interaction" pathway (http://www.genome.jp/kegg), and clustered with BioLayout Express^{3D} [51]. By confirming with real-time RT-PCR, 17 resistance gene candidates (**Table 3**) were selected for bioinformatics analysis, they have been proposed to be involved in plant-pathogen interaction pathway, biosynthetic pathway of plant hormones, autophagy and signal transduction prior to the induction of plant immune system [52].

To confirm the function of the identified genes from transcriptomic analysis, the SSR markers flanking in 5 cM region of these genes were retrieved from GRAME web site, screened for the polymorphic markers between TN1 (the susceptible parent) and SA0423 (the resistant parent), and then genotyping was performed in the F_2 population [53]. Simultaneously, the disease lesion of F_2 individuals was investigated to represent the resistance phenotype after the inoculation of *Xoo* XF89b. The linkage between genotype and phenotype was conducted using R/qtl software by the single marker regression model. The results displayed that only RM6838 adjacent to *Ankyrin* showed a significantly high LOD (6.86) (**Table 4**) indicting that *Ankyrin* (LOC_Os08g15840) has a high potential to be involved in the resistance of SA0423. The bioinformatics analysis shows that this Ankyrin protein shares 76% similarity with the *Arabidopsis* RING type ligase, XBAT32, of an XB3 family. In *Arabidopsis*, Ankyrin has been proposed to negatively regulate 1-aminocyclopropane-1-carboxylate synthase (ACS), a key enzyme involved in the ethylene biosynthesis pathway, and then compromised immune system [54]. The real-time RT-PCR displayed that the expression level of *Ankyrin* in SA0423 was lower than that of TNG67, and higher expression levels of *OsACS1* and *OsACS3* were found in the BBD resistance mutant, SA0423 (**Figure 8**). These findings showed that ethylene metabolism may involve in the disease resistance of SA0423. A total of 15 mutations in the coding region resulting two mutation residues, Ser280Pro and Thr381Ala, were discovered in the *Ankyrin* of SA0423 through cloning and sequencing (data not shown). At the same time, the transgenic rice plants with less expression of *ankyrin* showed a significant

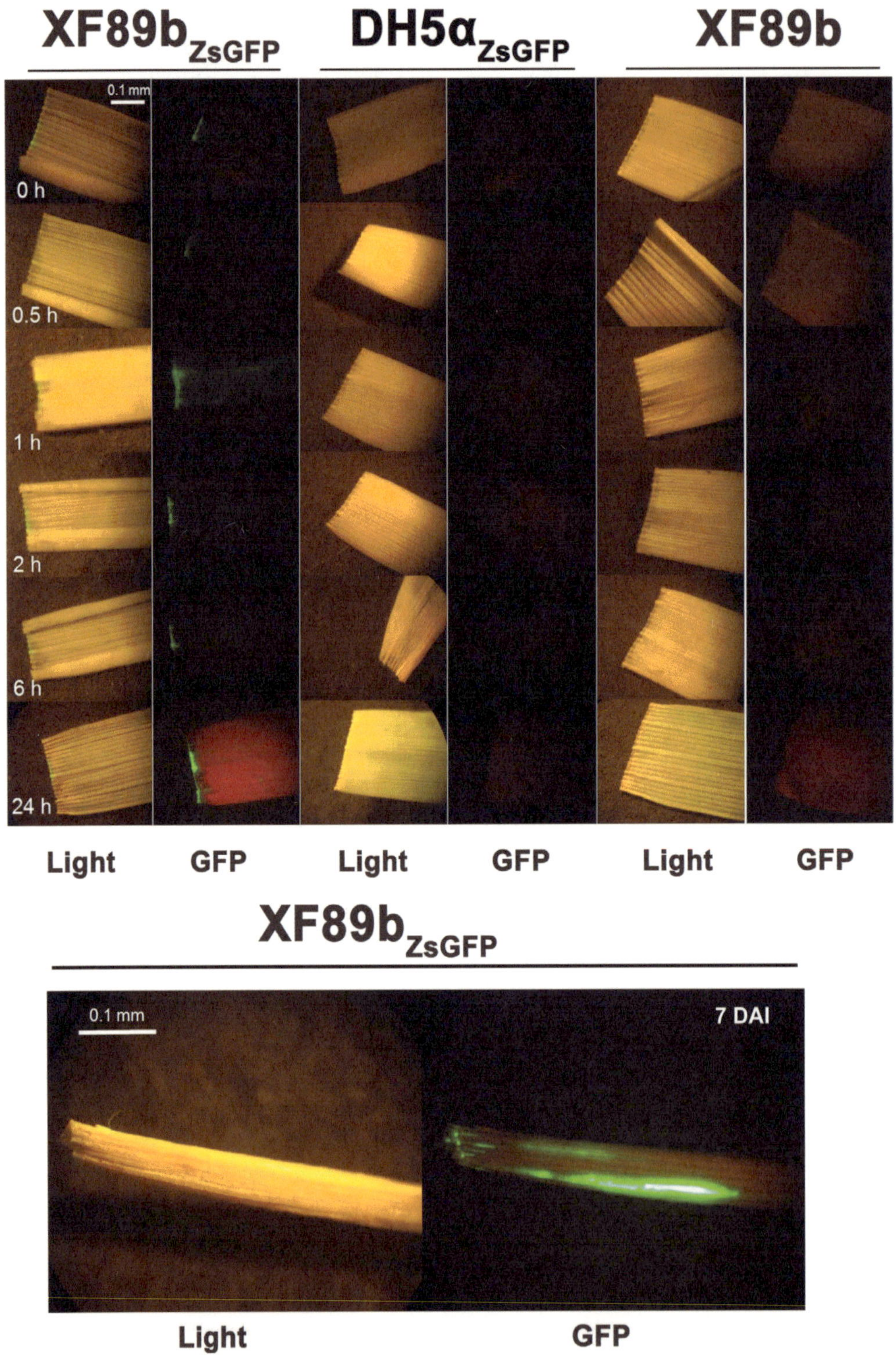

Figure 7. Visualization of *X. oryzae* pv. *oryzae* and *E. coli* expressing *GFP* in the dark-treated albino TNG67 seedlings.

Gene name	Gene ontology	
*Ankyrin*Ankyrin repeat-rich protein	BP	Cellular process, biosynthetic process, protein modification process, post-embryonic development, anatomical structure morphogenesis, response to endogenous stimulus
	MF	Binding, protein binding, catalytic activity,
*ATG1*ATG1	BP	Cellular process, cellular component organization, protein modification process
	CC	Plasma membrane
	MF	Molecular function
*CaM_Chr.1-1*OsCam1-3-Calmodulin	BP	Biological process, response to abiotic stimulus, post-embryonic development, signal transduction
	MF	Binding, protein binding
*CaM Chr.1-2*OsCam3-Calmodulin	BP	Biological process, response to abiotic stimulus, post-embryonic development, signal transduction
	MF	Binding, protein binding
*CaM_Chr.2*EF hand family protein	BP	Protein modification process, biosynthetic process
	CC	Cytoplasm
	MF	Binding
*CaM_Chr.5*OsCam2-Calmodulin	BP	Signal transduction
	CC	Plasma membrane
	MF	Signal transducer activity, binding, protein binding
*CMPG*Immediate-early fungal elicitor protein CMPG1	BP	Protein modification process, biological process
	CC	Intracellular
	MF	Catalytic activity, binding
*DUF26*Domain of Unknownfunction 26-lc	BP	Protein modification process, cellular process, metabolic process
	CC	Plasma membrane
	MF	kinase activity, protein binding, cellular process,
*FMO*Flavin-containing monooxygenase family protein	BP	Cell death, signal transduction, metabolic process, response to biotic stimulus, cellular process, response to stress
	CC	Endoplasmic reticulum, membrane, cell
	MF	Nucleotide binding, catalytic activity, binding
*JOM*JasmonateO-methyltransferase	BP	Multicellular organismal development, cellular process, metabolic process
	CC	Cellular component
	MF	Binding, protein binding, transferase activity
*PxMP*Peroxisomal membraneprotein	BP	Biological process
	CC	Peroxisome, membrane
	MF	Molecular function

Gene name	Gene ontology	
*SAM*SAM dependent carboxyl methyltransferase	BP	Biological process, cellular process, metabolic process
	CC	Cellular component
	MF	Transferase activity
*SNARE*SNARE associated Golgiprotein	CC	Cytosol
*Ubi*Ubiquitin family protein	MF	Molecular function
*Xa2*OsSAUR21—Auxin-responsive SAUR gene family member	BP	Response to endogenous stimulus
	MF	Molecular function
*Xa25*Nodulin MtN3 family protein	BP	Biological process, cellular process, transport
	CC	Plasma membrane, membrane, cell
	MF	Transporter activity
*xa5*Transcription initiation factor IIA gamma chain	BP	Biosynthetic process, nucleobase, nucleoside, nucleotide and nucleic acid metabolic process
	CC	Nucleoplasm

Note: BP, biological process; CC, cellular component; MF, molecular function.

Table 3. The resistance gene candidates identified from transcriptomic analysis in a bacterial blight-resistant mutant, SA0423.

Gene	Chromosome	Position (cM)	Marker	LOD[z]
CaM_Chr.1-1	1	50.8	RM6039	0.8941
CaM Chr.1-2	1	50.9	RM572	1.7618
CaM_Chr.2	2	25.3	RM6378	0.0886
CMPG	2	131	RM13938	0.5618
Xa2	4	107.4	RM17492	1.2685
JOM	4	120.3	RM17604	0.5006
xa5	5	3	RM17741	0.2725
CaM_Chr.5	5	104.7	RM6972	0.5717
FMO	6	19.1	RM19556	0.2034
SAM	6	33.5	RM276	0.1920
DUF26	7	73.2	RM3826	0.1547
SNARE	7	116.1	RM1362a	0.2325
Ankyrin	8	42.9	RM6838	6.8579
ATG	10	73.7	RM5471a	0.0479
Ubi	10	99.8	RM147	0.2880
Xa25	12	57.9	RM28157	0.4657
PxMP	12	69.6	RM519	0.2940

Note: [z] LOD, $\log_{10}$ of odds.

Table 4. Linkage analysis between the resistance gene candidates and the resistance trait of SA0423 by R/qtl.

resistance against *Xoo* XF89b isolate. Therefore, Ankyrin is considered to be one of the expression quantitative trait loci (eQTLs) involved in the bacterial blight resistance of SA0423.

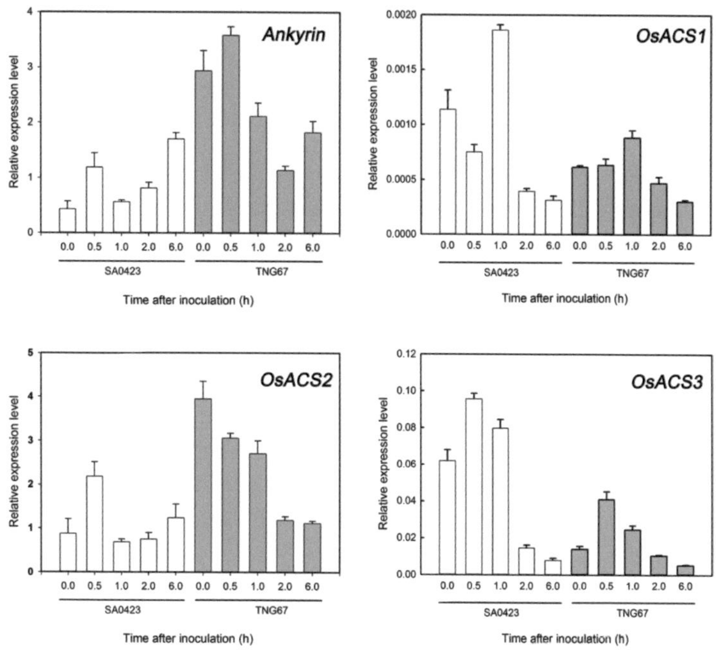

Figure 8. Quantitative analysis of mRNA expression of *Ankyrin* and *OsACS* homologs in TNG67 and SA0423 after inoculated with *Xoo* XF89b by using real-time RT-PCR.

5. Proteomics study

Proteomics technology provides a direct investigation of proteins which may participate in rice disease resistance. In previous studies, plasma membrane (PM) proteomic analysis of the genetically modified rice suspension cells with *Xa21* demonstrated that PM-associated ATPase, phosphatase, hypersensitive-induced response protein, prohibitin, zinc finger/C2 domain protein, universal stress protein and heat shock protein might be involved in the early immune response against compatible and incompatible *Xoos* [55]. A proteomic analysis of Java 14 seedling revealed that 20 differentially displayed proteins were responded to bacterial inoculation and categorized into energy, metabolism and defence pathways [56]. These proteomic studies were conducted at 0, 12, 24 even 72 hours after inoculation [55, 56] whereas considering the rapidity of defence observed in other plant-pathogen interactions [57] and the short life cycle of *Xoo*, it is expected that *Xoo* might induce rice reprogramming immediately after pathogen infection.

A comparative proteomics analysis was conducted to characterize the proteomic profiling in leaves of TN1 (as a susceptible control), TNG67 and SA0423 after the infection of *Xoo* XF89b at 0, 6, 48 and 72 hours after pathogen inoculation (**Figure 9**). There were 60, 38 and 96 differentially displayed protein spots identified only in SA0423, TNG67 and TN1, respectively, by the separation of two-dimensional gel electrophoresis (2-DE). Finally, a total of 150 disease resistance-related proteins were identified from these protein spots through the ESI-Q-TOF mass spectrometry (MS) analyses. Ten resistance protein candidates (**Table 5**) were then determined by bioinformatics approach including annotation of metabolic pathway, comparative mapping analysis with published resistance loci [refer to: http://www.nig.ac.jp/labs/PlantGen/english/oryzabase-e/; http://www.gramene.org/; http://www.ricedata.cn/and previous reviews] [29–31] as well as 'plant-pathogen interaction' pathway (http://www.genome.jp/kegg), and clustered with BioLayout Express^{3D} [51]. These candidates were proposed to be involved in ascorbate, glyoxylate and glutathione, and oxidative phosphorylation metabolisms.

The candidate genes identified from proteomics approach were genetically confirmed as previously described, the SSR markers flanking in 5 cM region of them were retrieved from GRAMENE web site, and screened for polymorphism TN1 (the susceptible parent) and SA0423 (the resistant parent). Genotyping analysis was performed in 94 TN1/SA0423 F_2 individuals using the polymorphic markers. The lesion of these F_2 individuals was investigated after the inoculation of *Xanthomonas oryzae* pv. *oryzae* XF89b as the resistance phenotype. The linkage between genotyping and resistance was analysed by MapDisto according to Lorieux's protocol [60]. The result displayed that only RM5970 adjacent to the putative 2,3-bisphosphoglycerate-independent phosphoglycerate mutase (BIPM) and RM14099 adjacent to aspartate aminotransferase (AST) showed significant association with the SA0423 resistance (**Table 5**). BIPM has been proposed to have some important roles in glycolysis, stomatal movement, vegetative growth and pollen production in *Arabidopsis* [61], but it was usually found to be differentially expressed under abiotic or biotic stress [62–64]. AST was found to be up-regulated in rice *spotted leaf 5* (*spl5*) mutant that showed spontaneous HR-like lesions on its leaves, and a broadly enhanced resistance against rice blast and bacterial blight pathogens [65]. Based on these findings, BIPM and AST are found to have high potential to participate in the resistance mechanism of SA0423. These results provide novel insights into the molecular mechanisms of rice response to *Xoo* infection and discovery of new resistance genes as the basis for application in molecular breeding. Therefore, both BIPM and AST are considered to be the proteomic quantitative trait loci (pQTLs) for the bacterial blight resistance in SA0423.

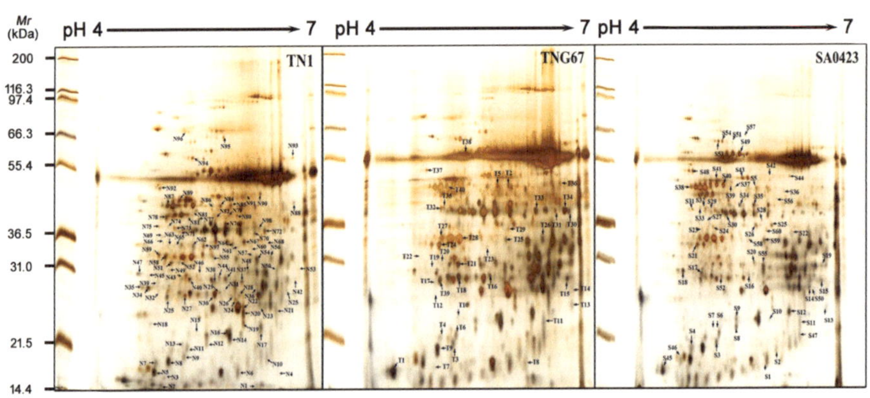

Figure 9. 2-DE image analysis of rice leaf proteome under *Xoo* XF89b infection. Total leaf proteins were extracted and separated by 2-DE then stained with sliver staining according to the previous protocol [58, 59]. An equal amount (200 μg) of the total proteins was loaded on each gel strip. The differentially expressed resistance-related proteins in TN1, TNG67 and SA0423 are marked as N, T and S, respectively.

http://dx.doi.org/10.5772/67331

Gene	Marker	hmzA	hmzB	htz	n	m(hmzA)	m(hmzB)	m(htz)	*R2*	*A*	*D*	|*D/A*|	*F*	*p*	
L-ascorbate peroxidase 1, cytosolic (APX1)	RM7197	30	35	29	94	6.56	8.28	7.57	0.05	0.862	0.15158	0.18	2.5	0.08795	
Putative 2,3-bisphosphoglycerate-independent phosphoglycerate mutase (BIPM)	RM5970	16	28	50	94	8.29	8.841	6.52	0.12	0.276	−2.04831	7.41	6.05	0.00337	**
Glyceraldehyde-3-phosphate dehydrogenase, putative, expressed (G3PD)	RM14336	23	23	40	86	6.75	7.833	7.4	0.02	0.539	0.10949	0.2	0.77	0.46538	
Aspartate aminotransferase (AST)	RM14099	42	22	14	78	6.61	9.433	7.29	0.14	1.412	−0.73185	0.52	6.34	0.00281	**
2,3-bisphosphoglycerate-independent phosphoglycerate mutase, putative, expressed (BIPME)	RM8084	24	52	18	94	6.74	8.134	6.73	0.05	0.695	−0.70982	1.02	2.34	0.10172	
Triosephosphate isomerase (TRI)	RM24714	16	33	45	94	7.43	8.23	7.01	0.03	0.401	−0.81535	2.03	1.44	0.24307	
30S ribosomal protein S4, chloroplastic (RP30S)	RM5579a	18	36	40	94	6.59	8.479	7.05	0.06	0.943	−0.48336	0.51	3	0.05439	
Fructose-bisphosphate aldolase, chloroplastic (FBPA)	RM26143	10	46	35	91	8.09	7.596	7.31	0.01	−0.247	−0.53336	2.16	0.24	0.78348	
Cysteine synthase (CYS1)	RM520	23	28	43	94	7.8	6.835	7.8	0.02	−0.483	0.47717	0.99	0.91	0.40443	

Table 5. Linkage analysis between the resistance protein candidates and the resistance trait of SA0423 by MapDisto. The "**" was indicated "statistical significance" ($p \leq 0.05$).

6. Conclusion

Breeding resistance variety is the best strategy to overcome the bacterial blight disease damage in rice and is a very challengeable work. Availability of resistant genotype is the major limitation to the resistance improvement. However, plant disease resistance is a complex trait usually regulated by QTLs, epistatic effect, and influenced by the interactions among pathogen, host and environment.

In this review, a durable resistance mutant, SA0423, was firstly obtained from screening a sodium azide-induced mutation pool on the genetic background of TNG67 rice variety. The genomic approaches and technologies were conducted according to the concept and flow of Central Dogma. In the genomic study, the inheritance and gene corresponding to the BBD resistance of SA0423 was conducted. Linkage maps were constructed, and three QTLs (qBBR06.1, qBBR08.1 and qBBR11.1) for resistance were identified from SA0423. Meanwhile, the linkage markers for each QTL were developed according to the linkage map for marker-assisted breeding.

The transcriptomics and proteomics technologies were applied to identify the expressed genes and proteins corresponding to the pathogen inoculation for BBD resistance on SA0423. The differential displayed genes (or proteins) were annotated by blast with the gene database (NCBI and GRAMENE websites), and then their putative biological functions or the participating pathways were predicted by GO analysis. Besides, they were compared with the published resistance genes in *Xa* locus or putative rice 'plant-pathogen interaction' pathway to confirm the resistance genes or pathway in SA0423. The results demonstrated that 17 candidate genes (eQTLs) and 10 candidate proteins (pQTLs) might be involved in SA0423 resistance mechanism. The association between these candidates and SA0423 resistance was further evaluated by integration of genotyping and phenotyping of TN1/SA0423 F_2 progeny through genetic approach. Both genomic and bioinformatics approaches were integrated to confirm the function and genetic relationship of the candidate genes with BBD resistance. The final results suggested that only one major expression QTLs (eQTLs) [53] and two protein QTLs (pQTLs) (Lin et al., 2017, paper in preparation) are confirmed to confer the resistance of SA0423. It is worth to note that both the eQTLs and pQTLs identified in this study are not identified in the genetic mapping approach, and the products of eQTLs were not found in the protein profiling (pQTLs), and vice versa. These results showed that the genomic approach alone cannot unravel all the genes involved in the disease resistance of SA0423.

Phenomics or phenotype can provide the solid evidence for gene function. Our previous findings were tested through transgenic approach as well as marker-assisted backcrossing (MABC). The transgenic rice plants with less expression of *ankyrin* and *BIPM* showed significant resistance against *Xoo* XF89b isolate, supporting that these two eQTLs are involved in BBD resistance in rice. The identified resistance QTL, qBBR11.1, of SA0423 was introduced and improved the BBD resistance in a very susceptible indica variety, TCS10, through MABC approach. These results demonstrated that the QTLs identified from genomic, transcriptomic and proteomic approaches can be practically applied to improve the BBD resistance in rice breeding program.

http://dx.doi.org/10.5772/67331

Author details

Chang-Sheng Wang[1]* and Da-Gin Lin[2]

*Address all correspondence to: wangchansen@nchu.edu.tw

1 Department of Agronomy, National Chung-Hsing University, Taichung, Taiwan

2 Division of Biotechnology, Taiwan Agricultural Research Institute, Taichung, Taiwan

References

[1] Rayas-Duarte P, McGlynn WG, Stoecker BJ. Cereal foods: A full serving of nutrition. Food Sci. 2004; **25**: 437–444.

[2] Mizukami T, Wakimoto S. Epidemiology and control of bacterial leaf blight of rice. Ann Rev Phytopathol. 1969; **7**: 51–72.

[3] Mew TW, Alvarez AM, Leach JE, Swings J. Focus on bacterial blight of rice. Plant Dis. 1993; **77**: 5–12.

[4] Adhikari TB, Mew TW, Teng PS. Phenotypic diversity of *Xanthomonas orzyae* pv. *orzyae* in Nepal. Plant Dis. 1994; **78**: 68–72.

[5] Gnanamanickam SS, Brindha Priyadarisini V, Narayanan NN, Vasudevan P, Kavitha S. An overview of bacterial blight disease of rice and strategies for its management. Curr Sci. 1999; **77**: 1435–1443.

[6] Hsieh SPY. Rice bacterial blight. In Cheng CH, editors. Plant Protection Illustrations 8: Rice Protection. 1st ed. Taipei: Bureau of Animal and Plant Health Inspection and Quarantine, Council of Agriculture, Executive Yuan; 2003. pp. 317–338.

[7] Wang CS, Wang AZ, Lin DG. The application of mutants in breeding disease resistance in rice. In Wang AZ, Wang CS, Ann PJ, Yu XZ, Wu HS, editors. Special Issue or the Symposium on Important Crop Pathogen Detection and Management. 1st ed. Taichung: Bureau of Animal and Plant Health Inspection and Quarantine, Council of Agriculture, Executive Yuan; 2013. 193 p.

[8] Goto M. Fundamentals of Bacterial Plant Pathology. 1st ed. San Diego, CA: Academic Press; 1992. 342 p.

[9] Lee KS, Rasabandith S, Angeles ER, Khush GS. Inheritance of resistance to bacterial blight in 21 cultivars of rice. Phytopathology. 2003; **93**: 147–152.

[10] Ou SH. A Handbook of Rice Diseases in the Tropics. 1st ed. Los Banos: International Rice Research Institute; 1973. 58 p.

[11] Lin TF. Influence of bacterial blight on the yield and quality of rice and the breeding of resistant lines. Bull Taichung Distr Agric Improv Station. 1990; **29**: 29–38.

[12] Lin CS, Chang SJ. Studies on the resistance to *Xanthomonas campestris* pv. *oryzae* of rice in Taiwan I. The reaction of newly developed rice lines to Taiwanese isolates. Bull Taichung Distr Agric Improv Station. 1992; **321**: 25–31.

[13] Lin CS, Chen CW, Tzeng DS. The effect of microorganism bioagent for biological control of rice leaf blight disease and influence of rice agriculture characters. Bull Taichung Distr Agric Improv Station. 2003; **81**: 65–77.

[14] Mew TW, Vera Cruz CM, Medalla ES. Changes in race frequency of *Xanthomonas oryza* pv. *oryza* in response to the planting of rice cultivars in Philippines. Plant Dis. 1992; **76**: 1029–1032.

[15] Huang N, Angeles ER, Domingo J, Magpantay G, Singh S, Zhang G, Kumaravadivel N, Bennett J, Khush GS. Pyramiding of bacterial blight resistance genes in rice: Marker-assisted selection using RFLP and PCR. Theor Appl Genet. 1997; **95**: 313–320.

[16] Wang AZ, Wang CS. Genomic breeding developing rice variety with durable resistance to bacterial leaf blight and blast disease. J Agric Forest. 2009; **58**: 11–24.

[17] Kottapalli KR, Sarla N, Kikuchi S. *In silico* insight into two rice chromosomal regions associated with submergence tolerance and resistance to bacterial leaf blight and gall midge. Biotechnol Adv. 2006; **24**: 561–589.

[18] Sama VS, Rawat N, Sundaram RM, Himabindu K, Naik BS, Viraktamath BC, Bentur JS. A putative candidate for the recessive gall midge resistance gene *gm3* in rice identified and validated. Theor Appl Genet. 2014; **127**: 113–124.

[19] Ezuka A, Sakaguchi S. Host-parasite relationship in bacterial blight of rice caused by *Xanthomonas oryzae*. Rev Plant Prot Res. 1978; **11**: 93–118.

[20] Leung H. Stressed genomics - bringing relief to rice fields. Curr Opin Plant Biol. 2008; **11**: 201–208.

[21] Leung H, Zhu Y, Revilla-Molina IM, Fan JX, Chen H, Pangga I, Vera Cruz C, Mew TW. Using genetic diversity to achieve sustainable rice disease management. Plant Dis. 2003; **87**: 1156–1169.

[22] Li ZK, Sanchez A, Angeles E, Singh S, Domingo J, Huang N, Khush GS. Are the dominant and recessive plant disease resistance genes similar? A case study of rice R genes and *Xanthomonas oryzae* pv. *oryzae* races. Genetics. 2001; **159**: 757–765.

[23] Davierwala AP, Reddy AP, Lagu MD, Ranjekar PK, Gupta VS. Marker assisted selection of bacterial blight resistance genes in rice. Biochem Genet. 2001; **39**: 261–278.

[24] Wang CS, Tseng TH, Lin CY. Rice biotech research at the Taiwan Agricultural Research Institute. Asia Pacific Biotech News. 2002; **6**: 950–956.

[25] Wang AZ, Hu TK, Wang CS. Whole genome analysis of disease resistance and defense-response gene of rice. Crop, Environ & Bioinformatics. 2005; **2**:267–281.

[26] Tseng TH, Cheng CH, Chern CG, Yen SM, Jeng TL, Jwo WS, Wang CS. Response and genetic analysis of brown planthopper resistance in rice mutants of the TNG67 variety. Plant Prot Bull. 2003; **45**: 211–223.

[27] Tseng TY, Ou JF, Wang CY. Role of ascorbate-glutathione cycle in paraquat tolerance of rice (*Oryza sativa*). Weed Sci. 2013; **61**: 361–373.

[28] Johnson R. A critical analysis of durable resistance. Ann Rev Phytopathol. 1984; **22**: 309–330.

[29] Lin XH, Zhang DP, Xie YF, Gao HP, Zhang Q. Identifying and mapping a new gene for bacterial blight resistance in rice based on RFLP markers. Phytopathology. 1996; **86**: 1156–1159.

[30] Nino-Liu DO, Ronald PC, Bogdanove AJ. *Xanthomonas oryzae* pathovars: model pathogens of a model crop. Mol Plant Pathol. 2006; **7**: 303–324.

[31] Shanti ML, Shenoy VV, Lalitha Devi G, Mohan Kumar V, Premalatha P, Naveen Kumar G, Shashidhar HE, Zehr UB, Freeman WH. Marker-assisted breeding for resistance to bacterial leaf blight in popular cultivar and parental lines of hybrid rice. J Plant Pathol. 2010; **92**: 495–501.

[32] Verdier V, Vera Cruz CM, Leach JE. Controlling rice bacterial blight in Africa: needs and prospects. J Biotechnol. 2012; **159**: 320–328.

[33] Chen S, Liu X, Zeng L, Ouyang D, Yang J, Zhu X. Genetic analysis and molecular mapping of a novel recessive gene *xa34(t)* for resistance against *Xanthomonas oryzae* pv. *oryzae*. Theor Appl Genet. 2011; **122**: 1331–1338.

[34] Liu Q, Yuan M, Zhou Y, Li X, Xiao J, Wang S. A paralog of the MtN3/saliva family recessively confers race-specific resistance to *Xanthomonas oryzae* in rice. Plant Cell Environ. 2011; **34:** 1958–1969.

[35] Chu Z, Fu B, Yang H, Xu C, Li Z, Sanchez A, Park YJ, Bennetzen JL, Zhang Q, Wang S. Targeting *xa13*, a recessive gene for bacterial blight resistance in rice. Theor Appl Genet. 2006; **112**: 455–461.

[36] Gu K, Yang B, Tian D, Wu L, Wang D, Sreekala C, Yang F, Chu Z, Wang GL, White FF. *R* gene expression induced by a type-III effector triggers disease resistance in rice. Nature. 2005; **435**: 1122–1125.

[37] Iyer AS, McCouch SR. The rice bacterial blight resistance gene *xa5* encodes a novel form of disease resistance. Mol Plant-Microbe Interact. 2004; **17**: 1348–1354.

[38] Jiang GH, Xia ZH, Zhou YL, Wan J, Li DY, Chen RS, Zhai WX, Zhu LH. Testifying the rice bacterial blight resistance gene *xa5* by genetic complementation and further analyzing xa5 (Xa5) in comparison with its homolog TFIIAγ1. Mol Genet Genom. 2006; **275**: 354–366.

[39] Song WY, Wang GL, Chen LL, Kim HS, Pi LY, Holsten T, Gardner J, Wang B, Zhai WX, Zhu LH. A receptor kinase-like protein encoded by the rice disease resistance gene, *Xa21*. Science. 1995; **270**: 1804–1806.

[40] Sun X, Cao Y, Yang Z, Xu C, Li X, Wang S, Zhang Q. *Xa26*, a gene conferring resistance to *Xanthomonas oryzae* pv. *oryzae* in rice, encodes an LRR receptor kinase-like protein. Plant J. 2004; **37**: 517–527.

[41] Xiang Y, Cao Y, Xu C, Li X, Wang S. *Xa3*, conferring resistance for rice bacterial blight and encoding a receptor kinase-like protein, is the same as *Xa26*. Theor Appl Genet. 2006; **113**: 1347–1355.

[42] Yoshimura S, Yamanouchi U, Katayose Y, Toki S, Wang ZX, Kono I, Kurata N, Yano M, Iwata N, Sasaki T. Expression of *Xa1*, a bacterial blight-resistance gene in rice, is induced by bacterial inoculation. Proc Natl Acad Sci USA. 1998; **95**: 1663–1668.

[43] Tian D, Wang J, Zheng X, Gu K, Qiu C, Yang X, Zhou Z, Goh M, Luo Y, Murata-Hori M. The rice TAL effector-dependent resistance protein Xa10 triggers cell death and calcium depletion in the endoplasmic reticulum. Plant Cell. 2014; **26**: 497–515.

[44] Wang C, Fan Y, Zheng C, Qin T, Zhang X, Zhao K. High-resolution genetic mapping of rice bacterial blight resistance gene *Xa23*. Mol Genet Genom. 2014; **289**: 745–753.

[45] Kim SM, Suh JP, Noh TH, Reinke RF, Jena KK. Identification and fine-mapping of a new resistance gene, *Xa40*, conferring resistance to bacterial blight races in rice (*Oryza sativa* L.). Theor Appl Genet. 2015; **128(10)**: 1933–1943. doi: 10.1007/s00122-015-2557-2.

[46] Broman KW, Wu H, Sen Ś, Churchill GA. R/qtl: QTL mapping in experimental crosses. Bioinformatics. 2003; **19**: 889–890.

[47] Broman KW, Sen S. A Guide to QTL Mapping with R/qtl. 1st ed. New York: Springer; 2009. 394 p.

[48] Tseng HY, Lin DG, Hsieh HY, Tseng YJ, Tseng WB, Chen CW, Wang CS. Genetic analysis and molecular mapping of QTLs associated with resistance to bacterial blight in a rice mutant, SA0423. Euphytica. 2015; **205**: 231–241.

[49] Lin DG, Lin YL, Chen CW, Lai HC, Wang YW, Chou SY, Wang CS. Development of an efficient fluorescent visualization method for the detection of *Xanthomonas oryzae* pv. *oryzae* infection in rice planta. J Taiwan Agric Res. 2015; **64**: 10–20.

[50] Huang CY, Chou SY, Lee YL, Tseng WB, Chen CW, Wang CS, Lin DG. Transcriptomic analysis of a resistant mutant in response to *Xanthomonas oryzae* pv. *oryzae* Infection. J Taiwan Agric Res. 2015; **64**: 145–158.

[51] Theocharidis A, van Dongen S, Enright AJ, Freeman TC. Network visualization and analysis of gene expression data using BioLayout Express(3D). Nat Proto. 2009; **4**: 1535–1550.

[52] Panstruga R, Parker JE, Schulze-Lefert P. SnapShot: plant immune response pathways. Cell. 2009; **136**: 978; e971–973.

[53] Lin DG, Lin YL, Huang CY, Tseng HY, Tseng WB, Chen CW, Wang CS. Study on the resistance eQTLs in the rice bacterial blight resistant mutant, SA0423. J Taiwan Agric Res. 2016; **65**: 54–69.

[54] Lyzenga WJ, Booth JK, Stone SL. The *Arabidopsis* RING-type E3 ligase XBAT32 mediates the proteasomal degradation of the ethylene biosynthetic enzyme, 1-aminocyclopropane-1-carboxylate synthase 7. Plant J. 2012; **71**: 23–34.

[55] Chen F, Yuan Y, Li Q, He Z. Proteomic analysis of rice plasma membrane reveals proteins involved in early defence response to bacterial blight. Proteomics. 2007; **7**: 1529–1539.

[56] Mahmood T, Jan A, Kakishima M, Komatsu S. Proteomic analysis of bacterial-blight defense-responsive proteins in rice leaf blades. Proteomics. 2006; **6**: 6053–6065.

[57] Zipfel C, Robatzek S, Navarro L, Oakeley EJ, Jones JD, Felix G, Boller T. Bacterial disease resistance in *Arabidopsis* through flagellin perception. Nature. 2004; **428**: 764–776.

[58] Lin DG, Chou SY, Wang AZ, Wang YW, Kuo SM, Lai CC, Chen LJ, Wang CS. A proteomic study of rice cultivar TNG67 and its high aroma mutant SA0420. Plant Sci. 2014; **214**: 20–28.

[59] Lin DG, Wang CS. Extraction of total proteins from rice. Plant Bio-protocol. 2015; **4**: e1277.

[60] Lorieux M. MapDisto: fast and efficient computation of genetic linkage maps. Mol Breed. 2012; **31(2)**: 1231–1235.

[61] Zhao Z, Assmann AM. The glycolytic enzyme, phosphoglycerate mutase, has critical roles in stomatal movement, vegetative growth, and pollen production in *Arabidopsis thaliana*. J Exp Bot. 2011; **62(14)**: 5179–5189.

[62] Ashikari M, Wu J, Yano M, Sasaki T, Yoshimura A. Rice gibberellin-insensitive dwarf mutant gene *Dwarf 1* encodes the alpha-subunit of GTP-binding protein. Proc Natl Acad Sci USA. 1999; **96(18)**: 10284–10289.

[63] Fujisawa Y, Kato T, Ohki S, Ishikawa A, Kitano H, Sasaki T, Asahi T, Iwasaki Y. Suppression of the heterotrimeric G protein causes abnormal morphology, including dwarfism, in rice. Proc Natl Acad Sci USA. 1999; **96(13)**: 7575–7580.

[64] Zuo J, Li J. Molecular genetic dissection of quantitative trait loci regulating rice grain size. Ann Rev Genet. 2014; **48**: 99–118.

[65] Chen X, Fu S, Zhang P, Gu Z, Liu J, Qian Q, Ma B. Proteomic analysis of a disease-resistance-enhanced lesion mimic mutant spotted leaf 5 in rice. Rice. 2013; **6**: 1–10.

[66] Kauffman HE, Reddy APK, Hsieh SPY, Merca SD. An improved technique for evaluation resistance of rice varieties to *Xanthomonas oryzae*. Plant Dis Rep. 1973; **57**: 537–541.

[67] McCouch SR, Cho YG, Yano M, Paul E, Blinstrub M, Morishima H, Kinoshita T. Report on QTL nomenclature. Rice Genet Newslett. 1997; 14: 11–13.

[68] Tabien R, Li Z, Paterson AH, Marchetti MA, Stansel JW, Pinson SRM. Mapping QTLs for field resistance to the rice blast pathogen and evaluating their individual and combined utility in improved varieties. Theor Appl Genet. 2002; **105**: 313–324.

Chapter 14

Rice Plant Height Monitoring from Space with Bistatic Interferometry

Cristian Rossi

Additional information is available at the end of the chapter

http://dx.doi.org/10.5772/65868

Abstract

This chapter provides an overview of the possibility to derive paddy rice plant heights with spaceborne bistatic SAR interferometry (InSAR). By using the only available interferometer in space, TanDEM-X, an investigation of rice crops located in Turkey is performed. Before analyzing the main outcomes, an introduction to the generation of elevation models with InSAR is provided, with a special focus on the agricultural land cover. The processing chain and the modifications foreseen to properly produce plant elevations and a roadmap for the quality assessment are described. The results obtained, with a very high interferometric coherence supporting an accurate estimation due to a limited electromagnetic wave penetration into the canopy, support a temporal change analysis on a field-by-field basis. For the purpose, an automatic approach to segment the fields without external auxiliary data is also provided. The study is concluded with an analysis of the impact of the wave polarization in the results.

Keywords: SAR, InSAR, DEM, TanDEM-X, agricultural remote sensing

1. Introduction

Remote sensing is a mature technology for the observation of natural environmental changes. In terms of agricultural monitoring applications, radar sensors differ from optical, multispectral, and thermal sensors for two main reasons: (1) radar systems can collect imagery independent of solar illumination and cloud cover. This is particularly relevant for countries affected by heavy precipitations during the plant growing stages. (2) The system measures amplitudes and phases of the backscattered signal, yielding the joint derivation of absolute ranging and backscattering coefficients. Both of them can be exploited to derive the plant height, as explained in the following.

The investigation presented in this chapter is performed for paddy-rice fields, even though in principle it can be generalized for other vertical-oriented vegetation crops. The relevance of the study comes from economical and geo-political aspects. According to the Food and Agricultural Organization (FAO), rice is one of the most valuable livestock products in the world, with a production of more than 700 million tons per year [1]. As a consequence, a big interest of international agencies, insurance companies, and governments are posed on this staple food. For instance, politicians and governments are particularly interested in the monitoring of farming practices and land control, e.g., to check for hidden and/or spoofed markets. Insurance companies are interested in forecasting coverage costs by knowing the status of crops at the moment of possible flooding. Agencies would like to regulate the product import/export based on the yield estimation and the current demand. The possibility to globally monitor paddies, by providing the growth status and field borders, is then very relevant.

This global monitoring can be ensured with the utilization of synthetic aperture radar (SAR) systems. SAR images have been already used for several campaigns for crop inspections (e.g., [2–6]). Many possible measures of rice growth such as canopy height, LAI, biomass, etc. are considered in the works cited above. Among them, canopy height is the most direct measurement and has direct relationship with growth rate, especially in the early growing stage. There are three techniques that can be employed to derive the rice plant height with SAR data: single-image backscatter analysis, SAR interferometry (InSAR), and Polarimetric SAR Interferometry (PolInSAR).

1.1. Single-image backscatter analysis

A practiced strategy relies in finding the correlation between canopy height and backscattering coefficients, although the scattering process is not a function depending only on crop height. In fact, an indirect relationship can be assessed. The electromagnetic scattering of the plant is a function of intricate interrelations among physical parameters of rice [7]. By taking into account the different scattering mechanisms involved in the acquisition process, the system parameters, and the physical properties of the plant, it is in principle possible to invert a complex model and derive, among other parameters, also the plant height. Nevertheless, only a few studies are reported in the literature for this purpose and they are based on experimental data sets and locally selected thresholds, thus limiting their accuracy and not being suitable for operational processors commanded to process various data sets [8].

1.2. SAR interferometry (InSAR)

Direct height information can be instead derived with the cost of two SAR images, by employing the interferometric technique [9]. In contrast to the single-image backscattering information, InSAR exploits the phase information embedded in the received signal. From an agricultural application point of view, in the literature, interferometric phase information has been employed by making use of the coherence as in Refs. [10, 11]. In these works, most of the attention has been given on the accuracy of the interferometric phase for the C-band European remote sensing (ERS) tandem data set. However, ERS data spatial resolution is very low, about 30 m, not being able to tackle the physical-based spatial heterogeneity problem in paddy-rice fields. Two other limiting factors are the wave penetration at C-band, causing an underestimation in volume deviations, and the nonzero temporal baseline, causing unreliable interferometric

http://dx.doi.org/10.5772/65868

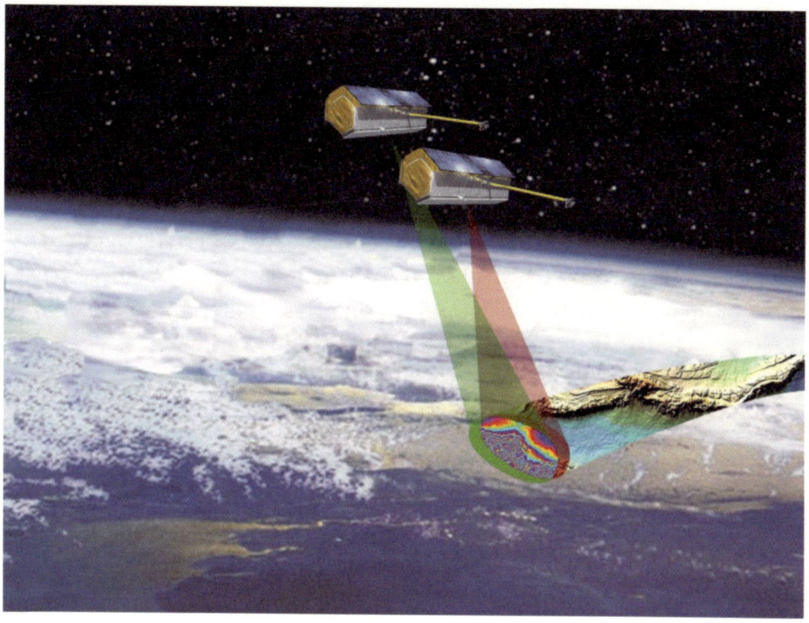

Figure 1. Artist's view of the TanDEM-X mission (©DLR).

phase information. A promising SAR concept to attenuate these limitations is TanDEM-X. An artist's view of the mission is sketched in **Figure 1**. TanDEM-X is an innovative mission, started in 2010 with the launch of a twin satellite (TDX-1) placed in close formation with the TerraSAR-X satellite (TSX-1). The main mission objective is the generation of a global digital elevation model (DEM) with HRTI-3 accuracy standards [12]. The mission acronym says just that: *TerraSAR-X add-on for Digital Elevation Measurement*. By definition, the DEM renders the height of what lays on the Earth at a given position, thus, also paddy-rice plants. A study about the accuracy of the DEM for crops is the main objective of this chapter.

The standard TanDEM-X mode of operation is bistatic, i.e., established on a single signal transmission and a dual reception. The chapter title term *bistatic interferometry* refers to this technique. The satellite transmitting and receiving the signal is also named *active* satellite, while the one only receiving the signal is named *passive* satellite. By doing so, strong DEM error sources for agricultural mapping such as atmospheric artifacts or temporal changes are avoided. Moreover, the wave penetration into the canopy is strongly limited with the employed wavelength of about 3.1 cm (X-band). Among other possible operation modes, it is worth mentioning the *monostatic* one, where the two satellites are run independently. This is the case of repeat-pass acquisitions, i.e., acquisitions taking place at different times. The potentials of TanDEM-X to render paddy-rice heights have been reported in [13, 14]. The flexible commanding yields the acquisition of several DEMs over the same area in a short revisit time, thus allowing a temporal study about the plant growth. This chapter takes inspiration from these works and revisits the results with an extended introduction about the uncertainty assessment of agricultural DEMs generated with bistatic interferometry.

1.3. Polarimetric SAR interferometry (PolInSAR)

The last technique taken into consideration for plant height derivation is Polarimetric InSAR (PolInSAR) [15]. PolInSAR requires multiple-polarized SAR images. Like the single-image backscatter image analysis, the PolInSAR height estimation is also based on scattering models. In particular, these models relate the crop height to the interferometric coherence, and they vary depending on the physical structure of the plant [16]. A limitation of this technique is the required geometrical configuration of the satellites. Indeed, to obtain the required sensitivity of a few centimeters for plants growing to about 1 m, a spatial separation between satellites (also called *baseline*) of some kilometers is required [16]. This limitation strongly impact on the applicability to spaceborne systems. The first demonstration of usage has been reported with an airborne system [17].

This chapter is organized in the following way: Section 2 presents the system employed for the height derivation and provides an overview of bistatic interferometry. Section 3 applies the technique to the mapping of paddy-rice and presents and discusses the results. Section 4 deals with the impact of the wave polarization in the results and Section 5 traces the conclusions.

2. DEM generation with bistatic interferometry

A digital elevation model is a model describing the topographical variations of the Earth. Terrain height is the main information. The elevation is generally given above a certain level, e.g., a geodetic datum. For instance, TanDEM-X elevations are over the WGS84 ellipsoid. DEMs can be generated with various sensors, such as optical, LiDAR (Light Detection and Ranging), and SAR.

Stereo photogrammetry is the standard technique to generate DEMs with optical data. It refers to the technique of measuring the position of Earth points from a set of photographs—minimum two [18]. LiDAR (Light Detection and Ranging) is another popular system to produce DEMs [19]. It is an active system based on a pulse/CW laser employed to determine the distance between sensor and target. This technology reached its maturity in the 1990s and nowadays several companies offer laser surveys with an airborne system. As for LiDAR, SAR is an active system, i.e., based on the transmission and reception of signals. The whole process is coherent, i.e., established on the use of both amplitude and phase information. Several studies have been reported in the literature. For instance, successful usage of photogrammetry and laser scanning for crop height monitoring can be found in Refs. [20, 21], respectively.

In contrast with LiDAR, which determines a 3D location from one range measurement and 2D pointing angles, the InSAR 3D positioning relies on two antenna locations and on the measure of the interferometric unwrapped phase. The processing from SAR raw data to DEM is shown in **Figure 2**.

A complete description of the processing steps is out of the scope of this chapter and can be found in several articles and books, e.g., [9, 22]. Instead, their main characteristics and modification adapted to the mapping of agricultural crops are outlined in Section 2.1.

http://dx.doi.org/10.5772/65868

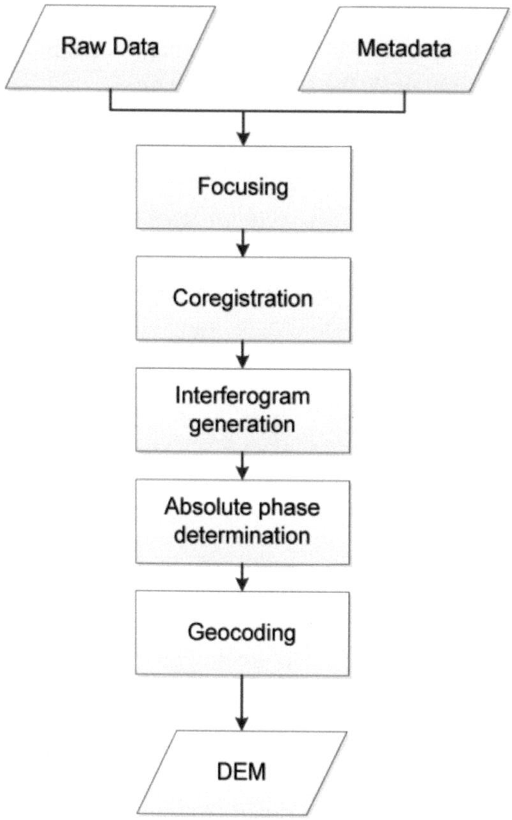

Figure 2. Flowchart of a typical InSAR processing chain finalized to DEM generation from SAR raw data.

2.1. InSAR processing steps

Agricultural crops are not a particularly difficult terrain to map and generally do not require dedicated processing solutions (see also Section 2.2) or modification to nominal InSAR processors. The processor used for the generation of the results presented in this chapter is the integrated TanDEM-X processor (ITP) [23, 24]. ITP is the operational processor employed in the German Aerospace Center (DLR) for the generation of TanDEM-X products. In the following, a brief description of the processing stages shown in **Figure 2** is provided with a special focus on the crop elevation modeling.

Focusing. Focusing is the process to form a SAR image from raw data [25]. The SAR image is a bidimensional complex array. The along-track dimension is named *azimuth,* while the across-track is named *range*. The conversion from pixel value to physical backscatter is also called radiometric calibration and is performed as:

$$\sigma^0 = (k|x|^2 - \beta_N)\sin\theta_i \tag{1}$$

where σ^0, or *Sigma Nought,* is the measure of the radar return, k is a sensor-dependent calibration factor, x is the pixel value after SAR focusing, β_N, or *Noise Equivalent Beta Naught,* equivalent beta naught, represents the noise contribution into the signal and it is usually annotated in the SAR product, and θ_i is the local incidence angle. Since rice paddies develop in locally flat terrain, θ_i is equal to θ_l, the radar looking angle.

In the bistatic interferometric scenario, the focusing operation is performed for the active channel, generating the *master* image, and for the passive channel, generating the *slave* image.

Coregistration. Coregistration has the objective to obtain a precise sample-overlap between two SAR images. A typical algorithm employed for coregistering SAR data is the crosscorrelation [23]. No peculiar algorithmic issues are expected for paddies.

Interferogram generation. The interferogram, generated by complex conjugate multiplication of the two coregistered images, is the main product for the DEM generation, since its phase is directly related to the terrain height. Typically, to reduce speckle noise, a multilooking process is implemented. For the considered agricultural scenario, an efficient moving-average 2D window is sufficient. The number of looks used in the processing defines an important DEM parameter, the horizontal resolution.

Horizontal resolution (Ω_r). Ω_r represents the minimum resolvable distance between two objects at different height. It is determined as:

$$\Omega_r = \frac{n_{az}\delta_{az}^{gr} + n_{rg}\delta_{rg}^{gr}}{2} \tag{2}$$

where n_{az} and n_{rg} are the azimuth and range independent number of looks and δ_{az}^{gr} and δ_{rg}^{gr} represent the single SAR pixel azimuth and range ground resolution. The independent number of range and azimuth looks is a function of the looks used in the multilooking process [12]. Ω_r represents the average of the range and azimuth interferogram resolutions.

Absolute phase determination. The SAR interferometric technique is based on the exploitation of the complex interferogram. The interferogram is defined through phase principal values, with values ranging into the interval $(-\pi,+\pi]$. A critical stage of the interferometric chain is the absolute phase retrieval given the wrapped interferogram phase. This process, named *phase unwrapping*, is one of the most delicate of the whole processing chains. It consists, for every interferogram pixel, in the estimation of the number of phase cycles to be added to the wrapped value. The topographic phase ϕ_{top}, also called absolute unwrapped phase, is sensitive to the terrain height h through the relation

$$\frac{\partial\phi_{top}}{\partial h} = \frac{2\pi B_\perp}{\lambda r \sin\theta_l} = \frac{2\pi}{h_a} \tag{3}$$

where $B_\perp$ is the perpendicular baseline between satellites, λ is the wavelength, r is the slant range, and h_a is a useful derived parameter called height of ambiguity. The phase unwrapping step defines the unwrapped phase from the (wrapped) interferometric phase by adding an estimated integer number of cycles. The accuracy of this operation depends on h_a. Indeed, large heights of ambiguity data-takes are less prone to phase unwrapping errors that manifest in the DEM as height discontinuities of multiples of h_a. In contrast, according to Eq. (3), small heights of ambiguity yield better results in terms of height sensitivity. For rice paddies, considering that the plant height is very small, growing up to 1–1.5 m, a small height of ambiguity would be preferred to obtain precise results. It has to be noticed that the unwrapping operation may even be not necessary for terrain height variation smaller than h_a, thus dramatically simplifying the overall InSAR processing. The nominal TanDEM-X ambiguity heights are around 40–60 m.

Finally, the unwrapped phase must be properly calibrated before the final geocoding step. The calibration involves the estimation of the *absolute phase offset*, which can be derived with external ground control points, with an external DEM, or with the DEM derived with the internal coregistration shifts, as in [24]. This DEM calibration is an important processing step for a multi-temporal elevation study like the one proposed in this chapter, since uncalibrated data provide misinterpretations of the geophysical outcomes. The method in [24], operationally employed for TanDEM-X production, should be actually discarded for multitemporal studies since every single absolute phase offset estimation is computed independently and is based on the local InSAR geometry. Error sources, such as baseline inaccuracies, may vary between geometries, thus producing absolute height differences between DEMs. For this reason, the calibration with a common reference is a more favorable solution. Obviously, the calibration points or region must be located outside the paddies and must consist of temporally stable elevations.

Geocoding. This processing step implies an absolute phase offset conversion in surface elevation and a georeference in a specific datum. It is a standard operation and no modifications are foreseen for agricultural mapping.

Figure 3 shows exemplary outputs from these processing stages for the test site considered in this chapter. Here, the master and slave amplitude channels in the top box reveal the changes in backscatter for the different land cover in the scene. The flattened interferogram in the second box, i.e., the interferometric phase compensated for the ellipsoidal height, shows the topographical variations.

One fringe represents a height variation equal to h_a, about 26 m in this case. The coherence gives a picture of the output quality, with very low values for low-backscatter areas (e.g., water) and high values in the central portion of the scene, covered by crops (see Section 2.2). The phase unwrapping, mandatory in this case due to the will to represent also the hilly portions of the scene at the upper and lower portions of the scene, is not creating artifacts, as can also be seen in the third box by the differential phase between the unwrapped phase and the equivalent phase generated with a reference elevation model, in this case represented by the Shuttle Radar Topography Mission (SRTM) [26] one. Finally, the generated DEM is displayed in a 3D view at the bottom of **Figure 3**.

2.2. DEM error sources and investigation

Although in principle every terrain can be mapped in elevation with InSAR, the obtained accuracy is strongly land cover dependent.

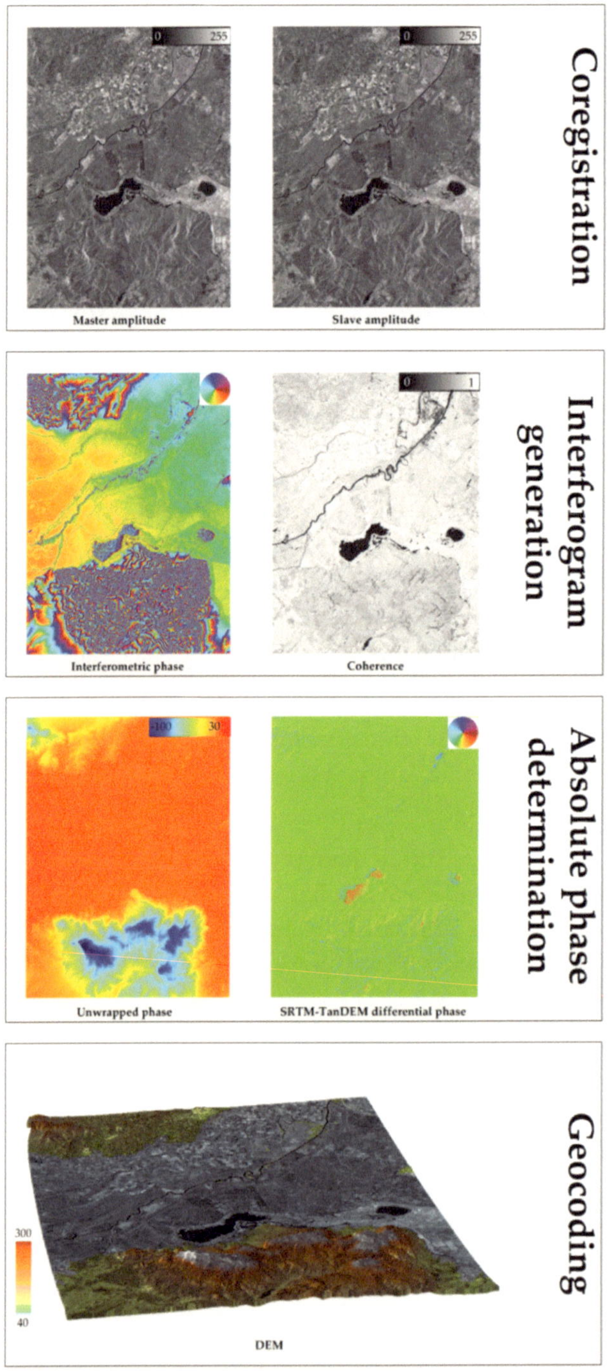

Figure 3. Interferometric processing example for the test site considered in the chapter. From the top, *coregistration stage*, with the master and slave amplitudes, *interferogram generation*, with the flattened interferometric phase, and the coherence, *absolute phase determination*, with the unwrapped phase and the differential phase between SRTM and TanDEM-X, and *geocoding*, with the final DEM.

2.2.1. *Local geomorphology impact*

Since SAR is a side-looking sensor, terrain slope impacts in the elevation model, with slopes that are even not representable due to the shadowing effect or to the multiple mapping in a single resolution cell (*layover*) [27]. Since agricultural crops are usually settled over flat or smooth terrains, the local geomorphology is not a source of error to take into account.

2.2.2. *Plant structure impact*

A relevant source of error for agricultural crops is instead the terrain itself. Being SAR an active system, i.e., transmitting and receiving energy, it is affected by wave propagation phenomena. Indeed, the wave propagates into the terrain depending on the material property [7, 31]. The measured height, i.e., the measured scattering phase center, depends on this property and in particular on the complex dielectric constant $\varepsilon_r = \varepsilon_r^{'} - j\varepsilon_r^{''}$. ε_r describes the medium characteristics in relationship to the electric field, i.e., how its power decreases in the medium where it travels. The loss of power density is described by the *penetration depth*

$$\delta_p = \frac{\lambda}{2\pi}\frac{\sqrt{\varepsilon_r^{'}}}{\varepsilon_r^{''}} \tag{4}$$

that is, the value for which the power density is reduced to 1/e. Deeper penetration is measured for low bandwidths and low moisture contents ($\varepsilon_r^{''}$ is proportional to moisture). The radar signal travels two times into the canopy, so that the equivalent penetration depth, or the scattering phase center location, is actually at $\delta_p/2\cos\theta_i$ below the top of the surface. In reality, the physical description of the electromagnetic interaction between the transmitted wave and the paddy-rice field is much more complex than that. For instance, also inhomogeneities of the inner portion of the plant and their integration into the SAR resolution cell contribute to the total signal extinction. This yields an overall loss of the interferometric coherence, which is also named *volume decorrelation* (see Section 2.2.3). Rather than inverting electromagnetic models and estimate the physical characteristics of the plants, this study aims to experimentally demonstrate the capabilities of the bistatic system in tracking the rice plant heights, thus indirectly deriving the impact of the signal extinction into the estimate.

2.2.3. *Interferometric coherence*

As aforementioned, in InSAR processors, the random error is measured by the coherence parameter. Coherence assesses the quantity of decorrelation that occurs between two SAR signals. It is defined as the crosscorrelation between two complex SAR images x_1 and x_2 and can be estimated as

$$\gamma = \left| \frac{\sum x_1 x_2 \exp\{-j\phi_{\text{known}}\}}{\sqrt{\sum |x_1|^2 \sum |x_2|^2}} \right|. \tag{5}$$

In Eq. (5), ϕ_{known} is a deterministic phase value, representing the topography and other known phase trends in the estimation window. This factor must be compensated to accomplish stationarity [28]. Given the coherence, the marginal probability density function for the interferometric phase ϕ can be first estimated and the standard deviation of the interferometric phase $\sigma_\phi(r,a)$ can be then derived by integrating it [9]. The DEM standard error for every range and azimuth samples (r,a) is then calculated, according to Eq. (3), as

$$h_{\mathrm{err}}(r,a) = \sigma_\phi(r,a)\frac{h_a}{2\pi}. \quad (6)$$

The error is proportional with the height of ambiguity: higher heights of ambiguity yield higher errors. To have an impression, for $h_a = 50$ m, a coherence value of 0.8 and 30 looks, the standard relative error is about 0.8 m. This error is only 0.3 m for $h_a = 20$ m.

Coherence can be decomposed in several factors [12], among which the volume decorrelation term, anticipated in Section 2.2.2, is the most relevant for rice paddies. To be noticed, coherence provides an estimate of the *relative* height error, as in Eq. (6). Relative height error refers to the error between two defined points in the elevation model, and sometimes it is specified as *point-to-point* error. It must not be confused with the *absolute* height error, i.e., how close the elevation cell is to the real height. A measure of the absolute error is described in the next section.

2.2.4. *Difference with reference*

The most straightforward way to evaluate the DEM quality is a direct comparison with references, in form of another DEM or in form of ground control points. It is clear that the reference must be originated from a different acquisition than the one under test. A typical solution for agricultural monitoring, as performed for the inspection presented in this chapter, is the setup of ground control points (GCPs) distributed in the field. More in detail, considering the current study, reference data has been collected in cooperation with the Istanbul Technical University (ITU). In particular, the state organization Trakya Agricultural Research Institute collected detailed ground truth in 8 fields with 4 independent samples per field during the growth cycle (May–October) of paddy-rice in 2013. Among the various gathered physical parameters, *height above ground* and *water* are the one of interest for the demonstration. The fieldwork dates are presented in **Figure 4** with the pictures taken during the campaign. To highlight the spatial variation in response to changes in agricultural practice, the first line in **Figure 4** shows the pictures taken from different fields on the same day. In this region, crops

Figure 4. Pictures taken for eight reference field during the ground truth data collection campaign [13]. The first line shows portion of the fields acquired on May 30, 2013, also illustrating the differences in agricultural practice. The second line shows the temporal evolution of field 8.

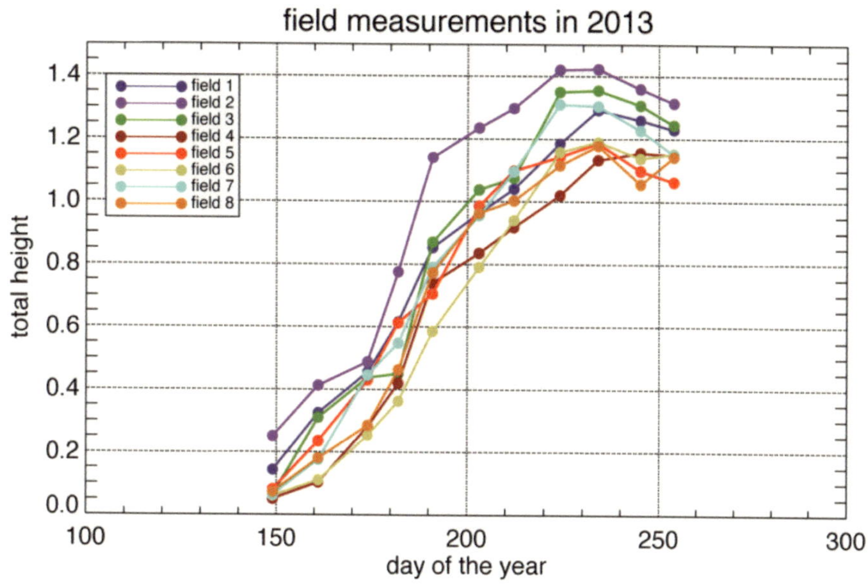

Figure 5. Relationship between the day of the year and the canopy height for the eight monitored fields [13].

are cultivated independently depending on the field owner's decision. Here, the sowing method is direct seeding by broadcasting, implying a random seeding instead of a regular straight-row one. This is a rather important point, since it highlights the expected randomness of the scattering. **Figure 5** shows the plots of the relationship between canopy height and day of the year obtained during the field works. Most fields were homogeneous and crops reached maximum height after flowering. Plant height ranges in between 0 and 140 cm.

The reference discrepancy needs quantification. In the mapping field, the standardized value for the vertical and horizontal positional accuracy is the root mean square error (RMSE). It is defined as

$$\mathrm{RMSE} = \sqrt{\frac{\sum_i^n (x_i - x_i^{\mathrm{REF}})^2}{n}} \tag{7}$$

where x_i and x_i^{REF} are the ith sample of the DEM and the reference, respectively. RMSE is of particular interest since it fully characterizes the error distribution, but just in case of normally distributed errors with zero-mean. Another used statistical descriptor of the DEM error is the standard deviation, which describes about the 68% of the normal population:

$$\hat{\sigma}_{\mathrm{err}} = \sqrt{\frac{\sum_i^n (x_i - x_i^{\mathrm{REF}} - \acute{e})^2}{n-1}} \tag{8}$$

where $\acute{e}$ is the mean error.

In this chapter, the validation is performed taking as reference the aforementioned ground campaign. The measures in Eqs. (5) and (6) present statistics of the absolute elevation error.

2.2.5. *TanDEM-X specifications*

DEM standards usually define a confidence interval, e.g., 90%, in order to discard outlier values. The positional accuracy is defined in the horizontal and vertical dimensions. The horizontal dimension determines the *absolute circular error*, i.e., the radius of a circle in which a specific feature must lie. The vertical dimension determines instead the *absolute linear error*, i. e., the elevation discrepancy between measure and ground truth. The TanDEM-X specification states a 90% absolute circular error of 10 m and a 90% absolute linear error of 10 m [12]. As for the absolute specification, the *relative circular error* describes how well the distance between two points in the model is represented. This horizontal error component has a 3 m specification for TanDEM-X at a 90% confidence. Similarly, the *relative linear error* describes the elevation error in between two points. For TanDEM-X, always at 90%, it shall be smaller than 2 m for slopes smaller than 20°, and smaller than 4 m for larger slopes.

3. Plant height derivation strategy and results

The test site chosen for the demonstration is the Lake Gala National Park, at the border between Greece and Turkey. The park is a particular wetland environment that consists of rivers, lakes, and agricultural fields (see **Figure 6**). In the last 50 years, topographical changes caused by heavy rain and debris flow were measured. More recently, the region is controlled by the Turkish government and made available for agricultural practice, in particular for paddy-rice. Considering the regional risk of debris flow, agricultural fields have to be monitored, controlling by this way the effect of flow. For instance, if the seeding has been affected from flow and irrigation, farmers can do transplanting again before it is too late for seeding. TanDEM-X monitoring is then particularly appealing for this test site.

3.1. Rice growth cycle

Before proceeding with technical details, the rice plant growth cycle shall be introduced. This cycle, from panicle initiation to maturing, lasts 110 to 250 days and can be divided in three stages: *vegetative*, *reproductive*, and *maturation*. Every stage is composed by different structural differences for the rice plant, described by a special scale called Biologische Bundesanstalt, bundessortenamt und CHemische industrie (BBCH) [29]. All the growing stages can be associated with the BBCH-scale, as shown in **Table 1**.

3.2. TanDEM-X dataset

Nine dual-pol TanDEM-X acquisitions have been acquired over the Lake Gala region in 2012 at an incidence angle of 36.8°. The data stack is processed with the integrated TanDEM-X processor [24]. The processor is commanded to generate HH (horizontal polarization in transmission and reception) and VV (vertical polarization in transmission and reception)

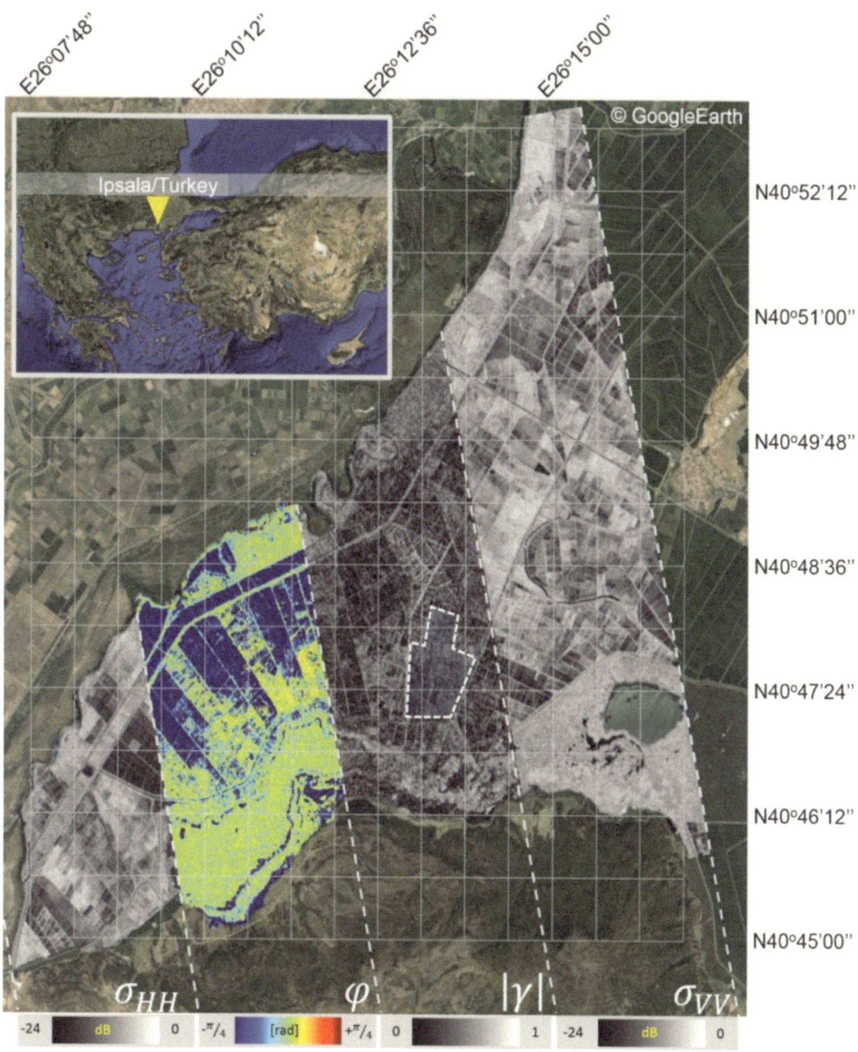

Figure 6. Agricultural study area in Ipsala, Turkey (top-left). Four features are highlighted in the picture [14]. From left to right, over the GoogleEarth image: backscatter in HH polarization, copolar phase difference, copolar coherence, and backscatter in HH polarization. The selected fields for the polarization study performed in Section 4 are highlighted in the coherence portion.

DEMs, for a total of 18 DEMs, with an output raster of 6 m. As shown in **Table 2**, all the rice growing stages in Turkey are covered (May–October), allowing then a temporal study. The height of ambiguity h_a is ranging between about 20 and 30 m. As briefly mentioned in Section 2.1, the relative error can be estimated given the number of looks used in the processing, the coherence, and the ambiguity height. Assuming a coherence value of 0.8 (a reasonable value at the crop locations, as explained in the following) and an independent number of looks of 30, the standard error varies in between 15 cm, as displayed in the last column of **Table 2**. To be noticed, these values refer to a single sample height estimate. The

Major stage	BBCH	Description
Vegetative	00	Germination
	10	Leafing
	20	Tillering
	30	Stem elongation
	40	Booting
Reproductive	50	Heading
	60	Flowering
	70	Fruiting
Maturation	80	Ripening
	90	Senescence

Table 1. BBCH-scale of the rice plant.

Acquisition date (DOY)	Perpendicular baseline [m]	Height of ambiguity [m]	Horizontal resolution [m]	Standard error [cm]
12.05.2012 (133)	253.7	23.1	10.2	36
14.06.2012 (166)	242.3	24.2	10.3	38
06.07.2012 (188)	234.3	25.1	10.2	40
17.07.2012 (199)	227.2	25.8	10.3	41
28.07.2012 (210)	222.7	26.3	10.2	42
19.08.2012 (232)	213.4	27.4	10.3	43
10.09.2012 (254)	204.4	28.7	10.3	46
13.10.2012 (297)	187.1	31.3	10.3	50
26.11.2012 (331)	181.3	32.3	10.3	51

Note: The standard error in the last column is computed for a fixed coherence value of 0.8 and an independent number of looks of 30.

Table 2. Main parameters of the TanDEM-X data set.

independent number of looks of 30 comes from the actual data processing, where a total of 45 looks have been used in the interferogram generation stage (9 in the range and 5 in the azimuth dimensions), and about 30% and 12% of the azimuth and range bandwidth has been filtered out after the spectral shift filter operation [9]. The acquisition mode of the imagery is the standard stripmap one, with a ground range pixel spacing of about 1.5 m and an azimuth one of about 2.5 m. The resulting horizontal resolution, according to Eq. (2), is displayed in the fourth column of **Table 2**. This multilooking operation is a necessary step to reduce the phase noise and the standard height error to a decimetric level for the single pixel estimate. As aforementioned, due to the relatively smooth topography of the scenes, phase unwrapping is not creating artifacts (even for small height of ambiguities), i.e., no unwrapping errors have been detected. To ensure a straightforward temporal analysis, all the DEMs have been generated using the same output grid and have been equally calibrated using a corrected version of SRTM with ICESat data.

3.3. Field segmentation

In the context of precise farming it is substantial to define field borders that are usually changing every cultivation period. Water management pattern is a further asset useful to supplier. Thus, crop segmentation is mandatory for a field-by-field uncertainty assessment, reasonably assuming a consistent growing within single fields. For this purpose, the interferometric coherence is an important subproduct to exploit, supporting the segmentation algorithm. The adopted strategy is to relate the field segmentation in a water detection problem. Indeed, flooded parcels of land characterize the first phenological phase of the plant. During this state, fields are covered by water and separated by a path network composed by soil or rare grass, as visible also in **Figure 7**, representing the May acquisition. A gravel road network is also present in the test site and separates parcel groups. This natural segmentation is visually detectable by inspecting master channel amplitude in **Figure 7(a)**, as well as the interferometric coherence in **Figure 7(b)**. This visibility relies on the water body dielectric properties. Nonmoving water behaves like a mirror, reflecting the incident signal wave in a specular direction, yielding a very low return to the SAR antenna. This phenomenon brings also a low interferometric coherence. Moreover, it is also known that a water body decorrelates within tens of milliseconds [9] (TanDEM-X small along-track time lags vary between 50 ms (equator) and 0 ms (poles)). The technique proposed by Wendleder et al. [30], operationally employed for the generation of water body mask as an auxiliary product of the official TanDEM-X DEM, is adopted. Specifically, a threshold value of 40 for the amplitude digital number (corresponding to σ^0 =−20 dB (Eq. (1)) and 0.23 for the coherence (Eq. (5)) were selected. In the study, this strategy is applied for scenes having flooded crops. In **Figure 7**, the May amplitude and coherence data show the flooded parcels for that date with low values. As visible, not all the fields were already flooded (see also **Figure 4**). To better cover the test site, additional information is retrieved by using also two complementary acquisitions taken in May 2013 over the same area.

The sole thresholding operation is not accurate enough to provide a precise segmentation, since fields that are closer than the image resolution (about 10 m, **Table 2**) may result in a single segmented field. Thus, a refinement is necessary. Among various filtering strategies, one of the most straightforward and fast, binary morphology, is chosen [13]. More in detail, erosion with a square (3×3) element is first performed to the binary water mask to remove artifacts, followed by a shape fill to remove holes within the detections. Afterward, the segmentation is performed. A total of more than 2000 fields are detected. The water detection, morphological filtering, and segmentation are performed in the geocoded (geographical coordinates) domain, in order to easily compare them with ground truth data.

(a) amplitude

(b) coherence

Figure 7. SAR master channel amplitude (a) and interferometric coherence (b) of the 12.05.2012 TanDEM-X acquisition, used to extract field shapes.

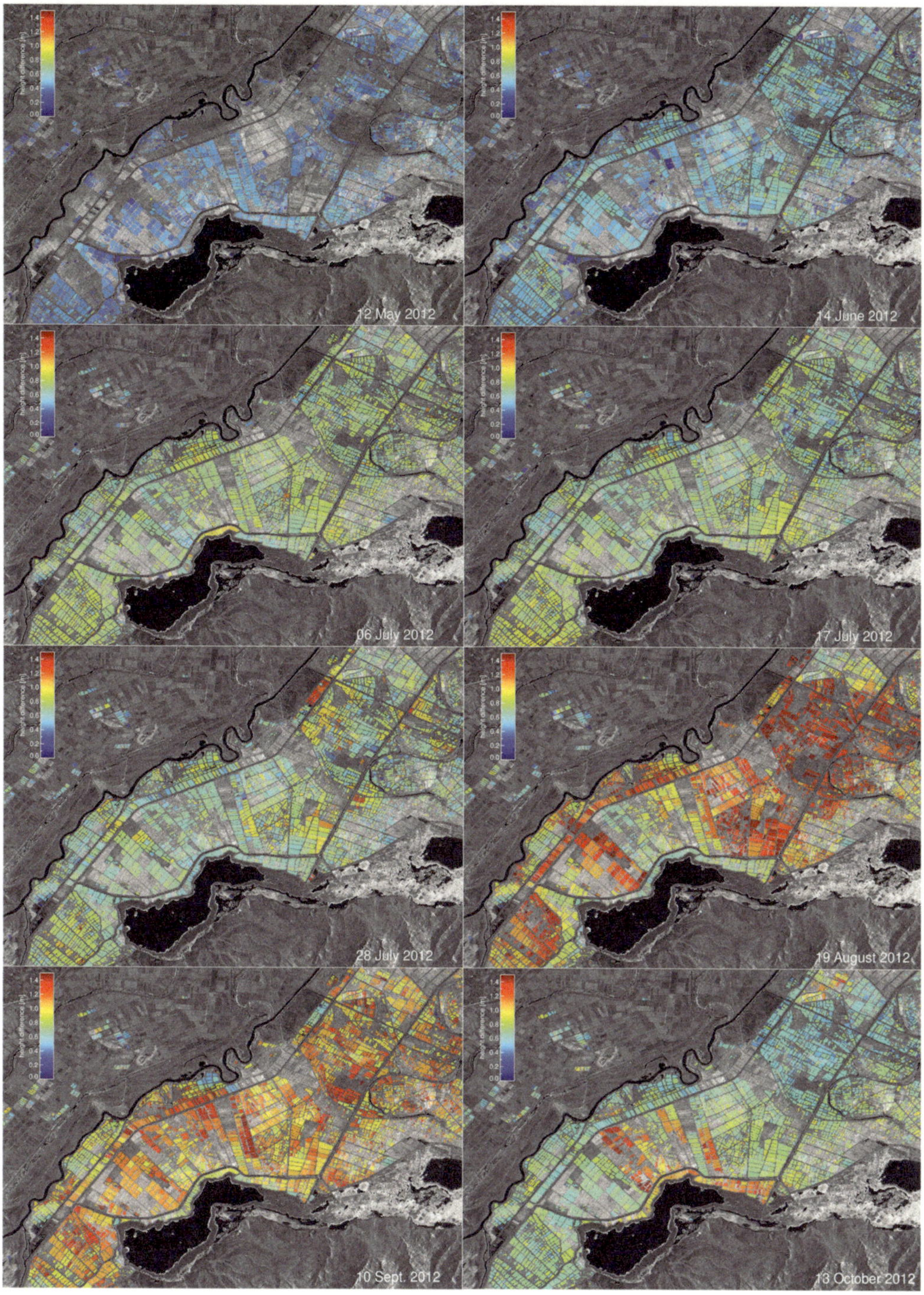

Figure 8. Temporal rice plant heights for the data stack derived with a difference between the DEM generated for the date annotated at the bottom-right and the reference one. The heights are shown in a field-by-field basis, for fields having a mean coherence value higher than 0.8 for both the analyzed and the reference acquisitions.

3.4. Temporal analysis

Since the analysis is on the plant elevation, and the generated DEM is defined over the WGS84 ellipsoid, a reference height corresponding to the plant base must be considered. For that purpose, the last acquisition, in late November, is taken as reference. Indeed, at this acquisition date, the fields have been harvested and the DEM can then be considered as a digital terrain model (DTM), i.e., representing the bare soil elevation. In the following, the November DEM is called for brevity DTM, although this is strictly true only at crop locations. The height difference between the DEMs and the DTM, i.e. the plant heights, is displayed in **Figure 8**, with a single average height value per field. The plant heights are here represented with an overlay between the SAR amplitude and the mean height difference for detected fields, which have an average coherence higher than 0.8 in both the analyzed and the reference acquisitions. A visual analysis of the maps allows the evaluation of the rice plant growth on a field-by-field basis. For instance, the first acquisition shows a limited number of crops since most of them were still flooded. The height of crops is around 20 cm. The numbers of detections increases starting from the second acquisition, i.e., the remaining fields are not flooded anymore, and a visual height growth is noticeable. The growth continues in early July, with a quite homogeneous result with plant heights around 70 cm. The following July maps reveal local changes among fields, e.g., crops located at the northern part grow faster than the ones located at the south. The August map reveals the growing of most of the plants, with doubled heights compared to late June/early July. For some of the fields, the higher maturation level is reached about a month later, as visible in the September map. The mid-October map shows the partial harvesting of some field (to be reminded: a single averaged value is displayed per segmented field), and the full growing for the fields located close to the lake northern coast. In general,

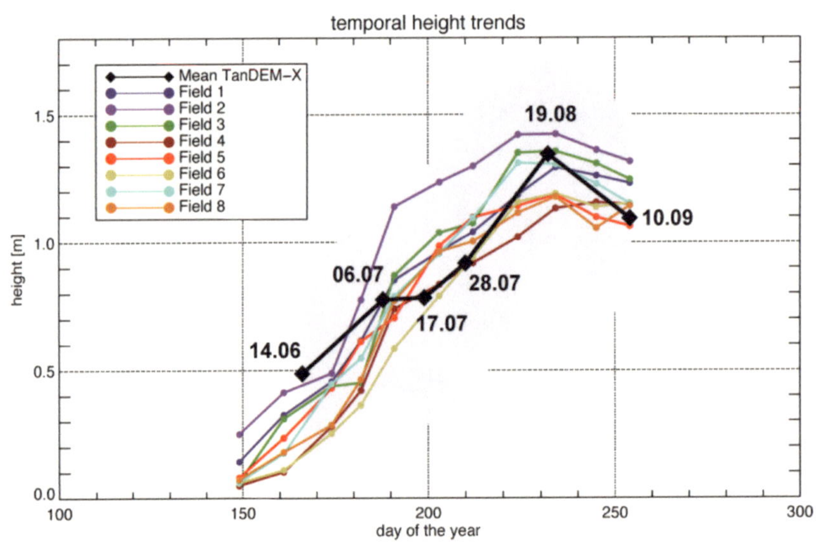

Figure 9. Mean temporal TanDEM-X elevation trend for all the 2012 detected fields over the specific date marked in the plot (black) and corresponding standard deviation (purple). The reference fields are overplotted with colors in the legend.

these maps can be used for the agricultural planning, in terms of production volume and outcomes.

This qualitative inspection already demonstrates the capability to reach the centimetric accuracy necessary to track rice plants. To further highlight it, in **Figure 9** the mean height for the detected fields is shown in black and the standard deviation highlighted in purple. Although crops exhibit variations due to the different seeding dates, the mean height trend exhibits a good accordance with the reference, overplotted in this figure. The height deviation for the late July acquisitions has to be linked to the different growing periods of the detected fields.

The quantitative inspection is performed for three out of eight fields (marked in **Figure 7(b)**. The analysis shall link the obtained accuracy derived through a comparison with reference data with the physical characteristics of the plant. The framework is the one delineated in Section 2.2.2. In particular, considering the interferometric analysis, the smaller the extinction, the lower the scattering center (Eq. (4)). Consequently, the retrieved plant elevation will be equal or smaller than the plant top depending on the actual effective dielectric constant of the canopy and the ground, since in the proposed approach the canopy height is retrieved with a difference between a plant growing phase and bare soil.

The differential-InSAR-based and the field-measurement-based canopy height are shown in **Figure 10** in form of scatterplot for three fields. Due to the growing height trend in time, this plot can be easily interpreted. Generally, the elevation trend is well detected by the interferometric measure for the late vegetative phase, reproductive, and maturation stages. Instead, the early vegetative phase represented by the May acquisition yields strongly biased elevation values due to the noisy values originated by the water reflection and is not considered. The plotted heights lie in between mid-June and mid-September (see second to seventh row in **Table 2**).

The June acquisition corresponds to the central vegetative stage (tillering, **Table 1**). At this phase, plants emerge from water (see the second and the third picture in the second row in **Figure 4**). In the SAR resolution cell different phenomena such as direct reflections from water, direct reflections from the surface, and double reflections water-surface (and vice versa) combine together. The interferometric elevation results underestimated due to this combination. The mean difference with reference data results of 7.7 cm for the eight fields taken into consideration in the ground truth campaign. A singular exception is measured for the field 5, marked with blue circles in the scatterplot in **Figure 10**, with an overestimation of about 10 cm. The overestimation has to be attributed to a low mean coherence value (about 0.5), yielding a high phase noise. During this stage, double bounces between growing vegetation and standing water should be the dominant part of the radar return. This implies a scattering phase center located at the water elevation for the cardinal effect on corners—in this case represented by quasi-vertical stems on calm water. However, the small measured height difference suggests the partial presence of the phenomenon due to the use of a short wavelength (3.1 cm) at a relatively high incidence angle (about 37°), yielding a limited penetration of the echo inside the fresh vegetated volume [13]. For the three July measurements the plant elevation exhibits the largest underestimation, with a mean difference of 26.5 cm. Also this discrepancy, at the end of the vegetative stage and beginning of the reproduction (BBCH scale 40–50, **Table 1**), can be explained with the radar wave interaction with the inner part of fresh canopy (see fourth

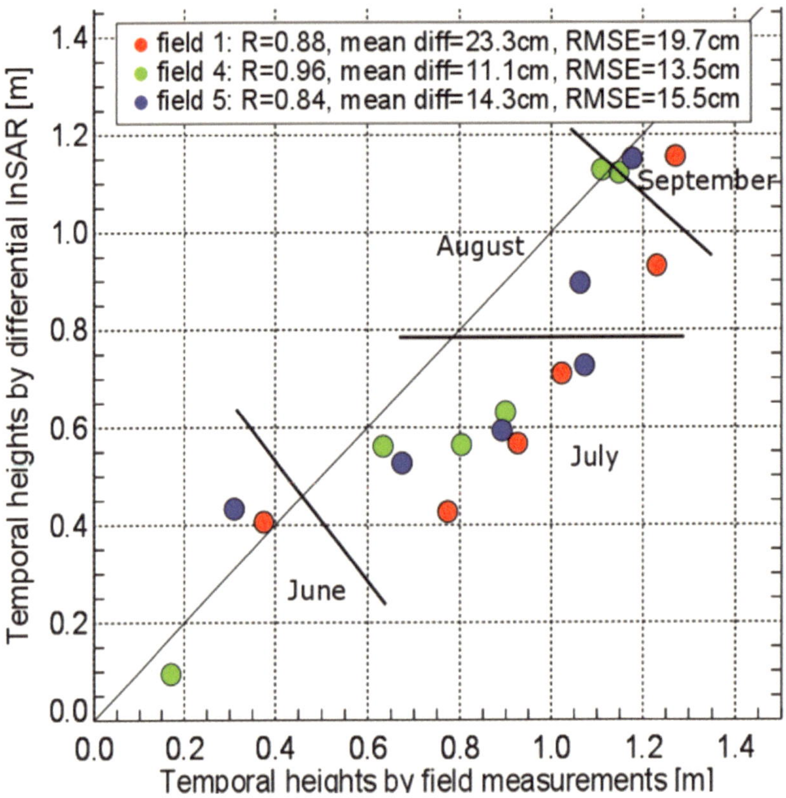

Figure 10. Comparison between interferometric height and ground truth in form of scatterplot for three fields.

picture in the second row in **Figure 4**) and a higher volume decorrelation. The difference in growing can be appreciated for the three fields in the scatterplot, with three different growing rates (higher for field 1, red circles). The August acquisition exhibits instead a generally good matching, with a mean underestimation of 4.8 cm. Being at the beginning of the maturation stage (see fifth picture in the second row in **Figure 4**), plants start to densely produce milky grains at their surface which are the main source of reflection of the signal at X-band. Again, every field should be considered independently due to the structural differences between crops. For instance, field 1 is still in its reproductive stage and shows a scattering phase center about 20 cm below the surface top. The last considered acquisition, in September, falls at the end of the maturation (see seventh picture in the second row in **Figure 4**). The grain is dry and mature, with a maximum height slightly smaller than the previous stage. On this date the interferometric elevations result again underestimated on average, with a mean difference of 16 cm. In principle, at this stage, plant elements are more randomly oriented and drier than in previous ones, hence making more similar the propagation for all polarizations. The aforementioned values represent average values for the eight fields. Just considering the three fields in the plot, the August acquisition reveals actually a higher mismatch than the mean one, while the September acquisition exhibits instead a better match for all the three fields. Once more, this is due to the discrepancies in seeding dates among the fields.

The best fit analysis in the form of $y = ax + b$ in **Figure 10** is used for calculating the offset between the two measurements [13]. As the data time sampling is not overlapping, a linear interpolation for the reference at the InSAR locations is performed. The two sources result highly correlated, with a correlation coefficient R equal to 0.88, 0.96, and 0.84 for the three fields under analysis. The mean differences and root mean square errors are in the decimetric level. In detail, the mean differences between reference and InSAR result 23.3, 11.1, and 14.3 cm and the RMSE (Eq. (5)) 19.7, 13.5, and 15.5 cm for field 1, 4, and 5, respectively. Even though the scattering analysis and the quantitative evaluation performed on this section are useful to understand the overall process, the focus shall be on the centimeter accuracy of the system for this application, and its capability of temporarily tracking the elevation through most of all the growing stages of paddy-rice fields.

Finally, the mean interferometric coherence, proportional to the relative height error (Section 2.2.3) is displayed in **Figure 11**. The mean coherence values for the selected fields are high, with values above 0.8 for all the dates. The only exceptions are for May, when fields are flooded, and for the late July acquisition, when the volume decorrelation reaches its maximum. This contribution linearly increases in July, but is strongly diminished in August, when reflections at the surface top dominates. As a reference, the last column of **Table 2** shows the height error for a fixed coherence value of 0.8.

To characterize the final relative height accuracy, one must consider that the estimated plant height is derived through a difference of two DEMs, implying a standard deviation equal to the sum in quadrature of the standard deviations of the studied DEM and the DTM. Thus, in principle, it is important to ingest an accurate as possible DTM for the algorithm proposed. Considering the chapter case, the DTM, representing bare soil at field locations, is highly accurate, with a mean coherence of about 0.9 (**Figure 11**) and a corresponding relative height

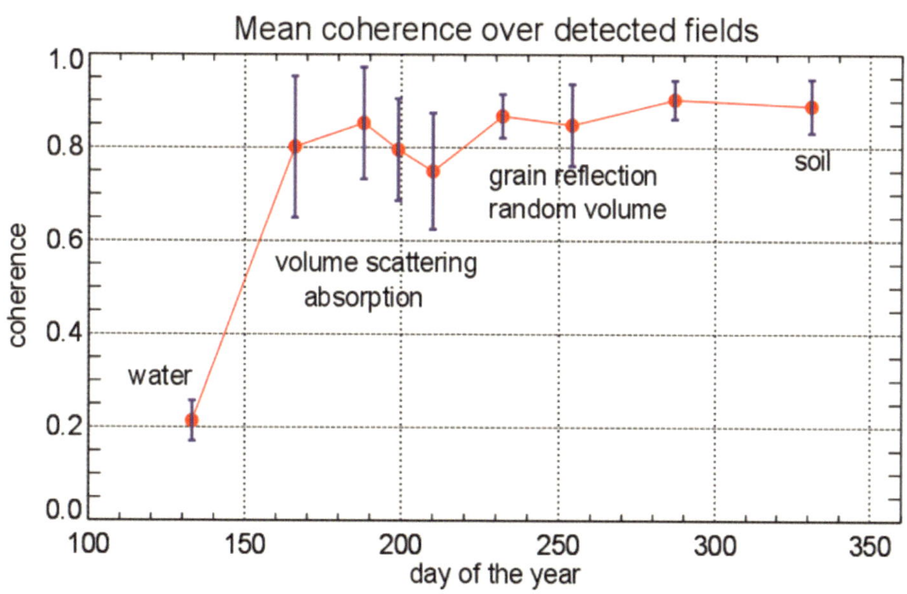

Figure 11. Interferometric mean coherence trend for the detected fields.

error of about 30 cm. Actually, a single height value is derived per field, thus dramatically reducing the overall relative height error of a factor depending on the number of samples composing the crop (fields may span more than 1000 SAR pixels).

4. Impact of wave polarization

The results and discussion provided in the previous section have been derived using the horizontal (HH) polarization and demonstrated the possibility of estimating the height (and derive the phenological stage) of the fields from TanDEM-X data with no additional ground measurements. In this section, the vertical (VV) polarization is studied, with the purpose to study the differences and possibly recommend the *best polarization* for crop monitoring.

In **Figure 12** the interferometric coherence is plotted for the HH and VV channels for the 30 randomly selected fields marked in **Figure 6**. An evident visual divergence appears for the late vegetative-early reproductive stage (around mid-July). Here, the HH elevation accuracy is larger than the VV one, since coherence values are higher (Section 2.2.3). Standard deviation is also smaller for the horizontal polarization. Thus, when considering assessing crop elevation with bistatic data for the central growing stage, it seems advantageous to privilege the HH channel. The two other stages perform similarly: early vegetative has very low coherence and poor elevation estimates for both channels, whereas late reproductive and maturation perform well.

To better characterize the polarization impact in the crop height estimates, the mean elevation difference between differently polarized DEMs for the sample fields is displayed in **Table 3**, together with the elevation standard deviation. For the first date, while fields are flooded, the copolar elevation difference measurement is very large because of the noisy phase information. For the other acquisitions the elevation differences are smaller, below 10 cm.

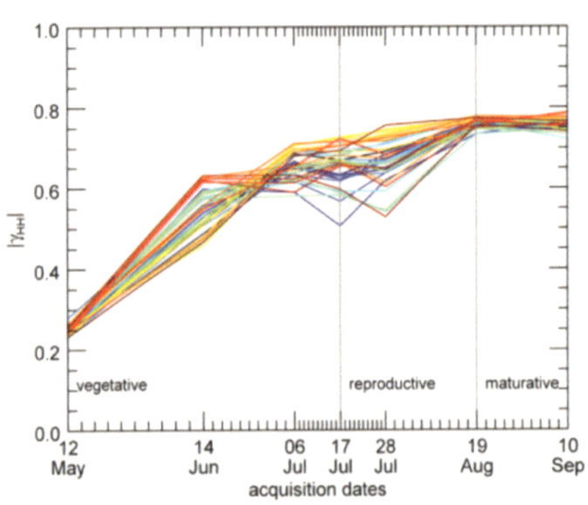

(a) HH coherence

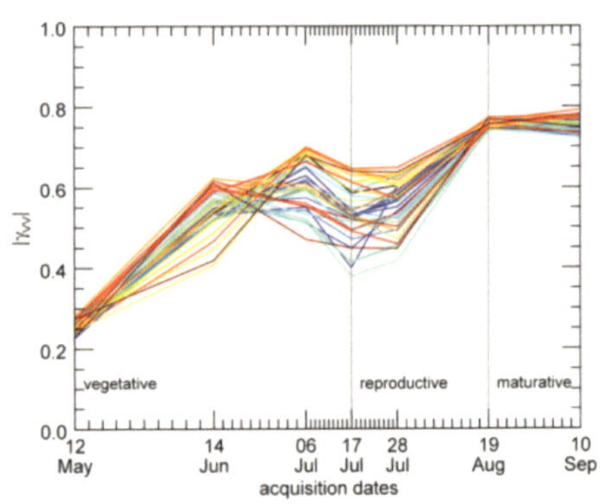

(b) VV coherence

Figure 12. Multitemporal coherence measurements from TanDEM-X HH (a) and VV (b) channels along the plan growth cycle for 30 different fields [14].

Acquisition date (DOY)	Sample mean [cm]	Sample STD [cm]
12.05.2012 (133)	+84	204
14.06.2012 (166)	−1	23
06.07.2012 (188)	−3	8
17.07.2012 (199)	−3	11
28.07.2012 (210)	−9	19
19.08.2012 (232)	−8	7
10.09.2012 (254)	−5	10
13.10.2012 (297)	+2	7
26.11.2012 (331)	0	6

Table 3. Copolar height difference statistics between HH and VV channels. The second column displays the mean height difference for 30 randomly selected fields, while the third column displays the standard deviation of the crop elevations for the two polarizations.

The analysis of **Table 3** allows an empirical evaluation of the effect of the extinction coefficient in the vertical channel through almost all the phenological stages. Excluding the first date, the temporal mean difference measurements increase monotonically until late July, i.e., the horizontally polarized signal penetrates more into the canopy compared to the vertically polarized one. The penetration discrepancy is in average of only 1 cm when the plant starts leafing, of 3 cm during tillering, and of 9 cm around the end of the vegetative stage. After, they decrease monotonically until when the plant starts to collapse and to lose its vertical structure. In particular, the measured discrepancy is still close to the maximum while reproduction and slowly decrease while flowering and finally maturing. The sample standard deviations for each acquisition date show the variability of the outcomes for each phenological stage. Values are nearly stable through maturation stage, but in vegetative and reproductive stages they are relatively high considering also the differences in growing rate.

Concluding, horizontal polarization yields digital elevation models with lower crop heights, up to about 10 cm differences. Vertical polarization yet yields higher elevation models, i.e., close to the true top canopy elevation. As aforementioned, horizontal polarization provides, on the average, more accurate elevation results for the central growing stage. So, which is the best polarization for crop elevation monitoring? Generally, if the objective is the determination of the crop elevation, local field coherence should be the final trigger. Nevertheless, for more reliable phenological stage estimation simply based on height, the VV channel can be preferred since it yields higher phase centers, therefore, better modeling the top of the canopy.

5. Conclusions

This chapter underlined the potential of TanDEM-X in paddy-rice elevation mapping. The outcomes can also be an input for the production estimation in terms of volumetric changes.

This is particularly remarkable, considering that the plant tracking requires a centimetric accuracy level and the TanDEM-X specifications are in the order of meters. The uncertainty study demonstrated three major points:

1. For the first time, plant growing has been *directly* measured from a spaceborne SAR system. As outlined in the introduction, previous demonstrations (e.g., ERS in TanDEM configuration) *indirectly* derived the elevation from coherence decomposition. The production of elevation models with InSAR has been reviewed with a special focus on the mapping of agricultural crops. An important point for the study is the presence of a DTM, in order to precisely derive plant heights. In this study, it has been shown how a postharvesting acquisition, and consequently a generated DEM, can serve for the purpose. A straightforward technique to derive field borders, with a simple thresholding operation followed by a refinement with morphological operators, has also been proposed. This refinement can be further improved for future works, for instance with more complex filtering strategies, such as unsupervised active contours techniques.

2. Also the impact of differential extinction on the crop height estimation by differential interferometry has been first experimentally studied with spaceborne SAR data. Although polarization differences are widely used for PolInSAR/PolSAR studies, and precise phenological stage derivation can be extracted by using different polarizations, it has been here demonstrated how the impact in the DEM is rather small, though still present in the DEMs.

3. Keeping the general view, it is important to carry out uncertainty studies in the temporal dimension. In particular, it has been demonstrated how the accuracy level varies depending on the plant phenological stage. Excluding the early stages, when the fields are flooded and the resulting DEM is not accurate, it has been shown how the accuracy level decreases for the late-vegetative stage, when the volume decorrelation is at its maximum, and increases for the following phenological stages, when reflections from the milky grains at the plant top dominates. An interesting and unexpected result comes from the early stages, when the plant can be assumed as a vertical and thin cylinder and the electromagnetic scattering should be dominated by double-bounces, thus with a scattering phase center at the water/soil level. Instead, the derived DEMs have showed a higher phase center, in between the plant top and the soil, suggesting a limited and not dominant double-bounce effect at X-band for rice paddies seeded by broadcasting, i.e., randomly and not in rows.

Acknowledgements

The rice and wetland monitoring in Turkey project was funded by the Scientific and Technological Research Council of Turkey (TUBITAK project no: 113Y446). TanDEM-X SAR data were supplied by the German Aerospace Center under the project no: XTILAND1476.

http://dx.doi.org/10.5772/65868

Author details

Cristian Rossi

Address all correspondence to: cristian.rossi@gmail.com

German Aerospace Center (DLR), Remote Sensing Technology Institute (IMF), Oberpfaffenhofen, Germany

References

[1] FAO. FAO Rice Market Monitor (RMM) [Internet]. [Accessed: http://www.fao.org/economic/est/publications/rice-publications/rice-market-monitor-rmm/en/]

[2] Shao, Y., Fan, X., Liu, H., Xiao, J., Ross, S., Brisco, et al. Rice monitoring and production estimation using multitemporal RADARSAT. Remote Sensing of Environment, 2001; 76 (3), 310–325.

[3] Chakraborty, M., Manjunath, K. R., Panigrahy, S., Kundu, N., & Parihar, J. S. Rice crop parameter retrieval using multi-temporal, multi-incidence angle Radarsat SAR data. ISPRS Journal of Photogrammetry and Remote Sensing, 2005; 59(5), 310–322.

[4] Wang, C., Wu, J., Zhang, Y., Pan, G., Qi, J., & Salas, W. A. Characterizing L-band scattering of paddy-rice in southeast China with radiative transfer model and multitemporal ALOS/PALSAR imagery. IEEE Transactions on Geoscience and Remote Sensing, 2009;47 (4), 988–998.

[5] Bouvet, A., Le Toan, T., & Lam-Dao, N. Monitoring of the rice cropping system in the Mekong Delta using ENVISAT/ASAR dual polarization data. IEEE Transactions on Geoscience and Remote Sensing, 2009; 47(2), 517–526.

[6] Inoue, Y., Sakaiya, E., & Wang, C. Capability of C-band backscattering coefficients from high-resolution satellite SAR sensors to assess biophysical variables in paddy-rice. Remote Sensing of Environment, 2014; 140, 257–266.

[7] Karam, M. A., Fung, A. K., & Antar, Y. M. Electromagnetic wave scattering from some vegetation samples. IEEE Transactions on Geoscience and Remote Sensing, 1988; 26(6), 799–808.

[8] Yuzugullu, O., Erten, E., & Hajnsek, I. Estimation of rice crop height from X-and C-Band PolSAR by metamodel-based optimization. IEEE Journal of Selected Topics in Applied Earth Observations and Remote Sensing, 2016; 1–11.

[9] Bamler, R., & Hartl, P. Synthetic aperture radar interferometry. Inverse Problems, 1998; 14 (4), R1.

[10] Wegmuller, U., & Werner, C. Retrieval of vegetation parameters with SAR interferometry. IEEE Transactions on Geoscience and Remote Sensing, 1997; 35(1), 18–24.

[11] Engdahl, M. E., Borgeaud, M., & Rast, M. The use of ERS-1/2 tandem interferometric coherence in the estimation of agricultural crop heights. IEEE Transactions on Geoscience and Remote Sensing, 2001; 39(8), 1799–1806.

[12] Krieger, G., Moreira, A., Fiedler, H., Hajnsek, I., Werner, M., Younis, M., & Zink, M. TanDEM-X: A satellite formation for high-resolution SAR interferometry. IEEE Transactions on Geoscience and Remote Sensing, 2007; 45(11), 3317–3341.

[13] Rossi, C.,& Erten, E. Paddy-rice monitoring using TanDEM-X. IEEE Transactions on Geoscience and Remote Sensing, 2015; 53(2), 900–910.

[14] Erten, E., Rossi, C., & Yüzügüllü, O. Polarization impact in TanDEM-X data over vertical-oriented vegetation: The paddy-rice case study. IEEE Geoscience and Remote Sensing Letters, 2015; 12(7), 1501–1505.

[15] Cloude, S. R., & Papathanassiou, K. P. Polarimetric SAR interferometry. IEEE Transactions on Geoscience and Remote Sensing, 1998; 36(5), 1551–1565.

[16] Lopez-Sanchez, J. M., & Ballester-Berman, J. D. Potentials of polarimetric SAR interferometry for agriculture monitoring. Radio Science, 2009; 44(2), 1–20.

[17] Lopez-Sanchez, J. M., Hajnsek, I., & Ballester-Berman, J. D. First demonstration of agriculture height retrieval with PolInSAR airborne data. IEEE Geoscience and Remote Sensing Letters, 2012; 9(2), 242–246.

[18] Szeliski, R. Computer vision: algorithms and applications. Springer Science & Business Media. 2010.

[19] Wehr, A., & Lohr, U. Airborne laser scanning—An introduction and overview. ISPRS Journal of Photogrammetry and Remote Sensing, 1999; 54(2), 68–82.

[20] Murakami, T., Yui, M., & Amaha, K. Canopy height measurement by photogrammetric analysis of aerial images: Application to buckwheat (Fagopyrum esculentum Moench) lodging evaluation. Computers and Electronics in Agriculture, 2012; 89, 70–75.

[21] Yu, X., Hyyppä, J., Kukko, A., Maltamo, M., & Kaartinen, H.Change detection techniques for canopy height growth measurements using airborne laser scanner data. Photogrammetric Engineering and Remote Sensing, 2006; 72(12), 1339–1348.

[22] Hanssen, R. F. Radar interferometry: data interpretation and error analysis (Vol. 2). Springer Science & Business Media. 2001.

[23] Fritz, T., Rossi, C., Yague-Martinez, N., Rodriguez-Gonzalez, F., Lachaise, M., & Breit, H. Interferometric processing of TanDEM-X data. In Geoscience and Remote Sensing Symposium (IGARSS), 2011 IEEE International (pp. 2428–2431). IEEE.

[24] Rossi, C., Gonzalez, F. R., Fritz, T., Yague-Martinez, N., & Eineder, M. TanDEM-X calibrated raw DEM generation. ISPRS Journal of Photogrammetry and Remote Sensing, 2012; 73, 12–20.

[25] Breit, H., Fritz, T., Balss, U., Niedermeier, A., Eineder, M., Yague-Martinez, N., & Rossi, C. Processing of bistatic TanDEM-X data. In Geoscience and Remote Sensing Symposium (IGARSS), 2010 IEEE International (pp. 2640–2643). IEEE.

[26] Rabus, B., Eineder, M., Roth, A., & Bamler, R. The shuttle radar topography mission—A new class of digital elevation models acquired by spaceborne radar. ISPRS Journal of Photogrammetry and Remote Sensing, 2003; 57(4), 241–262.

[27] Rossi, C., & Eineder, M. High-resolution InSAR building layovers detection and exploitation. IEEE Transactions on Geoscience and Remote Sensing, 2015;53(12), 6457–6468.

[28] Touzi, R., Lopes, A., Bruniquel, J., & Vachon, P. W. Coherence estimation for SAR imagery. IEEE Transactions on Geoscience and Remote Sensing, 1999; 37(1), 135–149.

[29] Meier, U., Bleiholder, H., Buhr, L., Feller, C., Hack, H., Heß, M.,& Weber, E. The BBCH system to coding the phenological growth stages of plants–history and publications. Journal of für Kulturpflanzen, 2009; 61(2), 41–52.

[30] Wendleder, A., Wessel, B., Roth, A., Breunig, M., Martin, K., & Wagenbrenner, S.TanDEM-X water indication mask: Generation and first evaluation results. IEEE Journal of Selected Topics in Applied Earth Observations and Remote Sensing, 2013; 6(1), 171–179.

[31] Rossi, C., Minet, C., Fritz, T., Eineder, M., & Bamler, R. Temporal monitoring of subglacial volcanoes with TanDEM-X—Application to the 2014–2015 eruption within the Bárðarbunga volcanic system, Iceland. Remote Sensing of Environment, 2016; 181, 186–197.